**IEE TELECOMMUNICATIONS SERIES 49**

Series Editors: Professor C. J. Hughes
 Professor J. O'Reilly
 Professor G. White

# Standard Codecs:
## Image Compression to Advanced Video Coding

# Contents

| | | |
|---|---|---|
| **Preface to first edition** | | xvii |
| **Preface to new edition** | | xix |
| **1 History of video coding** | | **1** |
| **2 Video basics** | | **9** |
| 2.1 | Analogue video | 9 |
| | 2.1.1 Scanning | 9 |
| | 2.1.2 Colour components | 10 |
| 2.2 | Digital video | 11 |
| 2.3 | Image format | 12 |
| | 2.3.1 SIF images | 12 |
| | 2.3.2 Conversion from SIF to CCIR-601 format | 15 |
| | 2.3.3 CIF image format | 16 |
| | 2.3.4 SubQCIF, QSIF, QCIF | 17 |
| | 2.3.5 HDTV | 18 |
| | 2.3.6 Conversion from film | 19 |
| | 2.3.7 Temporal resampling | 19 |
| 2.4 | Picture quality assessment | 20 |
| 2.5 | Problems | 22 |
| 2.6 | References | 23 |
| **3 Principles of video compression** | | **25** |
| 3.1 | Spatial redundancy reduction | 25 |
| | 3.1.1 Predictive coding | 25 |
| | 3.1.2 Transform coding | 27 |
| | 3.1.3 Mismatch control | 30 |
| | 3.1.4 Fast DCT transform | 31 |
| 3.2 | Quantisation of DCT coefficients | 31 |
| 3.3 | Temporal redundancy reduction | 35 |
| | 3.3.1 Motion estimation | 35 |

|       |       | 3.3.2 | Fast motion estimation | 37 |
|---|---|---|---|---|
|       |       | 3.3.3 | Hierarchical motion estimation | 39 |
|       | 3.4   | Variable length coding | | 41 |
|       |       | 3.4.1 | Huffman coding | 41 |
|       |       | 3.4.2 | Arithmetic coding | 43 |
|       |       |       | 3.4.2.1 Principles of arithmetic coding | 44 |
|       |       |       | 3.4.2.2 Binary arithmetic coding | 46 |
|       |       |       | 3.4.2.3 An example of binary arithmetic coding | 50 |
|       |       |       | 3.4.2.4 Adaptive arithmetic coding | 53 |
|       |       |       | 3.4.2.5 Context-based arithmetic coding | 53 |
|       | 3.5   | A generic interframe video codec | | 55 |
|       |       | 3.5.1 | Interframe loop | 55 |
|       |       | 3.5.2 | Motion estimator | 55 |
|       |       | 3.5.3 | Inter/intra switch | 55 |
|       |       | 3.5.4 | DCT | 56 |
|       |       | 3.5.5 | Quantiser | 56 |
|       |       | 3.5.6 | Variable length coding | 56 |
|       |       | 3.5.7 | IQ and IDCT | 56 |
|       |       | 3.5.8 | Buffer | 57 |
|       |       | 3.5.9 | Decoder | 57 |
|       | 3.6   | Constant and variable bit rates | | 57 |
|       | 3.7   | Problems | | 58 |
|       | 3.8   | References | | 60 |
| **4** | **Subband and wavelet** | | | **63** |
|       | 4.1   | Why wavelet transform? | | 63 |
|       | 4.2   | Subband coding | | 64 |
|       | 4.3   | Wavelet transform | | 69 |
|       |       | 4.3.1 | Discrete wavelet transform (DWT) | 71 |
|       |       | 4.3.2 | Multiresolution representation | 71 |
|       |       | 4.3.3 | Wavelet transform and filter banks | 74 |
|       |       | 4.3.4 | Higher order systems | 75 |
|       |       | 4.3.5 | Wavelet filter design | 76 |
|       | 4.4   | Coding of the wavelet subimages | | 79 |
|       |       | 4.4.1 | Quantisation by successive approximation | 80 |
|       |       | 4.4.2 | Similarities among the bands | 81 |
|       | 4.5   | Embedded zero tree wavelet (EZW) algorithm | | 81 |
|       |       | 4.5.1 | Analysis of the algorithm | 83 |
|       | 4.6   | Set partitioning in hierarchical trees (SPIHT) | | 84 |
|       |       | 4.6.1 | Coding algorithm | 86 |
|       | 4.7   | Embedded block coding with optimised truncation (EBCOT) | | 88 |
|       |       | 4.7.1 | Bit plane quantisation | 90 |
|       |       | 4.7.2 | Conditional arithmetic coding of bit planes (tier 1 coding) | 90 |
|       |       | 4.7.3 | Fractional bit plane coding | 91 |

|   |   | 4.7.3.1 | Significance propagation pass | 93 |
|---|---|---------|-------------------------------|----|
|   |   | 4.7.3.2 | Magnitude refinement pass | 95 |
|   |   | 4.7.3.3 | Clean up pass | 95 |
|   | 4.7.4 | Layer formation and bit stream organisation (tier 2 coding) | | 98 |
|   | 4.7.5 | Rate control | | 99 |
| 4.8 | Problems | | | 100 |
| 4.9 | References | | | 101 |

# 5 Coding of still pictures (JPEG and JPEG2000) — 103

| 5.1 | Lossless compression | | | 104 |
|-----|----------------------|---|---|-----|
| 5.2 | Lossy compression | | | 105 |
|     | 5.2.1 | Baseline sequential mode compression | | 105 |
|     | 5.2.2 | Run length coding | | 108 |
|     |       | 5.2.2.1 | Coding of DC coefficients | 108 |
|     |       | 5.2.2.2 | Coding of AC coefficients | 109 |
|     |       | 5.2.2.3 | Entropy coding | 111 |
|     | 5.2.3 | Extended DCT-based process | | 112 |
|     | 5.2.4 | Hierarchical mode | | 113 |
|     | 5.2.5 | Extra features | | 115 |
| 5.3 | JPEG2000 | | | 116 |
| 5.4 | JPEG2000 encoder | | | 117 |
|     | 5.4.1 | Preprocessor | | 118 |
|     |       | 5.4.1.1 | Tiling | 118 |
|     |       | 5.4.1.2 | DC level shifting | 118 |
|     |       | 5.4.1.3 | Colour transformation | 118 |
|     | 5.4.2 | Core encoder | | 119 |
|     |       | 5.4.2.1 | Discrete wavelet transform | 120 |
|     |       | 5.4.2.2 | Quantisation | 121 |
|     |       | 5.4.2.3 | Entropy coding | 121 |
|     | 5.4.3 | Postprocessing | | 123 |
| 5.5 | Some interesting features of JPEG2000 | | | 124 |
|     | 5.5.1 | Region of interest | | 124 |
|     | 5.5.2 | Scalability | | 124 |
|     |       | 5.5.2.1 | Spatial scalability | 126 |
|     |       | 5.5.2.2 | SNR scalability | 126 |
|     | 5.5.3 | Resilience | | 127 |
| 5.6 | Problems | | | 128 |
| 5.7 | References | | | 129 |

# 6 Coding for videoconferencing (H.261) — 131

| 6.1 | Video format and structure | | 132 |
|-----|----------------------------|---|-----|
| 6.2 | Video source coding algorithm | | 133 |
|     | 6.2.1 | Prediction | 135 |

|  |  |  |  | |
|---|---|---|---|---|
| | | 6.2.2 | MC/NO_MC decision | 135 |
| | | 6.2.3 | Inter/intra decision | 136 |
| | | 6.2.4 | Forced updating | 137 |
| | 6.3 | Other types of macroblock | | 137 |
| | | 6.3.1 | Addressing of macroblocks | 138 |
| | | 6.3.2 | Addressing of blocks | 139 |
| | | 6.3.3 | Addressing of motion vectors | 139 |
| | 6.4 | Quantisation and coding | | 140 |
| | | 6.4.1 | Two-dimensional variable length coding | 140 |
| | 6.5 | Loop filter | | 144 |
| | 6.6 | Rate control | | 146 |
| | 6.7 | Problems | | 148 |
| | 6.8 | References | | 149 |
| **7** | **Coding of moving pictures for digital storage media (MPEG-1)** | | | **151** |
| | 7.1 | Systems coding outline | | 152 |
| | | 7.1.1 | Multiplexing elementary streams | 153 |
| | | 7.1.2 | Synchronisation | 153 |
| | 7.2 | Preprocessing | | 153 |
| | | 7.2.1 | Picture reordering | 154 |
| | 7.3 | Video structure | | 155 |
| | | 7.3.1 | Group of pictures (GOP) | 155 |
| | | 7.3.2 | Picture | 156 |
| | | 7.3.3 | Slice | 157 |
| | | 7.3.4 | Macroblock | 158 |
| | | 7.3.5 | Block | 159 |
| | 7.4 | Encoder | | 159 |
| | 7.5 | Quantisation weighting matrix | | 161 |
| | 7.6 | Motion estimation | | 162 |
| | | 7.6.1 | Larger search range | 163 |
| | | 7.6.2 | Motion estimation with half pixel precision | 164 |
| | | 7.6.3 | Bidirectional motion estimation | 166 |
| | | 7.6.4 | Motion range | 166 |
| | 7.7 | Coding of pictures | | 167 |
| | | 7.7.1 | I-pictures | 167 |
| | | 7.7.2 | P-pictures | 168 |
| | | 7.7.3 | B-pictures | 169 |
| | | 7.7.4 | D-pictures | 170 |
| | 7.8 | Video buffer verifier | | 171 |
| | | 7.8.1 | Buffer size and delay | 172 |
| | | 7.8.2 | Rate control and adaptive quantisation | 173 |
| | 7.9 | Decoder | | 175 |
| | | 7.9.1 | Decoding for fast play | 176 |
| | | 7.9.2 | Decoding for pause and step mode | 176 |
| | | 7.9.3 | Decoding for reverse play | 176 |

| | | | |
|---|---|---|---|
| 7.10 | Postprocessing | | 177 |
| | 7.10.1 Editing | | 177 |
| | 7.10.2 Resampling and up conversion | | 178 |
| 7.11 | Problems | | 179 |
| 7.12 | References | | 180 |

## 8 Coding of high quality moving pictures (MPEG-2) — 181

| | | |
|---|---|---|
| 8.1 | MPEG-2 systems | 182 |
| 8.2 | Level and profile | 185 |
| 8.3 | How does the MPEG-2 video encoder differ from MPEG-1? | 187 |
| | 8.3.1 Major differences | 187 |
| | 8.3.2 Minor differences | 187 |
| | 8.3.3 MPEG-1 and MPEG-2 syntax differences | 188 |
| 8.4 | MPEG-2 nonscalable coding modes | 189 |
| | 8.4.1 Frame prediction for frame pictures | 189 |
| | 8.4.2 Field prediction for field pictures | 189 |
| | 8.4.3 Field prediction for frame pictures | 191 |
| | 8.4.4 Dual prime for P-pictures | 191 |
| | 8.4.5 $16 \times 8$ motion compensation for field pictures | 193 |
| | 8.4.6 Restrictions on field pictures | 193 |
| | 8.4.7 Motion vectors for chrominance components | 194 |
| | 8.4.8 Concealment motion vectors | 194 |
| 8.5 | Scalability | 194 |
| | 8.5.1 Layering *versus* scalability | 195 |
| | 8.5.2 Data partitioning | 197 |
| | 8.5.3 SNR scalability | 200 |
| | 8.5.4 Spatial scalability | 205 |
| | 8.5.5 Temporal scalability | 208 |
| | 8.5.6 Hybrid scalability | 211 |
| |     8.5.6.1 Spatial and temporal hybrid scalability | 211 |
| |     8.5.6.2 SNR and spatial hybrid scalability | 212 |
| |     8.5.6.3 SNR and temporal hybrid scalability | 212 |
| |     8.5.6.4 SNR, spatial and temporal hybrid scalability | 213 |
| | 8.5.7 Overhead due to scalability | 213 |
| | 8.5.8 Applications of scalability | 216 |
| 8.6 | Video broadcasting | 218 |
| 8.7 | Digital versatile disc (DVD) | 219 |
| 8.8 | Video over ATM networks | 220 |
| 8.9 | Problems | 224 |
| 8.10 | References | 226 |

## 9 Video coding for low bit rate communications (H.263) — 227

| | | |
|---|---|---|
| 9.1 | How does H.263 differ from H.261 and MPEG-1? | 228 |
| | 9.1.1 Coding of H.263 coefficients | 228 |
| | 9.1.2 Coding of motion vectors | 228 |

xii  *Contents*

|  |  |  |  |  |
|---|---|---|---|---|
| | 9.1.3 | Source pictures | | 230 |
| | 9.1.4 | Picture layer | | 231 |
| 9.2 | Switched multipoint | | | 231 |
| | 9.2.1 | Freeze picture request | | 232 |
| | 9.2.2 | Fast update request | | 232 |
| | 9.2.3 | Freeze picture release | | 232 |
| | 9.2.4 | Continuous presence multipoint | | 232 |
| 9.3 | Extensions of H.263 | | | 232 |
| | 9.3.1 | Scope and goals of H.263+ | | 233 |
| | 9.3.2 | Scopes and goals of H.26L | | 234 |
| | 9.3.3 | Optional modes of H.263 | | 234 |
| 9.4 | Advanced motion estimation/compensation | | | 234 |
| | 9.4.1 | Unrestricted motion vector | | 235 |
| | 9.4.2 | Advanced prediction | | 235 |
| | | 9.4.2.1 | Four motion vectors per macroblock | 236 |
| | | 9.4.2.2 | Overlapped motion compensation | 236 |
| | 9.4.3 | Importance of motion estimation | | 239 |
| | 9.4.4 | Deblocking filter | | 240 |
| | 9.4.5 | Motion estimation/compensation with spatial transforms | | 242 |
| 9.5 | Treatment of B-pictures | | | 247 |
| | 9.5.1 | PB frames mode | | 247 |
| | | 9.5.1.1 | Macroblock type | 247 |
| | | 9.5.1.2 | Motion vectors for B-pictures in PB frames | 248 |
| | | 9.5.1.3 | Prediction for a B-block in PB frames | 249 |
| | 9.5.2 | Improved PB frames | | 249 |
| | 9.5.3 | Quantisation of B-pictures | | 250 |
| 9.6 | Advanced variable length coding | | | 251 |
| | 9.6.1 | Syntax-based arithmetic coding | | 251 |
| | 9.6.2 | Reversible variable length coding | | 252 |
| | 9.6.3 | Resynchronisation markers | | 252 |
| | 9.6.4 | Advanced intra/inter VLC | | 253 |
| | | 9.6.4.1 | Advanced intra coding | 254 |
| | | 9.6.4.2 | Advanced inter coding with switching between two VLC tables | 256 |
| 9.7 | Protection against error | | | 256 |
| | 9.7.1 | Forward error correction | | 257 |
| | 9.7.2 | Back channel | | 258 |
| | 9.7.3 | Data partitioning | | 259 |
| | 9.7.4 | Error detection by postprocessing | | 263 |
| | 9.7.5 | Error concealment | | 266 |
| | | 9.7.5.1 | Intraframe error concealment | 266 |
| | | 9.7.5.2 | Interframe error concealment | 267 |
| | | 9.7.5.3 | Loss concealment | 271 |
| | | 9.7.5.4 | Selection of best estimated motion vector | 272 |

| | | | |
|---|---|---|---|
| 9.8 | Scalability | | 273 |
| | 9.8.1 | Temporal scalability | 274 |
| | 9.8.2 | SNR scalability | 274 |
| | 9.8.3 | Spatial scalability | 275 |
| | 9.8.4 | Multilayer scalability | 275 |
| | 9.8.5 | Transmission order of pictures | 277 |
| 9.9 | Buffer regulation | | 278 |
| 9.10 | Advanced video coding (H.26L) | | 280 |
| | 9.10.1 | How does H.26L (H.264) differ from H.263 | 280 |
| | 9.10.2 | Integer transform | 281 |
| | 9.10.3 | Intra coding | 282 |
| | 9.10.4 | Inter coding | 284 |
| | 9.10.5 | Multiple reference prediction | 285 |
| | 9.10.6 | Deblocking filter | 285 |
| | 9.10.7 | Quantisation and scanning | 285 |
| | 9.10.8 | Entropy coding | 286 |
| | 9.10.9 | Switching pictures | 287 |
| | 9.10.10 | Random access | 289 |
| | 9.10.11 | Network adaptation layer | 289 |
| 9.11 | Problems | | 291 |
| 9.12 | References | | 293 |

## 10  Content-based video coding (MPEG-4) — **295**

| | | | |
|---|---|---|---|
| 10.1 | Levels and profiles | | 296 |
| 10.2 | Video object plane (VOP) | | 297 |
| | 10.2.1 | Coding of objects | 299 |
| | 10.2.2 | Encoding of VOPs | 299 |
| | 10.2.3 | Formation of VOP | 300 |
| 10.3 | Image segmentation | | 301 |
| | 10.3.1 | Semiautomatic segmentation | 302 |
| | 10.3.2 | Automatic segmentation | 302 |
| | 10.3.3 | Image gradient | 303 |
| | | 10.3.3.1  Nonlinear diffusion | 303 |
| | | 10.3.3.2  Colour edge detection | 304 |
| | 10.3.4 | Watershed transform | 305 |
| | | 10.3.4.1  Immersion watershed flooding | 306 |
| | | 10.3.4.2  Topological distance watershed | 306 |
| | 10.3.5 | Colour similarity merging | 307 |
| | 10.3.6 | Region motion estimation | 307 |
| | 10.3.7 | Object mask creation | 307 |
| 10.4 | Shape coding | | 309 |
| | 10.4.1 | Coding of binary alpha planes | 309 |
| | 10.4.2 | Chain code | 310 |
| | 10.4.3 | Quad tree coding | 311 |
| | 10.4.4 | Modified modified Reed (MMR) | 314 |
| | 10.4.5 | Context-based arithmetic coding | 316 |

xiv *Contents*

|  |  |  |  |
|---|---|---|---|
|  |  | 10.4.5.1 Size conversion | 317 |
|  |  | 10.4.5.2 Generation of context index | 319 |
|  | 10.4.6 | Greyscale shape coding | 321 |
| 10.5 | Motion estimation and compensation |  | 321 |
| 10.6 | Texture coding |  | 322 |
|  | 10.6.1 | Shape adaptive DCT | 323 |
| 10.7 | Coding of the background |  | 324 |
| 10.8 | Coding of synthetic objects |  | 326 |
| 10.9 | Coding of still images |  | 327 |
|  | 10.9.1 | Coding of the lowest band | 328 |
|  | 10.9.2 | Coding of higher bands | 329 |
|  | 10.9.3 | Shape adaptive wavelet transform | 331 |
| 10.10 | Video coding with the wavelet transform |  | 332 |
|  | 10.10.1 | Virtual zero tree (VZT) algorithm | 333 |
|  | 10.10.2 | Coding of high resolution video | 334 |
|  | 10.10.3 | Coding of low resolution video | 336 |
| 10.11 | Scalability |  | 339 |
|  | 10.11.1 | Fine granularity scalability | 339 |
|  | 10.11.2 | Object-based scalability | 339 |
| 10.12 | MPEG-4 *versus* H.263 |  | 340 |
| 10.13 | Problems |  | 343 |
| 10.14 | References |  | 345 |

**11 Content description, search and delivery (MPEG-7 and MPEG-21)** — **347**

|  |  |  |  |
|---|---|---|---|
| 11.1 | MPEG-7: multimedia content description interface |  | 348 |
|  | 11.1.1 | Description levels | 348 |
|  | 11.1.2 | Application area | 350 |
|  | 11.1.3 | Indexing and query | 351 |
|  | 11.1.4 | Colour descriptors | 352 |
|  |  | 11.1.4.1 Colour space | 352 |
|  |  | 11.1.4.2 Colour quantisation | 352 |
|  |  | 11.1.4.3 Dominant colour(s) | 353 |
|  |  | 11.1.4.4 Scalable colour | 353 |
|  |  | 11.1.4.5 Colour structure | 353 |
|  |  | 11.1.4.6 Colour layout | 353 |
|  |  | 11.1.4.7 GOP colour | 353 |
|  | 11.1.5 | Texture descriptors | 354 |
|  |  | 11.1.5.1 Homogeneous texture | 354 |
|  |  | 11.1.5.2 Texture browsing | 354 |
|  |  | 11.1.5.3 Edge histogram | 355 |
|  | 11.1.6 | Shape descriptors | 355 |
|  |  | 11.1.6.1 Region-based shapes | 355 |
|  |  | 11.1.6.2 Contour-based shapes | 355 |
|  |  | 11.1.6.3 Three-dimensional shapes | 356 |

|  |  | 11.1.7 | Motion descriptors | 356 |
|---|---|---|---|---|
|  |  |  | 11.1.7.1 Camera motion | 356 |
|  |  |  | 11.1.7.2 Motion trajectory | 356 |
|  |  |  | 11.1.7.3 Parametric motion | 357 |
|  |  |  | 11.1.7.4 Motion activity | 357 |
|  |  | 11.1.8 | Localisation | 357 |
|  |  |  | 11.1.8.1 Region locator | 357 |
|  |  |  | 11.1.8.2 Spatio–temporal locator | 358 |
|  |  | 11.1.9 | Others | 358 |
|  |  |  | 11.1.9.1 Face recognition | 358 |
|  | 11.2 | Practical examples of image retrieval |  | 358 |
|  |  | 11.2.1 | Texture-based image retrieval | 358 |
|  |  | 11.2.2 | Shape-based retrieval | 360 |
|  |  | 11.2.3 | Sketch-based retrieval | 363 |
|  | 11.3 | MPEG-21: multimedia framework |  | 364 |
|  |  | 11.3.1 | Digital item declaration | 364 |
|  |  | 11.3.2 | Digital item identification and description | 365 |
|  |  | 11.3.3 | Content handling and usage | 365 |
|  |  | 11.3.4 | Intellectual property and management | 366 |
|  |  | 11.3.5 | Terminal and networks | 366 |
|  |  | 11.3.6 | Content representation | 366 |
|  |  | 11.3.7 | Event reporting | 367 |
|  | 11.4 | References |  | 367 |

**A  A C program for the fast discrete cosine transform** — **369**

**B  Huffman tables for the DC and AC coefficients of the JPEG baseline encoder** — **373**

**C  Huffman tables for quad tree shape coding** — **377**

**D  Frequency tables for the CAE encoding of binary shapes** — **379**

**E  Channel error/packet loss model** — **383**

**F  Solutions to the problems** — **387**

**G  Glossary of acronyms** — **399**

**Subject index** — **403**

# Preface to first edition

Television is an important part of our lives. In developing countries, where the TV sets can outnumber telephones by more than 50, its impact can be even greater. Advances in the multimedia industry and advent of digital TV and other services will only increase its influence in global terms.

There is a considerable and growing will to support these developments through global standards. As an example, the call for videoconferencing proposals in the 1980s, which led to the H.261 audiovisual codec, attracted 15 proposals; in February 1999 650 research proposals on MPEG-7 were submitted to the MPEG committee meeting in Lancaster, UK.

This book aims to address some of these exciting developments by looking at the fundamentals behind them. The intention is to provide material which is useful to a wide range of readers, from researchers in video coding to managers in the multimedia industry.

In writing this book I have made use of invaluable documents prepared by the working parties of the ISO/IEC and ITU. I am also in debt to the work of my former and current research students and assistants. In particular I would like to acknowledge the work by: Pedro Assuancao, Soroush Ghanbari, Ebroul Izquierdo, Fernando Lopes, Antonio Pinheiro, Eva Rosdiana, Vassilis Seferidis, Tamer Shanableh, Eduardo da Silva, Kuan Hui Tan, Kwee Teck Tan, Qi Wang, David Wilson and John Woods, who have directly or indirectly contributed to the realisation of this book.

Finally, I would like to express my deepest gratitude to my mentor Professor Charles Hughes, who with his great vision and tireless effort taught me how to be a competitive researcher. He encouraged me to write this book, and very patiently read every chapter of it and made numerous valuable comments. Charles, thanks for everything.

Mohammed Ghanbari
June 1999

Author (left) receiving the Reiyleigh prize for the first edition of this book that was awarded by IEE as the best book of the year 2000.

# Preface to new edition

The first edition of the book was published in August 1999. It was received with overwhelming worldwide support, such that it had to be reprinted in less than 18 months. The book was reviewed by distinguished video coding experts across the globe, from USA to Australia, and the comments were published in numerous prestigious international journals, such as *IEEE Signal Processing*, *IEE Review*, *EBU Review*, etc. Due to these successes, the book was recognised as the best book of the year 2000 by the IEE and was awarded the Reiyleigh prize.

Video coding is a dynamic field, and many changes have happened since the first edition. At that time JPEG2000 was under development, but now it is mature enough to be reported. In 1999 work on MPEG-7 had just started, but today we have more information on how video databases can be searched and the work has progressed into MPEG-21. At the turn of the millennium particular emphasis was put on mobile communications and video streaming over IP networks. Consequently the joint work of ISO/IEC and ITU on very low bit rate video coding has led to the development of the H.26L standard. In the second edition, the important and fundamental aspects of all these new and exciting events are explained. Of course, the remaining parts of the book have also undergone some amendments, either to clarify certain subjects, or to explain new ideas in video coding. Based on the comments of my colleagues at various universities round the world (who have used the book as a text book), I have designed a few problems at the end of each chapter. The model answers are given at the end of the book.

In addition to the material used in the first edition, I would like to acknowledge the work of several of my research associates, assistants and students, in particular: Randa Atta, Soroush Ghanbari, Mislav Grgic, Hiroshi Gunji, Ekram Khan, Fernando Lopes, Mike Nilsson, Antonio Pinheiro, Kai Sun and C.K. Tan.

Finally my deepest gratitude goes to Professor Charles Hughes, who has very passionately read all parts of the manuscript and made numerous valuable comments. Charles, once more, thank you very much.

Mohammed Ghanbari
February 2003

## Chapter 1
# History of video coding

Digital video compression techniques have played an important role in the world of telecommunication and multimedia systems where bandwidth is still a valuable commodity. Hence, video coding techniques are of prime importance for reducing the amount of information needed for a picture sequence without losing much of its quality, judged by the human viewers. Modern compression techniques involve very complex electronic circuits and the cost of these can only be kept to an acceptable level by high volume production of LSI chips. Standardisation of the video compression techniques is therefore essential.

Straightforward PCM coding of TV signals at 140 Mbit/s was introduced in the 1970s. It conformed with the digital hierarchy used mainly for multichannel telephony but the high bit rate restricted its application to TV programme distribution and studio editing. Digital TV operation for satellites became attractive since the signals were compatible with the time-division multiple access systems then coming into use. Experimental systems in 1980 used bit rates of about 45 Mbit/s for NTSC signals and 60 Mbit/s for the PAL standard.

An analogue videophone system tried out in the 1960s had not proved viable, but by the 1970s it was realised that visual speaker identification could substantially improve a multiparty discussion and videoconference services were considered. This provided the impetus for the development of low bit rate video coding. With the available technology in the 1980s, COST211 video codec, based on differential pulse code modulation (DPCM) was standardised by CCITT, under the H.120 standard. The codec's target bit rate was at 2 Mbit/s for Europe and 1.544 Mbit/s for North America, suitable for their respective first levels of digital hierarchy. However, the image quality, although having very good spatial resolution (due to the nature of DPCM working on pixel-by-pixel bases), had a very poor temporal quality. It was soon realised that in order to improve the image quality, without exceeding the target bit rate, less than one bit should be used to code each pixel. This was only possible if a group of pixels were coded together, such that the bit per pixel is fractional. This led to the design of so-called block-based codecs.

During the late 1980s study period, of the 15 block-based videoconferencing proposals submitted to the telecommunication standardisation sector of the International Telecommunication Union (ITU-T formerly CCITT), 14 were based on the discrete cosine transform (DCT) and only one on vector quantisation (VQ). The subjective quality of video sequences presented to the panel showed hardly any significant differences between the two coding techniques. In parallel to the ITU-T's investigation during 1984–1988, the Joint Photographic Experts Group (JPEG) was also interested in compression of static images. They chose the DCT as the main unit of compression, mainly due to the possibility of progressive image transmission. JPEG's decision undoubtedly influenced the ITU-T in favouring DCT over VQ. By now there was worldwide activity in implementing the DCT in chips and on DSPs.

By the late 1980s it was clear that the recommended ITU-T videoconferencing codec would use a combination of interframe DPCM for minimum coding delay and the DCT. The codec showed greatly improved picture quality over H.120. In fact, the image quality for videoconferencing applications was found reasonable at 384 kbit/s or higher and good quality was possible at significantly higher bit rates of around 1 Mbit/s. This effort, although originally directed at video coding at 384 kbit/s, was later extended to systems based on multiples of 64 kbit/s ($p \times 64$ kbit, where $p$ can take values from 1 to 30). The standard definition was completed in late 1989 and is officially called the H.261 standard (the coding method is often referred to as '$p \times 64$').

The success of H.261 was a milestone for low bit rate coding of video at reasonable quality. In the early 1990s, the Motion Picture Experts Group (MPEG) started investigating coding techniques for storage of video, such as CD-ROMs. The aim was to develop a video codec capable of compressing highly active video such as movies, on hard discs, with a performance comparable to that of VHS home video cassette recorders (VCRs). In fact, the basic framework of the H.261 standard was used as a starting point in the design of the codec. The first generation of MPEG, called the MPEG-1 standard, was capable of accomplishing this task at 1.5 Mbit/s. Since, for storage of video, encoding and decoding delays are not a major constraint, one can trade delay for compression efficiency. For example, in the temporal domain a DCT might be used rather than DPCM, or DPCM used but with much improved motion estimation, such that the motion compensation removes temporal correlation. This latter option was adopted within MPEG-1.

It is ironic that in the development of H.261, motion compensation was thought to be optional, since it was believed that after motion compensation little was left to be decorrelated by the DCT. However, later research showed that efficient motion compensation can reduce the bit rate. For example, it is difficult to compensate for the uncovered background, unless one looks ahead at the movement of the objects. This was the main principle in MPEG-1, where the motion in most picture frames is looked at from past and future, and this proved to be very effective.

These days, MPEG-1 decoders/players are becoming commonplace for multimedia on computers. MPEG-1 decoder plug-in hardware boards (e.g. MPEG magic cards) have been around for a few years, and now software MPEG-1 decoders are available with the release of new operating systems or multimedia extensions for PC and Mac platforms. Since in all standard video codecs only the decoders have to comply with proper syntax, software-based coding has added extra flexibility that might even improve the performance of MPEG-1 in the future.

Although MPEG-1 was optimised for typical applications using noninterlaced video of 25 frames/s (in European format) or 30 frames/s (in North America) at bit rates in the range of 1.2–1.5 Mbit/s (for image quality comparable to home VCRs), it can certainly be used at higher bit rates and resolutions. Early versions of MPEG-1 for interlaced video, such as those used in broadcast, were called MPEG-1+. Broadcasters, who were initially reluctant to use any compression on video, fairly soon adopted a new generation of MPEG, called MPEG-2, for coding of interlaced video at bit rates of 4–9 Mbit/s. MPEG-2 is now well on its way to making a significant impact in a range of applications such as digital terrestrial broadcasting, digital satellite TV, digital cable TV, digital versatile disc (DVD) and many others. In November 1998, OnDigital of the UK started terrestrial broadcasting of BBC and ITV programmes in MPEG-2 coded digital forms, and almost at the same time several satellite operators such as Sky-Digital launched MPEG-2 coded television pictures direct to homes.

Since in MPEG-2 the number of bidirectionally predicted pictures is at the discretion of the encoder, this number may be chosen for an acceptable coding delay. This technique may then be used for telecommunication systems. For this reason ITU-T has also adopted MPEG-2 under the generic name of H.262 for telecommunications. H.262/MPEG-2, apart from coding high resolution and higher bit rate video, also has the interesting property of scalability, such that from a single MPEG-2 bit stream two or more video images at various spatial, temporal or quality resolutions can be extracted. This scalability is very important for video networking applications. For example, in applications such as video on demand, multicasting etc., the client may wish to receive video of his/her own quality choice, or in networking applications during network congestion less essential parts of the bit stream can be discarded without significantly impairing the received video pictures.

Following the MPEG-2 standard, coding of high definition television (HDTV) was seen to be the next requirement. This became known as MPEG-3. However, the versatility of MPEG-2, being able to code video of any resolution, left no place for MPEG-3, and hence it was abandoned. Although Europe has been slow in deciding whether to use HDTV, broadcast of HDTV with MPEG-2 compression in the USA has already started. It is foreseen that in the USA by the year 2014 the existing transmission of analogue NTSC video will cease and HDTV/MPEG-2 will be the only terrestrial broadcasting format.

After so much development on MPEG-1 and 2, one might wonder what is next. Certainly we have not yet addressed the question of sending video at very low bit rates, such as 64 kbit/s or less. This of course depends on the demand for such services. However, there are signs that in the very near future such demands may arise. For example, currently, owing to a new generation of modems allowing bit rates of 56 kbit/s or so over public switched telephone networks (PSTN), videophones at such low bit rates are needed. In the near future there will be demands for sending video over mobile networks, where the channel capacity is very scarce. In fact, the wireless industry is the main driving force behind the low bit rate image/video compression.

For instance, during the months of June and July 2002, about two million picture-phone sets were sold in Japan alone. A picture-phone is a digital photo camera that grabs still pictures, compresses and sends them as a text file over the mobile network. On the video front, in October 2002 the Japanese company NTT DoCoMo announced the launch of her first handheld mobile video codec. The codec is the size of today's mobile phones, at a price of almost 350 US dollars.

To fulfil this goal, the MPEG group started working on a very low bit rate video codec, under the name of MPEG-4. Before achieving acceptable image quality at such bit rates, new demands arose. These were mainly caused by the requirements of multimedia, where there was a considerable demand for coding of multiviewpoint scenes, graphics and synthetic as well as natural scenes. Applications such as virtual studio and interactive video were the main driving forces. Ironically, critics say that since MPEG-4 could not deliver the very low bit rate codec that had been promised the goal posts have been moved.

Work on very low bit rate systems, due to the requirement of PSTN and mobile applications, was carried out by ITU-T, and a new video codec named H.263 has been devised to fulfil the goal of MPEG-4. This codec, which is an extension of H.261, but uses lessons learned from MPEG developments, is sophisticated enough to code small dimensioned video pictures at low frame rates within 10–64 kbit/s. Over the years the compression efficiency of this codec has been improved steadily through several iterations and amendments. Throughout its evolution the codec has then been renamed H.263+ and H.263++ to indicate the improvements. Due to a very effective coding strategy used in this codec, the recommendation even defines the application of this codec to very high resolution images such as HDTV, albeit at higher bit rates.

Before leaving the subject of MPEG-4, I should add that today's effort on MPEG-4 is on functionality, since this is what makes MPEG-4 distinct from other coders. In MPEG-4 images are coded as objects, and the generated bit stream is scalable. This provides the possibility of interacting with video, choosing the parts that are of interest. Moreover, natural images can be mixed with synthetic video, in what is called virtual studio. MPEG-4 defines a new coding method based on models of objects for coding synthetic objects. It also uses the wavelet transform for coding of still images. However, MPEG-4, as part of its functionality for coding of natural images, uses a similar technique to H.263; hence it is now equally capable of coding video at very low bit rates.

The fruitful outcome of the MPEG-2/H.262 video codec product under the joint effort of the MPEG and ITU encouraged the two standards bodies in further collaboration. In 1997 the ISO/IEC MPEG group joined the video coding experts group of the ITU-T and formed a Joint Video Team (JVT) to work on very low bit rate video. The project was called H.26L, with L standing for the long-term objectives. The JVT objective is to create a single video coding standard to outperform the most optimised H.263 and MPEG-4 video codecs. The H.26L development is an ongoing activity, with the first version of the standard finalised at the end of the year 2002. In the end the H.26L codec will be called H.264 by the ITU-U community and MPEG-4 version 10 by the ISO/IEC MPEG group.

*History of video coding* 5

As we see the video coding standards have evolved under the two brand names of H.26x and MPEG-x. The H.26x codecs are recommended by the telecommunication standardisation sector of the International Telecommunication Union (ITU-T). The ITU-T recommendations have been designed for telecommunications applications, such as videoconferencing and videotelephony. The MPEG-x products are the work of the International Standardisation Organisation and the International Electrotechnical Commission, Joint Technical Committee number 1 (ISO/IEC JTC1). The MPEG standards have been designed mostly to address the needs of video storage (e.g. CD-ROM, DVD), broadcast TV and video streaming (e.g. video Internet). For the most part the two standardisation committees have worked independently on different standards. However, there were exceptions, where their joint work resulted in standards such as H.262/MPEG-2 and the H.26L. Figure 1.1 summarises the evolution of video coding standards by the two organisations and their joint effort from the beginning of 1984 until now (2003). The Figure also shows the evolution of still image coding under the joint work of ITU and ISO/IEC, which is best known as the JPEG group.

It should be noted that MPEG activity is not just confined to the compression of audio-visual contents. The MPEG committee has also been active in the other aspects of audio-visual information. For example, work on object or content-based coding of MPEG-4 has brought new requirements, in particular searching for content in image databases. Currently, a working group under MPEG-7 has undertaken to study these requirements. The MPEG-7 standard builds on the other standards, such as MPEG-1, 2 and 4. Its main function is to define a set of descriptors for multimedia databases to look for specific image/video clips, using image characteristics such as colour, texture

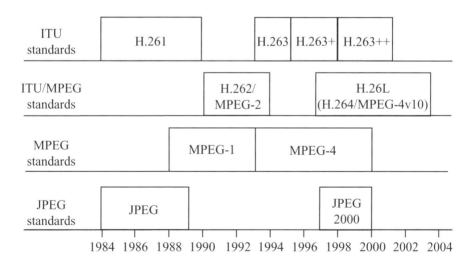

*Figure 1.1* *Evolution of video coding standards by the ITU-T and ISO/IEC committees*

and information about the shape of objects. These pictures may be coded by either of the standard video codecs, or even in analogue forms.

With the advances made on content description in MPEG-7 and coding and compression of contents under MPEG-4, it is now the time to provide the customers with efficient access to these contents. There is a need to produce specifications of standardised interfaces and protocols that allow customers to access the wide variety of content providers. This is the task undertaken by the MPEG-21, under the name of multimedia framework.

In this book, we start by reviewing briefly the basics of video, including scanning, formation of colour components at various video formats and quality evaluation of video. At the end of each chapter a few problems have been designed, either to cover some specific parts of the book in greater depth, or for a better appreciation of those parts. Principles of video compression techniques used in the standard codecs are given in Chapter 3. These include the three fundamental elements of compression: spatial, temporal and intersymbol redundancy reductions. The discrete cosine transform (DCT), as the core element of all the standard codecs, and its fast implementation are presented. Quantisation of the DCT coefficients for bit rate reduction is given. The most important element of temporal redundancy reduction, namely motion compensation, is discussed in this Chapter. Two variable length coding techniques for reduction of the entropy of the symbols, namely Huffman and arithmetic coding, are described. Special attention is paid to the arithmetic coding, because of its role and importance in most recent video codecs. The Chapter ends with an overview of a generic interframe video codec, which is used as a generic codec in the following chapters to describe various standard codecs.

Due to the importance of wavelet coding in the new generation of standard codecs, Chapter 4 is specifically devoted to the description of the basic principles of wavelet-based image coding. The three well known techniques for compression of wavelet-based image coding (EZW, SPIHT and EBCOT) are presented. Their relative compression efficiencies are compared against each other.

Coding of still pictures, under the Joint Photographic Experts Group (JPEG), is presented in Chapter 5. Lossless and lossy compression versions of JPEG are described, as is baseline JPEG and its extension with sequential and progressive modes. The Chapter also includes a new standard for still image coding, under JPEG2000. Potential for improving the picture quality under this new codec and its new functionalities are described.

Chapter 6 describes the H.261 video codec for teleconferencing applications. The structure of picture blocks and the concept of the macroblock as the basic unit of coding are defined. Selection of the best macroblock type for efficient coding is presented. The Chapter examines the efficiency of zigzag scanning of the DCT coefficients for coding. The efficiency of two-dimensional variable length coding of zigzag scanned DCT coefficients is compared with one-dimensional variable length codes.

Chapter 7 explains the MPEG-1 video coding technique for storage applications. The concept of group of pictures for flexible access to compressed video is explained. Differences between MPEG-1 and H.261, as well as the similarities, are highlighted.

These include the nature of motion compensation and various forms of coding of picture types used in this codec. Editing, pause, fast forward and fast reverse picture tricks are discussed.

Chapter 8 is devoted to coding of high quality moving pictures with the MPEG-2 standard. The concept of level and profile with its applications are defined. The two main concepts of interlacing and scalability, which discriminate this codec from MPEG-1, are given. The best predictions for the nonscalable codecs from the fields, frames and/or their combinations are discussed. On the scalability, the three fundamental scalable codecs, spatial, SNR and temporal scalable codecs are analysed, and the quality of some pictures coded using these methods is contrasted. Layered coding is contrasted against scalability and the additional overhead due to scalability/layering is also compared against the nonscalable encoder. The Chapter ends with transmission of MPEG-2 coded video for broadcast applications and video over ATM networks, as well as its storage on the digital versatile disc (DVD).

Chapter 9 discusses H.263 video coding for very low bit rate applications. The fundamental differences and similarities between this codec and H.261 and MPEG-1/2 are highlighted. Special interest is paid to the importance of motion compensation in this codec. Methods of improving the compression efficiency of this codec under various optional modes are discussed. Rather than describing all the annexes (optional modes) one by one, we have tried to group them into meaningful categories, and the importance of each category as well as the option itself is described. The relative compression performance of this codec with a limited option is compared with the other video codecs. Since H.263 is an attractive video encoding tool for mobile applications, special interest is paid to the transmission of H.263 coded video over unreliable channels. In this regard, error correction for transmission of video over mobile networks is discussed. Methods of improving the robustness of the codecs against channel errors are given, and methods for post processing and concealment of erroneous video are explained. Finally, the Chapter ends with the introduction of an improved successor of this codec under the name of H.26L. This codec is the joint work of the ITU-T and MPEG video coding expert groups and will be called H.264 by the ITU and MPEG-4 version 10 by the MPEG group.

In Chapter 10 a new method of video coding based on the image content is presented. The level and profiles set out for this codec are outlined. The concept of image plane that enables users to interact with the individual objects and change their characteristics is introduced. Methods for segmenting video frames into objects and their extractions are explained. Coding of arbitrary shaped objects with a particular emphasis on coding of their shapes is studied. The shape-adaptive DCT as a natural coding scheme for these objects is analysed.

Coding of synthetic objects with model-based coding and still images with the wavelet transform is introduced. It is shown how video can be coded with the wavelet transform, and its quality is compared against H.263. Performance of frame-based MPEG-4 is also compared against the H.263 for some channel error rates, using mobile and fixed network environments. The Chapter ends with scalability defined for content-based coding.

The book ends with a Chapter on content description, search and video browsing under the name MPEG-7. Various content search methods exploiting visual information such as colour, texture, shape and motion are described. Some practical examples for video search by textures and shapes are given. The Chapter ends with a brief look at the multimedia framework under MPEG-21, to define standards for easy and efficient use of contents by customers.

*Chapter 2*
# Video basics

Before discussing the fundamentals of video compression, let us look at how video signals are generated. Their characteristics will help us to understand how they can be exploited for bandwidth reduction without actually introducing perceptual distortions. In this regard, we will first look at image formation and colour video. Interlaced/progressive video is explained, and its impact on the signal bandwidth and display units is discussed. Representation of video in digital form and the need for bit rate reductions will be addressed. Finally, the image formats to be coded for various applications and their quality assessments will be analysed.

## 2.1 Analogue video

### 2.1.1 Scanning

Video signals are normally generated at the output of a camera by scanning a two-dimensional moving scene and converting it into a one-dimensional electric signal. A moving scene is a collection of individual pictures or images, where each scanned picture generates a frame of the picture. Scanning starts at the top left corner of the picture and ends at the bottom right.

The choice of number of scanned lines per picture is a trade-off between the bandwidth, flicker and resolution. Increasing the number of scanning lines per picture increases the spatial resolution. Similarly, increasing the number of pictures per second will increase the temporal resolution. There is a lower limit to the number of pictures per second, below which flicker becomes perceptible. Hence, flicker-free, high resolution video requires larger bandwidth.

If a frame is formed by the single scanning of a picture, it is called progressive scanning. Alternatively, two pictures may be scanned at two different times, with the lines interleaved, such that two consecutive lines of a frame belong to alternate fields to form a frame. In this case, each scanned picture is called a field, and the scanning is called interlaced. Figure 2.1 shows progressive and interlaced frames.

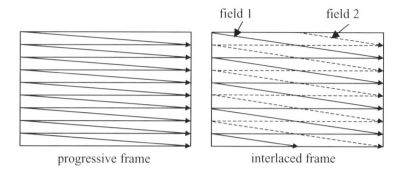

*Figure 2.1    Progressive and interlaced frames*

The concept behind interlaced scanning is to trade-off vertical–spatial resolution with that of the temporal. For instance, slow moving objects can be perceived with higher vertical resolution, since there are not many changes between the successive fields. At the same time, the human eye does not perceive flicker since the objects are displayed at field rates. For fast moving objects, although vertical resolution is reduced, the human eye is not sensitive to spatial resolutions at high display rates. Therefore, the bandwidth of television signals is halved without significant loss of picture resolution. Usually, in interlaced video, the number of lines per field is half the number of lines per frame, or the number of fields per second is twice the number of frames per second. Hence, the number of lines per second remains fixed.

It should be noted that if high spatio–temporal video is required, for example in high definition television (HDTV), then the progressive mode should be used. Although interlaced video is a good trade-off in television, it may not be suitable for computer displays, owing to the closeness of the screen to the viewer and the type of material normally displayed, such as text and graphs. If television pictures were to be used with computers, the result would be annoying interline flicker, line crawling etc. To avoid these problems, computers use noninterlaced (also called progressive or sequential) displays with refresh rates higher than 50/60 frames per second, typically 72 frames/s.

### 2.1.2    Colour components

During the scanning, a camera generates three primary colour signals called red, green and blue, the so-called RGB signals. These signals may be further processed for transmission and storage. For compatibility with black and white video and because of the fact that the three colour signals are highly correlated, a new set of signals at different colour space is generated. These are called colour systems, and the three standards are NTSC, PAL and SECAM [1]. We will concentrate on the PAL system as an example, although the basic principles involved in the other systems are very similar.

The colour space in PAL is represented by $YUV$, where $Y$ represents the luminance and $U$ and $V$ represent the two colour components. The basis $YUV$ colour space can be generated from gamma-corrected RGB (referred to in equations as $R'G'B'$) components as follows:

$$Y = 0.299R' + 0.587G' + 0.114B'$$
$$U = -0.147R' - 0.289G' + 0.436B' = 0.492(B' - Y) \quad (2.1)$$
$$V = 0.615R' - 0.515G' - 0.100B' = 0.877(R' - Y)$$

In the PAL system the luminance bandwidth is normally 5 MHz, although in PAL system-I, used in the UK, it is 5.5 MHz. The bandwidth of each colour component is only 1.5 MHz, because the human eye is less sensitive to colour resolution. For this reason, in most image processing applications, such as motion estimation, decisions on the types of block to be coded or not coded (see Chapter 6) are made on the luminance component only. The decision is then extended to the corresponding colour components. Note that for higher quality video, such as high definition television (HDTV), the luminance and chrominance components may have the same bandwidth, but nevertheless all the decisions are made on the luminance components. In some applications the chrominance bandwidth may be reduced much further than the ratio of 1.5/5 MHz.

## 2.2 Digital video

The process of digitising analogue video involves the three basic operations of filtering, sampling and quantisation. The filtering operation is employed to avoid the aliasing artefacts of the follow-up sampling process. The filtering applied to the luminance can be different to that for chrominance, owing to different bandwidth requirements.

Filtered luminance and chrominance signals are sampled to generate a discrete time signal. The minimum rate at which each component can be sampled is its Nyquist rate and corresponds to twice the signal bandwidth. For a PAL system this is in the range of 10–11 MHz. However, due to the requirement to make the sampling frequency a harmonic of the analogue signal line frequency, the sampling rate for broadcast quality signals has been recommended by CCIR to be 13.5 MHz, under recommendation CCIR-601 [2]. This is close to three times the PAL subcarrier frequency. The chrominance sampling frequency has also been defined to be half the luminance sampling frequency. Finally, sampled signals are quantised to eight-bit resolution, suitable for video broadcasting applications.

It should be noted that colour space recommended by CCIR-601 is very close to the PAL system. The precise luminance ($Y$) and chrominance ($C_b$ and $C_r$) equations under this recommendation are:

$$Y = 0.257R' + 0.504G' + 0.098B' + 16$$
$$C_b = -0.148R' - 0.291G' + 0.439B' + 128 \quad (2.2)$$
$$C_r = 0.439R' - 0.368G' - 0.071B' + 128$$

The slight departure from the PAL parameters is due to the requirement that, in the digital range, $Y$ should take values in the range of 16–235 quantum levels. Also, the normally AC chrominance components of $U$ and $V$ are centred on the grey level 128, and the range is defined from 16 to 240. The reasons for these modifications are:

(i) to reduce the granular noise of all three signals in later stages of processing
(ii) to make chrominance values positive to ease processing operations (e.g. storage).

Note that despite the unique definition for $Y$, $C_b$ and $C_r$, the CCIR-601 standard for European broadcasting is different from that for North America and the Far East. In the former, the number of lines per frame is 625 and the number of frames per second is 25. In the latter these values are 525 and 30, respectively. The number of samples per active line, called picture elements (pixels) is 720 for both systems. In the 625-line system, the total number of pixels per line, including the horizontal blanking, is 13.5 MHz times 64 μs, equal to 864 pixels. Note also that despite the differences in the number of lines and frames rates, the number of pixels generated per second under both CCIR-601/625 and CCIR-601/525 is the same. This is because in digital television we are interested in the active parts of the picture, and the number of active television lines per frame in CCIR-601/625 is 576 and the total number of pixels per second becomes equal to $720 \times 576 \times 25 = 10\,368\,000$. In CCIR-601/525 the number of active lines is 480, and the total number of pixels per second is $720 \times 480 \times 30 = 10\,368\,000$.

The total bit rate is then calculated by considering that there are half the luminance pixels for each of the chrominance pixels, and with eight bits per pixel, the total bit rate becomes $10\,368\,000 \times 2 \times 8 = 165\,888\,000$ bits/s. Had we included all the horizontal and vertical blanking, then the total bandwidth would be $13.5 \times 10^6 \times 2 \times 8 = 216$ Mbit/s. Either of these values is much greater than the equivalent analogue bandwidth, hence the video compression to reduce the digital bit rate is very demanding. In the following chapters we will show how such a huge bit rate can be compressed down to less than 10 Mbit/s, without noticeable effect on picture quality.

## 2.3 Image format

CCIR-601 is based on an image format for studio quality. For other applications, images with various degrees of resolution and dimensions might be preferred. For example, in videoconferencing or videotelephony, small image sizes with lower resolutions require much less bandwidth than the studio or broadcast video, and at the same time the resultant image quality is quite acceptable for the application. On the other hand, for HDTV, larger image sizes with improved luminance and chrominance resolutions are preferred.

### 2.3.1 SIF images

In most cases the video sources to be coded by standard video codecs are produced by CCIR-601 digitised video signals direct from the camera. It is then logical to relate

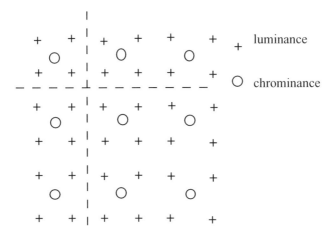

*Figure 2.2   Positioning of luminance and chrominance samples (dotted lines indicate macroblock boundaries)*

picture resolutions and dimensions of various applications to those of CCIR-601. The first sets of images related to CCIR-601 are the lower resolution images for storage applications.

A lower resolution to CCIR-601 would be an image sequence with half the CCIR-601 resolutions in each direction. That is, in each CCIR-601 standard, active parts of the image in the horizontal, vertical and temporal dimensions are halved. For this reason it is called the source input format, or SIF [3]. The resultant picture is noninterlaced (progressive). The positions of the chrominance samples share the same block boundaries with those of the luminance samples, as shown in Figure 2.2. For every four luminance samples, $Y$, there will be one pair of chrominance components, $C_b$ and $C_r$.

Thus, for the European standard, the SIF picture resolution becomes 360 pixels per line, 288 lines per picture and 25 pictures per second. For North America and the Far East, these values are 360, 240 and 30, respectively.

One way of converting the source video rate (temporal resolution) is to use only odd or even fields. Another method is to take the average values of the two fields. Discarding one field normally introduces aliasing artefacts, but simple averaging blurs the picture. For better SIF picture quality more sophisticated methods of rate conversion are required, which inevitably demand more processing power. The horizontal and vertical resolutions are halved after filtering and subsampling of the video source.

Considering that in CCIR-601 the chrominance bandwidth is half of the luminance, then the number of each chrominance pixel per line is half of the luminance pixels, but their frame rates and the number of lines per frame are equal. This is normally referred to as 4:2:2 image format. Figure 2.3 shows the luminance and chrominance components for the 4:2:2 image format. As the Figure shows, in the scanning direction (horizontal) there is a pair of chrominance samples for every

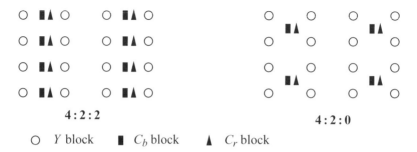

*Figure 2.3* Sampling pattern for 4 : 2 : 2 (CCIR 601) and 4 : 2 : 0 SIF

alternate luminance sample, but the chrominance components are present in every line. For SIF pictures, there is a pair of chrominance samples for every four luminance pixels as shown in the Figure.

Thus, in SIF, the horizontal and vertical resolutions of luminance will be half of the source resolutions, but for the chrominance, although horizontal resolution is halved, the vertical resolution has to be one quarter. This is called 4 : 2 : 0 format.

The lowpass filters used for filtering the source video are different for luminance and chrominance coefficients. The luminance filter coefficient is a seven-tap filter with characteristics:

$$[-29 \quad 0 \quad 88 \quad 138 \quad 88 \quad 0 \quad -29]//256 \tag{2.3}$$

Use of a power of two for the devisor allows a simple hardware implementation.

For the chrominance the filter characteristic is a four-tap filter of the type:

$$[1 \quad 3 \quad 3 \quad 1]//8 \tag{2.4}$$

Hence, the chrominance samples have to be placed at a horizontal position in the middle of the luminance samples, with a phase shift of half a sample. These filters are not part of the international standard, and other filters may be used. Figure 2.4 illustrates the subsampling and lowpass filtering of the CCIR-601 format video into SIF format.

Note that the number of luminance pixels per line of CCIR-601 is 720. Hence the horizontal resolutions of SIF luminance and chrominance should be 360 and 180, respectively. Since in the standard codecs the coding unit is based on macroblocks of 16 × 16 pixels, 360 is not divisible by 16. Therefore from each of the leftmost and rightmost sides of SIF four pixels are removed.

The preprocessing into SIF format is not normative, and other preprocessing steps and other resolutions may be used. The picture size need not even be a multiple of 16. In this case a video coder adds padding pixels to the right or bottom edges of the picture. For example, a horizontal resolution of 360 pixels could be coded by adding eight pixels to the right edge of each horizontal row, bringing the total to 368.

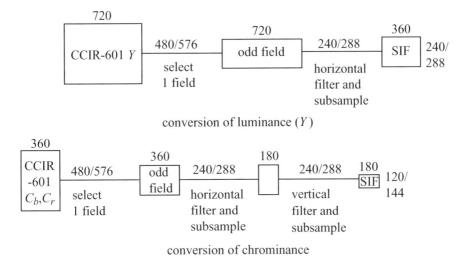

*Figure 2.4   Conversion of CCIR-601 to SIF*

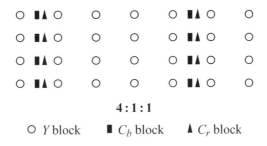

*Figure 2.5   Sampling pattern of 4 : 1 : 1 image format*

Now 23 macroblocks would be coded in each row. The decoder would discard the extra padding pixels after decoding, giving the final decoded resolution of 360 pixels.

The sampling format of 4 : 2 : 0 should not be confused with that of the 4 : 1 : 1 format used in some digital VCRs. In this format chrominance has the same vertical resolution as luminance, but horizontal resolution is one quarter. This can be represented with the sampling pattern shown in Figure 2.5. Note that 4 : 1 : 1 has the same number of pixels as 4 : 2 : 0!

### 2.3.2  Conversion from SIF to CCIR-601 format

A SIF is converted to its corresponding CCIR-601 format by spatial upsampling as shown in Figure 2.6. A linear phase finite impulse response (FIR) is applied after the insertion of zeros between samples [3]. A filter that can be used for upsampling the luminance is a seven-tap FIR filter with the impulse response of:

$$[-12 \quad 0 \quad 140 \quad 256 \quad 140 \quad 0 \quad -12]//256 \qquad (2.5)$$

At the end of the lines some special techniques such as replicating the last pixel must be used. Note that the DC response of this filter has a gain of two. This is due to the inserted alternate zeros in the upsampled samples, such that the upsampled values retain their maximum nominal value of 255.

According to CCIR recommendation 601, the chrominance samples need to be cosited with the luminance samples 1, 3, 5... In order to achieve the proper location, the upsampling filter should have an even number of taps, as given by:

$$[1 \quad 3 \quad 3 \quad 1]//4 \qquad (2.6)$$

Note again, the filter has a gain of two.

The SIF may be reconstructed by inserting four black pixels into each end of the horizontal luminance line in the decoded bitmap, and two grey pixels (value of 128) to each of the horizontal chrominance lines. The luminance SIF may then be upsampled horizontally and vertically. The chrominance SIF should be upsampled once horizontally and twice vertically, as shown in Figure 2.6b.

### 2.3.3 CIF image format

For worldwide videoconferencing, a video codec has to cope with the CCIR-601 of both European (625 line, 50 Hz) and North America and Far East (525 line, 60 Hz) video formats. Hence, CCIR-601 video sources from these two different formats have to be converted to a common format. The picture resolutions also need to be reduced, to be able to code them at lower bit rates.

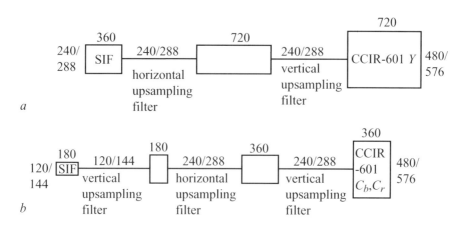

Figure 2.6  Upsampling and filtering from SIF to CCIR-601 format
       a  luminance signals
       b  chrominance signals

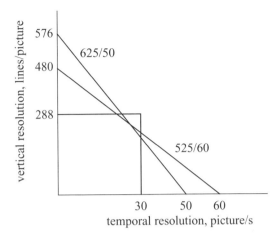

*Figure 2.7  Spatio–temporal relation in CIF format*

Considering that in CCIR-601 the number of pixels per line in both the 625/50 and 525/60 standards is 720 pixels per line, then half of this value, 360 pixels/line, was chosen as the horizontal resolution. For the vertical and temporal resolutions, a value intermediate between the two standards was chosen such that the combined vertical × temporal resolutions were one quarter of that of CCIR-601. The 625/50 system has the greater vertical resolution. Since the active picture area is 576 lines, half of this value is 288 lines. On the other hand, the 525/60 system has the greater temporal resolution, so that the half rate is 30 Hz. The combination of 288 lines and 30 Hz gives the required vertical × temporal resolution. This is illustrated in Figure 2.7.

Such an intermediate selection of vertical resolution from one standard and temporal resolution from the other leads to the adopted name common intermediate format (CIF). Therefore a CIF picture has a luminance with 360 pixels per lines, 288 lines per picture and 30 (precisely 29.97) pictures per second [4]. The colour components are at half the spatial resolution of luminance, with 180 pixels per line and 144 lines per picture. Temporal resolutions of colour components are the same as for the luminance at 29.97 Hz.

In CIF format, like SIF, pictures are progressive (noninterlaced), and the positions of the chrominance samples share the same block boundaries with those of the luminance samples, as shown in Figure 2.2. Also like SIF, the image format is also 4 : 2 : 0 and similar down-conversion and up-conversion filters to those shown in Figures 2.4 and 2.6 can also be applied to CIF images. Note the difference between SIF-625 and CIF and SIF-525 and CIF. In the former the only difference is in the number of pictures per second, while in the latter they differ in the number of lines per picture.

## 2.3.4  SubQCIF, QSIF, QCIF

For certain applications, such as video over mobile networks or videotelephony, it is possible to reduce the frame rate. Known reduced frame rates for CIF and SIF-525 are

15, 10 and 7.5 frames/s. These rates for SIF-625 are 12.5 and 8.3 frames/s. To balance the spatio–temporal resolutions, the spatial resolutions of the images are normally reduced, nominally by halving in each direction. These are called quarter-SIF (QSIF) and quarter-CIF (QCIF) for SIF and CIF formats, respectively. Conversion of SIF or CIF to QSIF and QCIF (or *vice versa*) can be carried out with a similar method to converting CCIR-601 to SIF and CIF, respectively, using the same filter banks shown in Figures 2.4 and 2.6. Lower frame rate QSIF and QCIF images are normally used for very low bit rate video.

Certain applications, such as video over mobile networks, demand even smaller image sizes. SubQCIF is the smallest standard image size, with the horizontal and vertical picture resolutions of 128 pixels by 96 pixels, respectively. The frame rate can be very low (e.g. five frames/s) to suit the channel rate. The image format in this case is 4 : 2 : 0, and hence the chrominance resolution is half the luminance resolution in each direction.

### 2.3.5 HDTV

Currently there is no European standard for HDTV. The North American and Far Eastern HDTV has a nominal resolution of twice the 525-line CCIR-601 format. Hence, the filter banks of Figures 2.4 and 2.6 can also be used for image size conversion. Also, since in HDTV higher chrominance bandwidth is desired, it can be made equal to the luminance, or half of it. Hence there will be upto a pair of chrominance pixels for every luminance pixel, and the image format can be made 4 : 2 : 2 or even 4 : 4 : 4. In most cases HDTV is progressive, to improve vertical resolution.

It is common practice to define image format in terms of relations between $8 \times 8$ pixel blocks with a macroblock of $16 \times 16$ pixels. The concept of macroblock and block will be explained in Chapter 6. Figure 2.8 shows how blocks of luminance and chrominance in various 4 : 2 : 0, 4 : 2 : 2 and 4 : 4 : 4 image formats are defined.

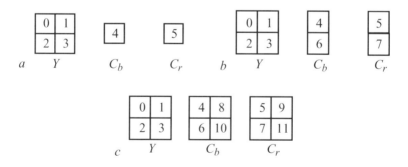

*Figure 2.8*  Macroblock structures
     *a*  4 : 2 : 0
     *b*  4 : 2 : 2
     *c*  4 : 4 : 4

*Table 2.1  Percentage of each chrominance component resolution with respect to luminance in the horizontal and vertical directions*

| Image format | Horizontal [%] | Vertical [%] |
|---|---|---|
| 4:4:4 | 100 | 100 |
| 4:2:2 | 50 | 100 |
| 4:2:0 | 50 | 50 |
| 4:1:1 | 25 | 100 |

For a more detailed representation of image formats, especially discriminating 4:2:0 from 4:1:1, one can relate the horizontal and vertical resolutions of the chrominance components to those of luminance as shown in Table 2.1. Note, the luminance resolution is the same as the number of pixels in each scanning direction.

### 2.3.6  Conversion from film

Sometimes sources available for compression consist of film material, which has a nominal frame rate of 24 pictures per second. This rate can be converted to 30 pictures per second by the pulldown technique [3]. In this mode digitised pictures are shown alternately for three and two television field times, generating 60 fields per second. This alteration may not be exact, since the actual frame rate in the 525/60 system is 29.97 frames per second. Editing and splicing of compressed video after the conversion might also have changed the pulldown timing. A sophisticated encoder might detect the duplicated fields, average them to reduce digitisation noise and code the result at the original 24 pictures per second rate. This should give a significant improvement in quality over coding at 30 pictures per second. This is because, first of all, when coding at 24 pictures per second the bit rate budget per frame is larger than that for 30 pictures per second. Secondly, direct coding of 30 pictures per second destroys the 3:2 pulldown timing and gives a jerky appearance to the final decoded video.

### 2.3.7  Temporal resampling

Since the picture rates are limited to those commonly used in the television industry, the same techniques may be applied. For example, conversion from 24 pictures per second to 60 fields can be achieved by the technique of 3:2 pulldown. Video coded at 25 pictures per second can be converted to 50 fields per second by displaying the original decoded lines in the odd CCIR-601 fields, and the interpolated lines in the even fields. Video coded at 29.97 or 30 pictures per second may be converted to a field rate twice as large using the same method.

## 2.4 Picture quality assessment

Conversion of digital pictures from one format to another, as well as their compression for bit rate reduction, introduces some distortions. It is of great importance to know whether the introduced distortion is acceptable to the viewers. Traditionally this has been done by subjective assessments, where the degraded pictures are shown to a group of subjects and their views on the perceived quality or distortions are sought.

Over the years, many subjective assessment methodologies have been developed and validated. Among them are: the double stimulus impairment scale (DSIS), where the subjects are asked to rate the impairment of the processed picture with respect to the reference unimpaired picture, and the double stimulus continuous quality scale (DSCQS), where the order of the presentation of the reference and processed pictures is unknown to the subjects. The subjects will then give a score between 1 and 100 containing adjectival guidelines placed at 20 point intervals (1–20 = bad, 21–40 = poor, 41–60 = fair, 61–80 = good and 81–100 = excellent) for each picture, and their difference is an indication of the quality [5]. Pictures are presented to the viewers for about ten seconds and the average of the viewers' scores, defined as the mean opinion score (MOS), is a measure of video quality. At least 20–25 nonexpert viewers are required to give a reliable MOS, excluding the outliers.

These methods are usually used in assessment of still images. For video evaluation single stimulus continuous quality evaluation (SSCQE) is preferred, where the time-varying picture quality of the processed video without reference is evaluated by the subjects [5]. In this method subjects are asked to continuously evaluate the video quality of a set of video scenes. The judgement criteria are the five scales used in the DSCQS above. Since video sequences are long, they are segmented into ten seconds shots, and for each video segment an MOS is calculated.

Although these methods give reliable indications of the perceived image quality, they are unfortunately time consuming and expensive. An alternative is objective measurements, or video quality metrics, which employ some mathematical models to mimic human visual systems behaviour.

In 1997 the Video Quality Experts Group (VQEG) formed from experts of ITU-T study group 6 and ITU-T study group 9 undertook this task [6]. They are considering three methods for the development of the video quality metric. In the first method, called the full reference (FR-TV) model, both the processed and the reference video segments are fed to the model and the outcome is a quantitative indicator of the video quality. In the second method, called the reduced reference (RR-TV) model, some features extracted from the spatio–temporal regions of the reference picture (e.g. mean and variance of pixels, colour histograms etc.) are made available to the model. The processed video is then required to generate similar statistics in those regions. In the third model, called no reference (NR-TV), or single ended, the processed video without any information from the reference picture excites the model. All these models should be validated with the SSCQE methods for various video segments. Early results indicate that these methods compared with the SSCQE perform satisfactorily, with a correlation coefficient of 0.8–0.9 [7].

Until any of these quality metrics become standards, it is customary to use the simplest form of objective measurement, which is the ratio of the peak-to-peak signal to the root-mean-squared processing noise. This is referred to as the peak-to-peak signal-to-noise ratio (PSNR) and defined as:

$$\text{PSNR} = 10 \log_{10} \left[ \frac{255^2}{(1/N) \sum_i \sum_j (Y_{ref}(i, j) - Y_{prc}(i, j))^2} \right] \quad (2.7)$$

where $Y_{ref}(i, j)$ and $Y_{prc}(i, j)$ are the pixel values of the reference and processed images, respectively, and $N$ is the total number of pixels in the image. In this equation, the peak signal with an eight-bit resolution is 255, and the noise is the square of the pixel-to-pixel difference (error) between the reference image and the image under study. Although it has been claimed that in some cases the PSNR's accuracy is doubtful, its relative simplicity makes it a very popular choice.

Perhaps the main criticism against the PSNR is that the human interpretation of the distortions at different parts of the video can be different. Although it is hoped that the variety of interpretations can be included in the objective models, there are still some issues that not only the simple PSNR but also more sophisticated objective models may fail to address. For example, if a small part of a picture in a video is severely degraded, this hardly affects the PSNR or any objective model parameters (depending on the area of distortion), but this distortion attracts the observers' attention, and the video looks as bad as if a larger part of the picture was distorted. This type of distortion is very common in video, where due to a single bit error, blocks of $16 \times 16$ pixels might be erroneously decoded. This has almost no significant effect on PSNR, but can be viewed as an annoying artefact. In this case there will be a large discrepancy between the objective and subjective test results.

On the other hand one may argue that under similar conditions if one system has a better PSNR than the other, then the subjective quality can be better but not worse. This is the main reason that PSNR is still used in comparing the performance of various video codecs. However, in comparing codecs, PSNR or any objective measure should be used with great care, to ensure that the types of coding distortion are not significantly different from each other. For instance, objective results from the blockiness distortion produced by the block-based video codecs can be different from the picture smearing distortion introduced by the filter-based codecs. The fact is that even the subjects may interpret these distortions differently. It appears that expert viewers prefer blockiness distortion to smearing, and nonexperts' views are opposite!

In addition to the abovementioned problems of subjective and objective measurements of video quality, the impact of people's expectation of video quality cannot be ignored. As technology progresses and viewers become more familiar with digital video, their level of expectation of video quality can grow. Hence, a quality that today might be regarded 'good', may be rated as 'fair' or 'poor' tomorrow. For instance, watching a head-and-shoulders video coded at 64 kbit/s by the early prototypes of the video codecs in the mid 1980s was very fascinating. This was despite the fact that pictures were coded at one or two frames per second, and waiving hands in front of the camera would freeze the picture for a few seconds, or cause a complete picture break-up. But today, even 64 kbit/s coded video at 4–5 frames per seconds, without picture freeze, does not look attractive. As another example, most people might be quite satisfied with the quality of the broadcast TV at home, both analogue and digital, but if they watch football spectators on a broadcast TV side by side with an HDTV video, they then realise how much information they are missing. These are all indications that people's expectations of video quality in the future will be higher. Thus video codecs either have to be more sophisticated or more channel bandwidth should be assigned to them. Fortunately, with the advances in digital technology and growth in network bandwidth, both are feasible and in the future we will witness better quality video services.

## 2.5 Problems

1 In a PAL system, determine the values of the three colour primaries R, G and B for the following colours: red, green, blue, yellow, cyan, magenta and white.

2 Calculate the luminance and chrominance values of the colours in problem 1, if they are digitised into eight bits, according to CCIR-601 specification.

3 Calculate the horizontal scanning line frequency for CCIR-601/625 and CCIR-601/525 line systems and hence their periods.

4 CCIR-601/625 video is normally digitised at 13.5 MHz sampling rate. Find the number of pixels per scanning line. If there are 720 pixels in the active part of the horizontal scanning, find the duration of horizontal scanning fly-back (i.e. horizontal blanking interval).

5 Repeat problem 4 for CCIR-601/525.

6 Find the bit rate per second of the following video formats (only active pixels are considered):

   a  CCIR-601/625; 4 : 2 : 2
   b  CCIR-601/525; 4 : 2 : 2
   c  SIF/625; 4 : 2 : 0
   d  SIF/525; 4 : 2 : 0
   e  CIF
   f  SIF/625; 4 : 1 : 1
   g  SIF/525; 4 : 1 : 1
   h  QCIF (15 Hz)
   i  subQCIF (10 Hz)

7 The luminance values of a set of pixels in a CCIR-601 video are: 128; 128; 128; 120; 60; 50; 180; 154; 198; 205; 105; 61; 93; 208; 250; 190; 128; 128; 128. They

are filtered and downsampled by 2:1 into SIF format. Find the luminance values of seven SIF samples, starting from the fourth pixel of CCIR-601.

8  The luminance values of the SIF pixels in problem 7 are upsampled and filtered into CCIR-601 format. Find the reconstructed CCIR-601 format pixel values. Calculate the PSNR of the reconstructed samples.

9  A pure sinusoid is linearly quantised into $n$ bits:

   $a$  show that the signal-to-quantisation noise ratio (SNR) in dB is given by SNR = $6n + 1.78$,
   $b$  find such an expression, for peak SNR (PSNR),
   $c$  calculate the minimum bit per pixel required for quantising video, such that PSNR is better than 58 dB.

   (hint: the mean-squared quantisation error of a uniformly quanitised waveform with step size $\Delta$ is $\Delta^2/12$).

## 2.6  References

1  NETRAVALI, A.N., and HASKELL, B.G.: 'Digital pictures, representation and compression and standards' (Plenum Press, New York, 1995, 2nd edn)
2  CCIR Recommendation 601: 'Digital methods of transmitting television information'. Recommendation 601, encoding parameters of digital television for studios
3  MPEG-1: 'Coding of moving pictures and associated audio for digital storage media at up to about 1.5 Mbit/s'. ISO/IEC 11172-2: video, November 1991
4  OKUBO, S.: 'Video codec standardisation in CCITT study group XV', *Signal Process. Image Commun.*, 1989, pp.45–54
5  Recommendation ITU-R BT.500 (revised): 'Methodology for the subjective assessment of the quality of television pictures'
6  VQEG: The Video Quality Experts Group, RRNR-TV Group Test Plan, draft version 1.4, 2000
7  TAN, K.T, GHANBARI, M., and PEARSON, D.E.: 'An objective measurement tool for MPEG video quality', *Signal Process.*, 1998, **7**, pp.279–294

*Chapter 3*

# Principles of video compression

The statistical analysis of video signals indicates that there is a strong correlation both between successive picture frames and within the picture elements themselves. Theoretically, decorrelation of these signals can lead to bandwidth compression without significantly affecting image resolution. Moreover, the insensitivity of the human visual system to loss of certain spatio–temporal visual information can be exploited for further reduction. Hence, subjectively lossy compression techniques can be used to reduce video bit rates while maintaining an acceptable image quality.

For coding still images, only the spatial correlation is exploited. Such a coding technique is called intraframe coding and is the basis for JPEG coding. If temporal correlation is exploited as well, then it is called interframe coding. Interframe predictive coding is the main coding principle that is used in all standard video codecs, such as H.261, H.263, MPEG-1, 2 and 4. It is based on three fundamental redundancy reduction principles:

1. Spatial redundancy reduction: to reduce spatial redundancy among the pixels within a picture (similarity of pixels, within the frames), by employing some data compressors, such as transform coding.
2. Temporal redundancy reduction: to remove similarities between the successive pictures, by coding their differences.
3. Entropy coding: to reduce the redundancy between the compressed data symbols, using variable length coding techniques.

A detailed description of these redundancy reduction techniques is given in the following sections.

## 3.1 Spatial redundancy reduction

### *3.1.1 Predictive coding*

In the early days of image compression, both signal processing tools and storage devices were scarce resources. At the time, a simple method for redundancy

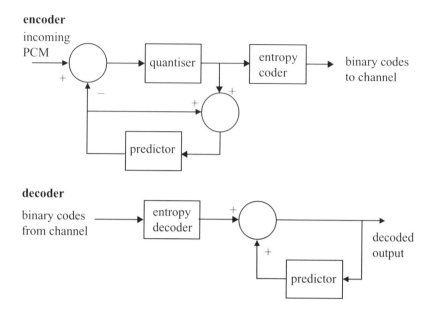

*Figure 3.1    Block diagram of a DPCM codec*

reduction was to predict the value of pixels based on the values previously coded, and code the prediction error. This method is called differential pulse code modulation (DPCM). Figure 3.1 shows a block diagram of a DPCM codec, where the differences between the incoming pixels from the predictions in the predictor are quantised and coded for transmission. At the decoder the received error signal is added to the prediction to reconstruct the signal. If the quantiser is not used it is called lossless coding, and the compression relies on the entropy coder, which will be explained later.

Best predictions are those from the neighbouring pixels, either from the same frame or pixels from the previous frame, or their combinations. The former is called intraframe predictive coding and the latter is interframe predictive coding. Their combination is called hybrid predictive coding.

It should be noted that, no matter what prediction method is used, every pixel is predictively coded. The minimum number of bits that can be assigned to each prediction error is one bit. Hence, this type of coding is not suitable for low bit rate video coding. Lower bit rates can be achieved if a group of pixels are coded together, such that the average bit per pixel can be less than one bit. Block transform coding is most suitable for this purpose, but despite this DPCM is still used in video compression. For example, interframe DPCM has lower coding latency than interframe block coding. Also, DPCM might be used in coding of motion vectors, or block addresses. If motion vectors in a moving object move in the same direction, coding of their differences will reduce the motion vector information. Of course, the coding would be lossless.

## 3.1.2 Transform coding

Transform domain coding is mainly used to remove the spatial redundancies in images by mapping the pixels into a transform domain prior to data reduction. The strength of transform coding in achieving data compression is that the image energy of most natural scenes is mainly concentrated in the low frequency region, and hence into a few transform coefficients. These coefficients can then be quantised with the aim of discarding insignificant coefficients, without significantly affecting the reconstructed image quality. This quantisation process is, however, lossy in that the original values cannot be retained.

To see how transform coding can lead to data compression, consider joint occurrences of two pixels as shown in Figure 3.2.

Although each pixel $x_1$ or $x_2$ may take any value uniformly between 0 (black) to its maximum value 255 (white), since there is a high correlation (similarity) between them, then it is most likely that their joint occurrences lie mainly on a 45 degrees line, as shown in the Figure. Now if we rotate the $x_1 x_2$ coordinates by 45 degrees, to a new position $y_1 y_2$, then the joint occurrences on the new coordinates have a uniform distribution along the $y_1$ axes, but are highly peaked around zero on the $y_2$ axes. Certainly, the bits required to represent the new parameter $y_1$ can be as large as any of $x_1$ or $x_2$, but that of the other parameter $y_2$ is much less. Hence, on average, $y_1$ and $y_2$ can be represented at a lower bit rate than $x_1$ and $x_2$.

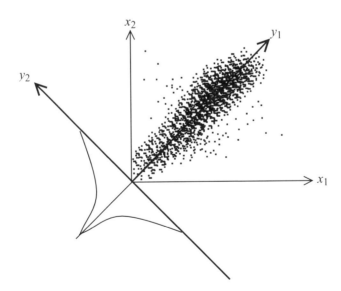

*Figure 3.2   Joint occurrences of a pair of pixels*

Rotation of $x_1 x_2$ coordinates by 45 degrees is a transformation of vector $[x_1, x_2]$ by a transformation matrix $T$:

$$T = \begin{bmatrix} \cos 45 & \sin 45 \\ \sin 45 & -\cos 45 \end{bmatrix} = \frac{1}{\sqrt{2}} \begin{bmatrix} 1 & 1 \\ 1 & -1 \end{bmatrix} \tag{3.1}$$

Thus, in this example the transform coefficients $[y_1, y_2]$ become:

$$[y_1, y_2] = \frac{1}{\sqrt{2}} \begin{bmatrix} 1 & 1 \\ 1 & -1 \end{bmatrix} \begin{bmatrix} x_1 \\ x_2 \end{bmatrix}$$

or

$$y_1 = \frac{1}{\sqrt{2}}(x_1 + x_2) \quad \text{and} \quad y_2 = \frac{1}{\sqrt{2}}(x_1 - x_2) \tag{3.2}$$

$y_1$ is called the average or DC value of $x_1$ and $x_2$ and $y_2$ represents their residual differences. The normalisation factor of $\frac{1}{\sqrt{2}}$ makes sure that the signal energy due to transformation is not changed (Parseval theorem). This means that the signal energy in the pixel domain, $x_1^2 + x_2^2$, is equal to the signal energy in the transform domain, $y_1^2 + y_2^2$. Hence the transformation matrix is orthonormal.

Now, if instead of two pixels we take $N$ correlated pixels, then by transforming the coordinates such that $y_1$ lies on the main diagonal of the sphere, then only the $y_1$ coefficient becomes significant, and the remaining $N-1$ coefficients, $y_2, y_3, \ldots, y_N$, only carry the residual information. Thus, compared with the two pixels case, larger dimensions of transformation can lead to higher compression. Exactly how large the dimensions should be depends on how far pixels can still be correlated to each other. Also, the elements of the transformation matrix, called the basis vectors, have an important role on the compression efficiency. They should be such that only one of the transform coefficients, or at most a few of them, becomes significant and the remaining ones are small.

An ideal choice for the transformation matrix is the one that completely decorrelates the transform coefficients. Thus, if $R_{xx}$ is the covariance matrix of the input source (pixels), $x$, then the elements of the transformation matrix $T$ are calculated such that the covariance of the coefficients $R_{yy} = TR_{xx}T^T$ is a diagonal matrix (zero off diagonal elements). A transform that is derived on this basis is the well known Karhunen–Loève transform (KLT) [1]. However, although this transform is optimum, and hence it can give the maximum compression efficiency, it is not suitable for image compression. This is because, as the image statistics change, the elements of the transform need to be recalculated. Thus, in addition to extra computational complexity, these elements need to be transmitted to the decoder. The extra overhead involved in the transmission significantly restricts the overall compression efficiency. Despite this, the KLT is still useful and can be used as a benchmark for evaluating the compression efficiency of other transforms.

A better choice for the transformation matrix is that of the discrete cosine transform (DCT). The reason for this is that it has well defined (fixed) and smoothly varying basis vectors which resemble the intensity variations of most natural images, such that image energy is matched to a few coefficients. For this reason its rate distortion performance closely follows that of the Karhunen–Loève transform, and results in almost identical compression gain [1]. Equally important is the availability of efficient fast DCT transformation algorithms that can be used, especially, in software-based image coding applications [2].

Since in natural image sequences pixels are correlated in the horizontal and vertical directions as well as in the temporal direction of the image sequence, a natural choice for the DCT is a three-dimensional one. However, any transformation in the temporal domain requires storage of several picture frames, introducing a long delay, which restricts application of transform coding in telecommunications. Hence transformation is confined to two dimensions.

A two-dimensional DCT is a separable process that is implemented using two one-dimensional DCTs: one in the horizontal direction followed by one in the vertical. For a block of $M \times N$ pixels, the forward one-dimensional transform of $N$ pixels is given by:

$$F(u) = \sqrt{\frac{2}{N}} C(u) \sum_{x=0}^{N-1} f(x) \cos\left(\frac{\pi(2x+1)u}{2N}\right) \quad u = 0, 1, \ldots, N-1 \quad (3.3)$$

where

$$C(u) = \sqrt{\frac{1}{2}} \quad \text{for } u = 0$$

$$C(u) = 1 \quad \text{otherwise}$$

$f(x)$ represents the intensity of the $x$th pixel, and $F(u)$ represents the $N$ one-dimensional transform coefficients. The inverse one-dimensional transform is thus defined as:

$$f(x) = \sqrt{\frac{2}{N}} \sum_{u=0}^{N-1} C(u) F(u) \cos\left(\frac{\pi(2x+1)u}{2N}\right) \quad x = 0, 1, \ldots, N-1 \quad (3.4)$$

Note that the $\sqrt{1/N}$ normalisation factor is used to make the transformation orthonormal. That is, the energy in both pixel and transform domains is to be equal. In the standard codecs the normalisation factor for the two-dimensional DCT is defined as $\frac{1}{2}$. This gives DCT coefficients in the range of $-2047$ to $+2047$. The normalisation factor in the pixel domain is then adjusted accordingly (e.g. it becomes $2/N$).

To derive the final two-dimensional transform coefficients, $N$ sets of one-dimensional transforms of length $M$ are taken over the one-dimensional transform coefficients of similar frequency in the vertical direction:

$$F(u, v) = \sqrt{\frac{2}{M}} C(v) \sum_{y=0}^{M-1} F(u, y) \cos\left(\frac{\pi(2y+1)v}{2M}\right) \quad v = 0, 1, \ldots, M-1$$

(3.5)

where $C(v)$ is defined similarly to $C(u)$.

Thus a block of $MN$ pixels is transformed into $MN$ coefficients. The $F(0, 0)$ coefficient represents the DC value of the block. Coefficient $F(0, 1)$, which is the DC value of all the first one-dimensional AC coefficients, represents the first AC coefficient in the horizontal direction of the block. Similarly, $F(1, 0)$, which is the first AC coefficient of all one-dimensional DC values, represents the first AC coefficient in the vertical direction, and so on.

In practice $M = N = 8$, such that a two-dimensional transform of $8 \times 8 = 64$ pixels results in 64 transform coefficients. The choice of such a block size is a compromise between the compression efficiency and the blocking artefacts of coarsely quantised coefficients. Although larger block sizes have good compression efficiency, the blocking artefacts are subjectively very annoying. At the early stage of standardisation of video codecs, the block sizes were made optional at $4 \times 4$, $8 \times 8$ and $16 \times 16$. Now the block size in the standard codecs is $8 \times 8$.

### 3.1.3 Mismatch control

Implementation of both forward and inverse transforms (e.g. eqns 3.3 and 3.4) requires that the cos elements be approximated with finite numbers. Due to this approximation the reconstructed signal, even without any quantisation, cannot be an exact replica of the input signal to the forward transform. For image and video coding applications this mismatch needs to be controlled, otherwise the accumulated error due to approximation can grow out of control resulting in an annoying picture artefact.

One way of preventing error accumulation is to let the error oscillate between two small levels. This guarantees that the accumulated error never exceeds its limit. The approach taken in the standard codecs is to say (e.g. MPEG-2), at the decoder, that the sum of all the values of the $8 \times 8 = 64$ transform coefficients should be an odd number (no matter whether they are quantised or not). In case the sum is an even number, the value of the highest frequency coefficient, F(7, 7), is either incremented or decremented by 1, depending whether its value itself is odd or even, respectively. This, of course, introduces a very small error, but it cannot be noticed on images for two reasons. First, at the inverse transform, the reconstructed pixels are divided by a large value of the order of $N^2$. Second, since error is introduced by the highest frequency coefficient, it appears as a very high frequency, small amplitude dither-like noise, which is not perceivable at all (the human eye is very tolerant to high frequency noise).

*Principles of video compression* 31

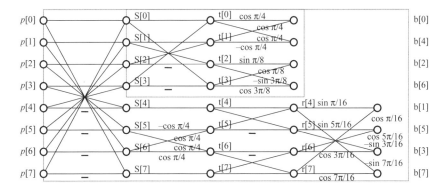

*Figure 3.3 A fast DCT flow chart*

### 3.1.4 Fast DCT transform

To calculate transform coefficients, every one-dimensional forward or inverse transformation requires eight multiplications and seven additions. This process is repeated for 64 coefficients, both in the horizontal and vertical directions. Since software-based video compression is highly desirable, methods of reducing such a huge computational burden are also highly desirable.

The fact that DCT is a type of discrete Fourier transform, with the advantage of all-real coefficients, means that one can use a fast transform, similar to the fast Fourier transform, to calculate transform coefficients with complexity proportional to $N \log_2 N$, rather than $N^2$. Figure 3.3 shows a butterfly representation of the fast DCT [2]. Intermediate nodes share some of the computational burden, hence reducing the overall complexity. In the Figure, p[0]–p[7] are the inputs to the forward DCT and b[0]–b[7] are the transform coefficients. The inputs can be either the eight pixels for the source image, or eight transform coefficients of the first stage of the one-dimensional transform. Similarly, for inverse transformation, b[0]–b[7] are the inputs to the IDCT, and p[0]–p[7] are the outputs. A C language programme for fast forward DCT is given in Appendix A. In this program, some modifications to the butterfly matrices are made to trade-off the number of additions for multiplications, since multiplications are more computationally intensive than additions. A similar program can be written for the inverse transform.

## 3.2 Quantisation of DCT coefficients

The domain transformation of the pixels does not actually yield any compression. A block of 64 pixels is transformed into 64 coefficients. Due to the orthonormality of the transformation, the energy in both the pixel and the transform domains are equal, hence no compression is achieved. However, transformation causes the significant part of the image energy to be concentrated at the lower frequency components,

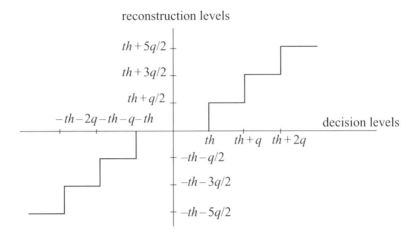

Figure 3.4    Quantisation characteristics

with the majority of the coefficients having little energy. It is the quantisation and variable length coding of the DCT coefficients that lead to bit rate reduction. Moreover, by exploiting the human eye's characteristics, which are less sensitive to picture distortions at higher frequencies, one can apply even coarser quantisation at these frequencies, to give greater compression. Coarser quantisation step sizes force more coefficients to zero and as a result more compression is gained, but of course the picture quality deteriorates accordingly.

The class of quantiser that has been used in all standard video codecs is based around the so-called uniform threshold quantiser (UTQ). It has equal step sizes with reconstruction values pegged to the centriod of the steps. This is illustrated in Figure 3.4.

The two key parameters that define a UTQ are the threshold value, $th$, and the step size, $q$. The centroid value is typically defined mid way between quantisation intervals. Note that, although AC transform coefficients have nonuniform characteristics, and hence can be better quantised with nonuniform quantiser step sizes (the DC coefficient has a fairly uniform distribution), bit rate control would be easier if they were quantised linearly. Hence, a key property of UTQ is that the step sizes can be easily adapted to facilitate rate control.

A further two subclasses of UTQ can be identified within the standard codecs, namely those with and without a dead zone. These are illustrated in Figure 3.5 and will be hereafter abbreviated as UTQ-DZ and UTQ, respectively. The term dead zone commonly refers to the central region of the quantiser, whereby the coefficients are quantised to zero.

Typically, UTQ is used for quantising intraframe DC, $F(0, 0)$, coefficients, and UTQ-DZ is used for the AC and the DC coefficients of interframe prediction error. This is intended primarily to cause more nonsignificant AC coefficients to become

*Principles of video compression* 33

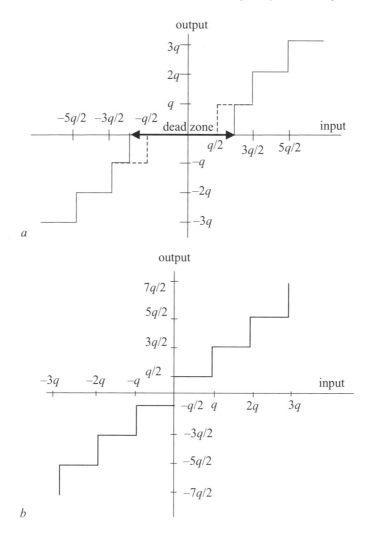

*Figure 3.5* Uniform quantisers
    *a* with dead zone
    *b* without dead zone

zero, so increasing the compression. Both quantisers are derived from the generic quantiser of Figure 3.4, where in UTQ $th$ is set to zero but in UTQ-DZ it is set to $q/2$ and in the most inner region it is allowed to vary between $q/2$ and $q$ just to increase the number of zero-valued outputs, as shown in Figure 3.5. Thus the dead zone length can be from $q$ to $2q$. In some implementations (e.g. H.263 or MPEG-4), the decision and/or the reconstruction levels of the UTQ-DZ quantiser might be shifted by $q/4$ or $q/2$.

In practice, rather than transmitting a quantised coefficient to the decoder, its ratio to the quantiser step size, called the quantisation index, $I$

$$I(u, v) = \left\lfloor \frac{F(u, v)}{q} \right\rfloor \tag{3.6}$$

is transmitted. (In eqn. 3.6 the symbol $\lfloor . \rfloor$ stands for rounding to the nearest integer.) The reason for defining the quantisation index is that it has a much smaller entropy than the quantised coefficient. At the decoder, the reconstructed coefficients, $F^q(u, v)$, after inverse quantisation are given by:

$$F^q(u, v) = \left\{ I(u, v) \pm \tfrac{1}{2} \right\} \times q \tag{3.7}$$

If required, depending on the polarity of the index, an addition or subtraction of half the quantisation step is required to deliver the centroid representation, reflecting the quantisation characteristics of Figure 3.5.

It is worth noting that, for the standard codecs, the quantiser step size $q$ is fixed at 8 for UTQ but varies from 2–62, in even step sizes, for the UTQ-DZ. Hence the entire quantiser range, or the quantiser parameter $Q_p$ (half the quantiser step size), can be defined with five bits (1–31).

Uniform quantisers with and without a dead zone can also be used in DPCM coding of pixels (section 3.1). Here, the threshold is set to zero, $th = 0$, and the quantisers are usually identified with even and odd numbers of levels, respectively.

One of the main problems of linear quantisers in DPCM is that for lower bit rates the number of quantisation levels is limited and hence the quantiser step size is large. In coding of plain areas of the picture, if a quantiser with an even number of levels is used, then the reconstructed pixels oscillate between $-q/2$ and $q/2$. This type of noise at these areas, in particular at low luminance levels, is visible and is called granular noise.

Larger quantiser step sizes with an odd number of levels (dead zone) reduce the granular noise, but cause loss of pixel resolution at the plain areas. This type of noise when the quantiser step size is relatively large is annoying and is called contouring noise.

To reduce granular and contouring noises, the quantiser step size should be reduced. This of course for a limited number of quantisation levels (low bit rate) reduces the outmost reconstruction level. In this case large pixel transitions such as sharp edges cannot be coded with good fidelity. It might take several cycles for the encoder to code one large sharp edge. Hence, edges appear smeared and this type of noise is known as slope overload noise.

In order to reduce the slope overload noise without increasing the granular or contouring noise, the quantiser step size can change adaptively. For example, a lower step size quantiser is used at the plain areas and a larger step size is employed at the edges and high texture areas. Note that the overhead of adaptation can be very costly (e.g. one bit per pixel).

*Principles of video compression* 35

The other method is to use a nonlinear quantiser with small step sizes at the inner levels and larger step sizes at the outer levels. This suits DPCM video better than the linear quantiser. Nonlinear quantisers reduce the entropy of the data more than linear quantisers. Hence data is less dependent on the variable length codes (VLC), increasing the robustness of the DPCM video to channel errors.

## 3.3 Temporal redundancy reduction

By using the differences between successive images, temporal redundancy is reduced. This is called interframe coding. For static parts of the image sequence, temporal differences will be close to zero, and hence are not coded. Those parts that change between the frames, either due to illumination variation or to motion of the objects, result in significant image error, which needs to be coded. Image changes due to motion can be significantly reduced if the motion of the object can be estimated, and the difference is taken on the motion compensated image.

Figure 3.6 shows the interframe error between successive frames of the Claire test image sequence and its motion compensated counterpart. It is clear that motion compensation can substantially reduce the interframe error.

### 3.3.1 Motion estimation

To carry out motion compensation, the motion of the moving objects has to be estimated first. This is called motion estimation. The commonly used motion estimation technique in all the standard video codecs is the block matching algorithm (BMA). In a typical BMA, a frame is divided into blocks of $M \times N$ pixels or, more usually, square blocks of $N^2$ pixels [3]. Then, for a maximum motion displacement of $w$ pixels per frame, the current block of pixels is matched against a corresponding block at the same coordinates but in the previous frame, within the square window

*a*  
*b*

*Figure 3.6*  *a*  Interframe  
*b*  motion compensated interframe

36  Standard codecs: image compression to advanced video coding

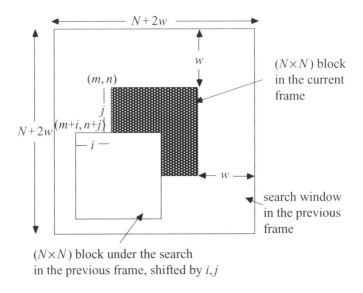

Figure 3.7  The current and previous frames in a search window

of width $N + 2w$ (Figure 3.7). The best match on the basis of a matching criterion yields the displacement.

Various measures such as the cross correlation function (CCF), mean-squared error (MSE) and mean absolute error (MAE) can be used in the matching criterion [4–6]. For the best match, in the CCF the correlation has to be maximised, whereas in the latter two measures the distortion must be minimised. In practical coders both MSE and MAE are used, since it is believed that CCF would not give good motion tracking, especially when the displacement is not large [6]. The matching functions of the type MSE and MAE are defined as, for MSE:

$$M(i, j) = \frac{1}{N^2} \sum_{m=1}^{N} \sum_{n=1}^{N} (f(m, n) - g(m + i, n + j))^2, \quad -w \leq i, j \leq w \quad (3.8)$$

and for MAE:

$$M(i, j) = \frac{1}{N^2} \sum_{m=1}^{N} \sum_{n=1}^{N} |f(m, n) - g(m + i, n + j)|, \quad -w \leq i, j \leq w$$

where $f(m, n)$ represents the current block of $N^2$ pixels at coordinates $(m, n)$ and $g(m + i, n + j)$ represents the corresponding block in the previous frame at new coordinates $(m + i, n + j)$. At the best matched position of $i = a$ and $j = b$, the motion vector $MV(a, b)$, represents the displacement of all the pixels within the block.

*Table 3.1    Percentage of processing time required to carry out motion estimation in an MPEG-1 encoder*

| Category | Fast DCT | | Brute force DCT | |
|---|---|---|---|---|
| Picture type | Mobile | Claire | Mobile | Claire |
| P-frame ME | 66.1 % | 68.4 % | 53.3 % | 56.1 % |
| B-frame ME | 58.2 % | 60.9 % | 46.2 % | 48.7 % |

To locate the best match by full search, $(2w+1)^2$ evaluations of the matching criterion are required. To reduce processing cost, MAE is preferred to MSE and hence is used in all the video codecs. However, for each block of $N^2$ pixels we still need to carry out $(2w+1)^2$ tests, each with almost $2N^2$ additions and subtractions. This is still far from being suitable for the implementation of BMA in software-based codecs. Measurements of the video encoders' complexity show that motion estimations comprise almost 50–70 per cent of the overall encoder's complexity [7]. This of course depends on the motion activity in the scene and whether a fast DCT is used in deriving the transform coefficients. For example, the percentage of the processing time required to calculate the motion vectors of the Mobile and Claire test image sequences in an MPEG-1 software-based encoder is given in Table 3.1. Note that although more processing time is required for motion estimation in B-pictures than P-pictures, since the search ranges in P-pictures are much larger than for B-pictures, the overall processing time for motion estimation in P-pictures can be larger than that of the B-pictures, as shown in the Table. The reason for these will be dealt with in Chapter 7, when we talk about different picture types in the MPEG-1 encoder. In any case, as motion estimation is a costly process, fast motion estimation techniques are highly desirable.

### 3.3.2    Fast motion estimation

In the past two decades a number of fast search methods for motion estimation have been introduced to reduce the computational complexity of BMA. The basic principle of these methods is that the number of search points can be reduced, by selectively checking only a small number of specific points, assuming that the distortion measure monotonically decreases towards the best matched point. Jain and Jain [6] were the first to use a two-dimensional logarithmic (TDL) search method to track the direction of a minimum mean-squared error distortion measure. In their method, the distortion for the five initial positions, one at the centre of the coordinate and four at coordinates $(\pm w/2, \pm w/2)$ of the search window, are computed first. In the next step, three more positions with the same step size in the direction of the previous minimum position are searched. The step size is then halved and the above procedure is continued until the step size becomes unity. Finally, all the nine positions are searched.

With this method, for $w = 5$ pixels/frame, 21 positions are searched as opposed to 121 positions required in the full search method.

Koga et al. [8] use a three-step search (TSS) method to compute motion displacements up to six pixels/frame. In their method all eight positions surrounding the coordinate with a step size of $w/2$ are searched first. At each minimum position the search step size is halved and the next eight new positions are searched. This method, for $w = 6$ pixels/frame, searches 25 positions to locate the best match. The technique is the recommended method for the test of software-based H.261 [9] for videophone applications.

In Kappagantula and Rao's [4] modified motion estimation algorithm (MMEA), prior to halving the step sizes, two more positions are also searched. With this method for $w = 7$ pixels/frame, only 19 MAE computations are required. In Srinivasan and Rao's [10] conjugate direction search (CDS) method, at every iteration of the direction search two conjugate directions with a step size of one pixel, centred at the minimum position, are searched. Thus, for $w = 5$ pixels/frame, there will be only 13 searches at most.

Another method of fast BMA is the cross search algorithm (CSA) [11]. In this method, the basic idea is still a logarithmic step search, which has also been exploited in [4,6,8], but with some differences, which lead to fewer computational search points. The main difference is that at each iteration there are four search locations which are the end points of a cross ($\times$) rather than ($+$). Also, at the final stage, the search points can be either the end points of ($\times$) or ($+$) crosses, as shown in Figure 3.8. For a maximum motion displacement of $w$ pixels/frame, the total number of computations becomes $5 + 4\log_2 w$.

Puri et al. [12] have introduced the orthogonal search algorithm (OSA) in which, with a logarithmic step size, at each iteration four new locations are searched. This is the fastest method of all known fast MBAs. In this method, at every step, two

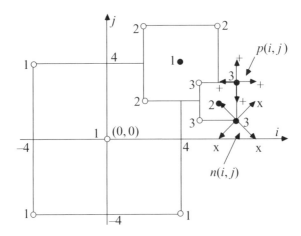

*Figure 3.8* An example of the CSA search for $w = 8$ pixels/frame

Table 3.2  Computational complexity

| Algorithm | Maximum number of search points | $w$ | | |
|---|---|---|---|---|
| | | 4 | 8 | 16 |
| FSM | $(2w+1)^2$ | 81 | 289 | 1089 |
| TDL | $2 + 7\log_2 w$ | 16 | 23 | 30 |
| TSS | $1 + 8\log_2 w$ | 17 | 25 | 33 |
| MMEA | $1 + 6\log_2 w$ | 13 | 19 | 25 |
| CDS | $3 + 2w$ | 11 | 19 | 35 |
| OSA | $1 + 4\log_2 w$ | 9 | 13 | 17 |
| CSA | $5 + 4\log_2 w$ | 13 | 17 | 21 |

Table 3.3  Compensation efficiency

| Algorithm | Split screen | | Trevor white | |
|---|---|---|---|---|
| | entropy (bits/pel) | standard deviation | entropy (bits/pel) | standard deviation |
| FSM | 4.57 | 7.39 | 4.41 | 6.07 |
| TDL | 4.74 | 8.23 | 4.60 | 6.92 |
| TSS | 4.74 | 8.19 | 4.58 | 6.86 |
| MMEA | 4.81 | 8.56 | 4.69 | 7.46 |
| CDS | 4.84 | 8.86 | 4.74 | 7.54 |
| OSA | 4.85 | 8.81 | 4.72 | 7.51 |
| CSA | 4.82 | 8.65 | 4.68 | 7.42 |

positions are searched alternately in the vertical and horizontal directions. The total number of test points is $1 + 4\log_2 w$.

Table 3.2 shows the computational complexity of various fast search methods, for a range of motion speeds from 4 to 16 pixels/frame. The motion compensation efficiency of these algorithms for a motion speed of $w = 8$ pixels/frame for two test image sequences is tabulated in Table 3.3.

It can be seen that, although fast search methods reduce the computational complexity of the full search method (FSM) significantly, their motion estimation accuracy (compensation efficiency) has not been degraded noticeably.

### 3.3.3  Hierarchical motion estimation

The assumption of monotonic variation of image intensity employed in the fast BMAs often causes false estimations, especially for larger picture displacements. These

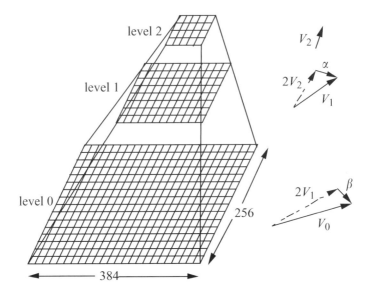

*Figure 3.9   A three-level image pyramid*

methods perform well for slow moving objects, such as those in video conferencing. However, for higher motion speeds, due to the intrinsic selective nature of these methods, they often converge to a local minimum of distortion.

One method of alleviating this problem is to subsample the image to smaller sizes, such that the motion speed is reduced by the sampling ratio. The process is done on a multilevel image pyramid, known as the hierarchical block matching algorithm (HBMA) [13]. In this technique, pyramids of the image frames are reconstructed by successive two-dimensional filtering and subsampling of the current and past image frames. Figure 3.9 shows a three-level pyramid, where for simplicity each level of the upper level of the pyramid is taken as the average of four adjacent pixels of one level below. Effectively this is a form of lowpass filtering.

Conventional block matching, either full search or any fast method, is first applied to the highest level of the pyramid (level 2 in Figure 3.9). This motion vector is then doubled in size, and further refinement within one pixel search is carried out in the following level. The process is repeated to the lowest level. Therefore, with an $n$-level pyramid the maximum motion speed of $w$ at the highest level is reduced to $w/2^{n-1}$.

For example, a maximum motion speed of 32 pixels/frame with a three-level pyramid is reduced to eight pixels/frame, which is quite manageable by any fast search method. Note also that this method can be regarded as another type of fast search with a performance very close to the full search irrespective of the motion speed, but the computational complexity can be very close to the fast logarithmic methods.

As an example, for a maximum motion speed of 32 pixels/frame, which is very common in high definition video or most TV sports programmes (particularly in the P-pictures of the standard codecs, which can be several frames apart from each other) although the nonhierarchical full search BMA requires $(2 \times 32 + 1)^2 = 4225$ operations, a four-level hierarchy, where the motion speed at the top level is $32/2^{4-1} = 4$ pixels/frame, only requires $(2 \times 4 + 1)^2 + 3 \times 9 = 108$ operations. Here with the full search method 81 operations are carried out at the top level, and at each lower level nine new positions are searched.

## 3.4 Variable length coding

For further bit rate reduction, the transform coefficients and the coordinates of the motion vectors are variable length coded (VLC). In VLC, short code words are assigned to the highly probable values and long code words to the less probable ones. The lengths of the codes should vary inversely with the probability of occurrences of the various symbols in VLC. The bit rate required to code these symbols is the inverse of the logarithm of probability, $p$, at base 2 (bits), i.e. $\log_2 p$. Hence, the entropy of the symbols which is the minimum average bits required to code the symbols can be calculated as:

$$H(x) = -\sum_{i=1}^{n} p_i \log_2 p_i \qquad (3.9)$$

There are two types of VLC, which are employed in the standard video codecs. They are Huffman coding and arithmetic coding. It is noted that Huffman coding is a simple VLC code, but its compression can never reach as low as the entropy due to the constraint that the assigned symbols must have an integral number of bits. However, arithmetic coding can approach the entropy since the symbols are not coded individually [14]. Huffman coding is employed in all standard codecs to encode the quantised DCT coefficients as well as motion vectors. Arithmetic coding is used, for example, in JPEG, JPEG2000, H.263 and shape and still image coding of MPEG-4 [15–17], where extra compression is demanded.

### 3.4.1 Huffman coding

Huffman coding is the most commonly known variable length coding method based on probability statistics. Huffman coding assigns an output code to each symbol with the output codes being as short as 1 bit, or considerably longer than the input symbols, depending on their probability. The optimal number of bits to be used for each symbol is $-\log_2 p$, where $p$ is the probability of a given symbol.

However, since the assigned code words have to consist of an integral number of bits, this makes Huffman coding suboptimum. For example, if the probability of a symbol is 0.33, the optimum number of bits to code that symbol is around 1.6 bits, but the Huffman coding scheme has to assign either one or two bits to the code. In either case, on average it will lead to more bits compared to its entropy. As the probability of a symbol becomes very high, Huffman coding becomes very nonoptimal. For example, for a symbol with a probability of 0.9, the optimal code size should be 0.15 bits, but Huffman coding assigns a minimum value of one bit code to the symbol, which is six times larger than necessary. Hence, it can be seen that resources are wasted.

To generate the Huffman code for symbols with a known probability of occurrence, the following steps are carried out:

- rank all the symbols in the order of their probability of occurrence
- successively merge every two symbols with the least probability to form a new composite symbol, and rerank order them; this will generate a tree, where each node is the probability of all nodes beneath it
- trace a path to each leaf, noting the direction at each node.

Figure 3.10 shows an example of Huffman coding of seven symbols, A–G. Their probabilities in descending order are shown in the third column. In the next column the two smallest probabilities are added and the combined probability is included in the new order. The procedure continues to the last column, where a single probability of 1 is reached. Starting from the last column, for every branch of probability a 0 is assigned on the top and a 1 in the bottom, shown in bold digits in the Figure. The corresponding codeword (shown in the first column) is read off by following the sequence from right to left. Although with fixed word length each sample is represented by three bits, they are represented in VLC from two to four bits.

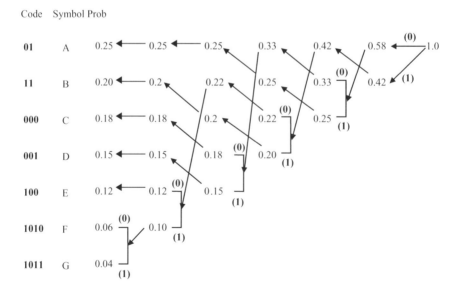

Figure 3.10   An example of Huffman code for seven symbols

The average bit per symbol is then:

$$0.25 \times 2 + 0.20 \times 2 + 0.18 \times 3 + 0.15 \times 3 + 0.12 \times 3 + 0.06 \times 4 + 0.04 \times 4 = 2.65 \text{ bits}$$

which is very close to the entropy given by:

$$-\begin{pmatrix} 0.25 \log_2 0.25 + 0.2 \log_2 0.2 + 0.18 \log_2 0.18 + 0.15 \log_2 0.15 \\ + 0.12 \log_2 0.12 + 0.06 \log_2 0.06 + 0.04 \log_2 0.04 \end{pmatrix}$$
$$= 2.62 \text{ bits}$$

It should be noted that for a large number of symbols, such as the values of DCT coefficients, such a method can lead to a long string of bits for the very rarely occurring values, and is impractical. In such cases normally a group of symbols is represented by their aggregate probabilities and the combined probabilities are Huffman coded, the so-called modified Huffman code. This method is used in JPEG. Another method, which is used in H.261 and MPEG, is two-dimensional Huffman, or three-dimensional Huffman in H.263 [9,16].

### 3.4.2 Arithmetic coding

Huffman coding can be optimum if the symbol probability is an integer power of $\frac{1}{2}$, which is usually not the case. Arithmetic coding is a data compression technique that encodes data by creating a code string, which represents a fractional value on the number line between 0 and 1 [14]. It encourages clear separation between the model for representing data and the encoding of information with respect to that model. Another advantage of arithmetic coding is that it dispenses with the restriction that each symbol must translate into an integral number of bits, thereby coding more efficiently. It actually achieves the theoretical entropy bound to compression efficiency for any source. In other words, arithmetic coding is a practical way of implementing entropy coding.

There are two types of modelling used in arithmetic coding: the fixed model and the adaptive model. Modelling is a way of calculating, in any given context, the distribution of probabilities for the next symbol to be coded. It must be possible for the decoder to produce exactly the same probability distribution in the same context. Note that probabilities in the model are represented as integer frequency counts. Unlike the Huffman-type, arithmetic coding accommodates adaptive models easily and is computationally efficient. The reason why data compression requires adaptive coding is that the input data source may change during encoding, due to motion and texture.

In the fixed model, both encoder and decoder know the assigned probability to each symbol. These probabilities can be determined by measuring frequencies in representative samples to be coded and the symbol frequencies remain fixed. Fixed models are effective when the characteristics of the data source are close to the model and have little fluctuation.

In the adaptive model, the assigned probabilities may change as each symbol is coded, based on the symbol frequencies seen so far. Each symbol is treated as an individual unit and hence there is no need for a representative sample of text. Initially, all the counts might be the same, but they update, as each symbol is seen, to approximate the observed frequencies. The model updates the inherent distribution so the prediction of the next symbol should be close to the real distribution mean, making the path from the symbol to the root shorter.

Due to the important role of arithmetic coding in the advanced video coding techniques, in the following sections a more detailed description of this coding technique is given.

### 3.4.2.1 Principles of arithmetic coding

The fundamental idea of arithmetic coding is to use a scale in which the coding intervals of real numbers between 0 and 1 are represented. This is in fact the cumulative probability density function of all the symbols which add up to 1. The interval needed to represent the message becomes smaller as the message becomes longer, and the number of bits needed to specify that interval is increased. According to the symbol probabilities generated by the model, the size of the interval is reduced by successive symbols of the message. The more likely symbols reduce the range less than the less likely ones and hence they contribute fewer bits to the message.

To explain how arithmetic coding works, a fixed model arithmetic code is used in the example for easy illustration. Suppose the alphabet is {**a**, **e**, **i**, **o**, **u**, **!**} and the fixed model is used with the probabilities shown in Table 3.4.

Once the symbol probability is known, each individual symbol needs to be assigned a portion of the [0, 1) range that corresponds to its probability of appearance in the cumulative density function. Note also that the character with a range of [lower, upper) owns everything from lower up to, but not including, the upper value. So, the alphabet **u** with probability 0.1, defined in the cumulative range of [0.8, 0.9), can take any value from 0.8 to 0.8999 . . . .

The most significant portion of an arithmetic coded message is the first symbol to be encoded. Using an example that a message **eaii!** is to be coded, the first symbol to be coded is **e**. Hence, the final coded message has to be a number greater than or equal to 0.2 and less than 0.5. After the first character is encoded, we know that the

Table 3.4  Example: fixed model for alphabet {*a, e, i, o, u, !*}

| Symbol | Probability | Range |
| --- | --- | --- |
| a | 0.2 | [0.0, 0.2) |
| e | 0.3 | [0.2, 0.5) |
| i | 0.1 | [0.5, 0.6) |
| o | 0.2 | [0.6, 0.8) |
| u | 0.1 | [0.8, 0.9) |
| ! | 0.1 | [0.9, 1.0) |

lower number and the upper number now bound our range for the output. Each new symbol to be encoded will further restrict the possible range of the output number during the rest of the encoding process.

The next character to be encoded, **a**, is in the range of 0–0.2 in the new interval. It is not the first number to be encoded, so it belongs to the range corresponding to 0–0.2, but in the new subrange of [0.2, 0.5). This means that the number is now restricted to the range of [0.2, 0.26), since the previous range was 0.3 (0.5 − 0.2 = 0.3) units long and one-fifth of that is 0.06. The next symbol to be encoded, **i**, is in the range of [0.5, 0.6), that corresponds to 0.5–0.6 in the new subrange of [0.2, 0.26) and gives the smaller range [0.23, 0.236). Applying this rule for coding of successive characters, Table 3.5 shows the successive build up of the range of the message coded so far.

Figure 3.11 shows another representation of the encoding process. The range is expanded to fill the whole range at every stage and marked with a scale that gives the end points as a number. The final range, [0.23354, 0.2336) represents the message

*Table 3.5 Representation of arithmetic coding process*

|  | New character | Range |
|---|---|---|
| Initially: |  | [0, 1) |
| After seeing a symbol: | e | [0.2, 0.5) |
|  | a | [0.2, 0.26) |
|  | i | [0.23, 0.236) |
|  | i | [0.233, 0.2336) |
|  | ! | [0.23354, 0.2336) |

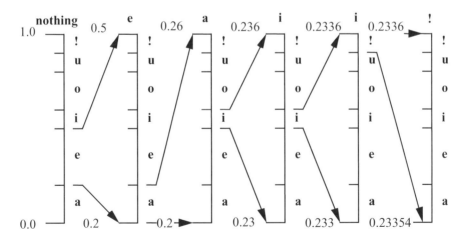

*Figure 3.11 Representation of arithmetic coding process with the interval scaled up at each stage for the message **eaii!***

Table 3.6 Representation of decoding process of arithmetic coding

| Encoded number | Output symbol | Range |
| --- | --- | --- |
| 0.23355 | e | [0.2, 0.5) |
| 0.11185 | a | [0.0, 0.2) |
| 0.55925 | i | [0.5, 0.6) |
| 0.59250 | i | [0.5, 0.6) |
| 0.92500 | ! | [0.9, 1.0) |

**eaii!**. This means that if we transmit any number in the range of $0.23354 \leq x < 0.2336$, that number represents the whole message of **eaii!**.

Given this encoding scheme, it is relatively easy to see how during the decoding the individual elements of the **eaii!** message are decoded. To verify this, suppose a number $x = 0.23355$ in the range of $0.23354 \leq x < 0.2336$ is transmitted. The decoder, using the same probability intervals as the encoder, performs a similar procedure. Starting with the initial interval [0, 1), only the interval [0.2, 0.5) of **e** envelops the transmitted code of 0.23355. So the first symbol can only be **e**. Similar to the encoding process, the symbol intervals are then defined in the new interval [0.2, 0.5). This is equivalent to defining the code within the initial range [0, 1), but offsetting the code by the lower value and then scaling up within its original range. That is, the new code will be $(0.23355 - 0.2)/(0.5 - 0.2) = 0.11185$, which is enveloped by the interval [0.0, 0.2) of symbol **a**. Thus the second decoded symbol is **a**. To find the third symbol, a new code within the range of **a** should be found, i.e. $(0.11185 - 0.0)/(0.2 - 0.0) = 0.55925$. This code is enveloped by the range of [0.5, 0.6) of symbol **i** and the resulting new code after decoding the third symbol will be $(0.55925 - 0.5)/(0.6 - 0.5) = 0.5925$, which is again enveloped by [0.5, 0.6). Hence the fourth symbol will be **i**. Repeating this procedure will yield a new code of $(0.5925 - 0.5)/(0.6 - 0.5) = 0.925$. This code is enveloped by [0.9, 1), which decodes symbol **!**, the end of decoding symbol, and the decoding process is terminated. Table 3.6 shows the whole decoding process of the message **eaii!**.

In general, the decoding process can be formulated as:

$$R_{n+1} = \frac{R_n - L_n}{U_n - L_n} \tag{3.10}$$

where $R_n$ is a code within the range of lower value $L_n$ and upper value $U_n$ of the $n$th symbol and $R_{n+1}$ is the code for the next symbol.

#### 3.4.2.2 Binary arithmetic coding

In the preceding section we saw that, as the number of symbols in the message increases, the range of the message becomes smaller. If we continue coding more symbols then the final range may even become smaller than the precision of any computer to define such a range. To resolve this problem we can work with binary arithmetic coding.

In Figure 3.11 we saw that after each stage of coding, if we expand the range to its full range of [0, 1), the apparent range is increased. However, the values of the lower and upper numbers are still small. Therefore, if the initial range of [0, 1) is replaced by a larger range of [0, *MAX_VAL*), where *MAX_VAL* is the largest integer number that a computer can handle, then the precision problem is resolved. If we use 16-bit integer numbers, then $MAX\_VAL = 2^{16} - 1$. Hence, rather than defining the cumulative probability in the range of [0,1), we define their cumulative frequencies scaled up within the range of $[0, 2^{16} - 1)$.

At the start, the coding interval [lower, upper) is initialised to the whole scale [0, *MAX_VAL*). The model's frequencies, representing the probability of each symbol in this range, are also set to their initial values in the range. To encode a new symbol element $e_k$ assuming that symbols $e_1 \ldots e_{k-1}$ have already being coded, we project the model scale to the interval resulting from the sequence of events. The new interval [*lower'*, *upper'*) for the sequence $e_1 \ldots e_k$ is calculated from the old interval [*lower*, *upper*) of the sequence $e_1 \ldots e_{k-1}$ as follows:

$$lower' = lower + width * low/maxfreq$$
$$width' = width * symb\_width/maxfreq$$
$$upper' = lower' + width'$$
$$= lower + width * (low + symb\_width)/maxfreq$$
$$= lower + width * up/maxfreq$$

with:

$$width = upper - lower \text{ (old interval)}$$
$$width' = upper' - lower' \text{ (new interval)}$$
$$symb\_width = up - low \text{ (model's frequency)}$$

At this stage of the coding process, we do not need to keep the previous interval [*lower*, *upper*) in the memory, so we allocate the new values [*lower'*, *upper'*). We then compare the new interval with the scale [0, *MAX_VAL*) to determine whether there is any bit for transmission down the channel. These bits are due to the redundancy in the binary representation of lower and upper values. For example, if values of both lower and upper are less than half the [0, *MAX_VAL*) range, then their most significant number in binary form is 0. Similarly, if both belong to the upper half range, their most significant number is 1. Hence we can make a general rule:

- if lower and upper levels belong to the first half of the scale, the most significant bit for both will be 0
- if lower and upper belong to the second half of the scale, their most significant bit will be 1

- otherwise, when lower and upper belong to the different halves of the scale their most significant bits will be different (0 and 1).

Thus for the cases where lower and upper values have the same most significant bit, we send this bit down the channel and calculate the new interval as follows:

- sending a 0 corresponds to removal of the second half of the scale and keeping its first half only; the new scale is expanded by a factor 2 to obtain its representation in the whole scale of [0, *MAX_VAL*) again, as shown in Figure 3.12
- sending a 1 corresponds to the shift of the second half of the scale to its first half; that is subtracting half of the scale value and multiplying the result by a factor 2 to obtain its representation in the whole scale again, as shown in Figure 3.13.

If the interval always remains in either half of the scale after a bit has been sent, the operation is repeated as many times as necessary to obtain an interval occupying both halves of the scale. The complete procedure is called the interval testing loop.

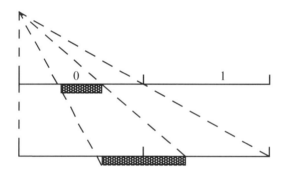

if lower and upper belong to the first half of the scale, their higher bits will both be 0

*Figure 3.12   Both lower and upper values in the first half*

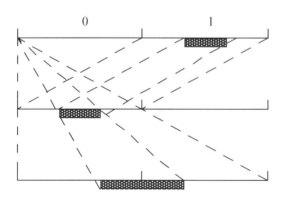

if lower and upper belong to the second half of the scale, their higher bits will both be 1

*Figure 3.13   Both lower and upper values in the second half*

*Principles of video compression* 49

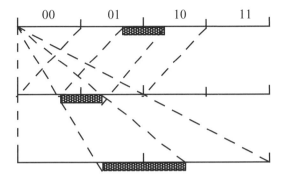

if lower and upper belong to the second & third quarter of the scale, an unknown bit is sent to a waiting buffer for later transmission, and the interval shifted by a quarter

*Figure 3.14   Lower and upper levels in the second and third quarter, respectively*

Now we go back to the case where both of the lower and upper values are not in either of the half intervals. Here we identify two cases: first the lower value is in the second quarter of the scale and the upper value is in the third quarter. Hence the range of frequency is less than 0.5. In this case, in the binary representation, the two most significant bits are different, 01 for the lower and 10 for the upper, but we can deduce some information from them. That is, in both cases, the second bit is the complementary bit to the first. Hence, if we can find out the second bit, the previous bit is its complement. Therefore, if the second bit is 1, then the previous value should be 0, and we are in the second quarter, meaning the lower value. Similar conclusions can be drawn for the upper value.

Thus we need to divide the interval within the scale of [0, *MAX_VAL*) into four segments instead of two. Then the scaling and shifting for this case will be done only on the portion of the scale containing second and third quarters, as shown in Figure 3.14.

Thus the general rule for this case is that:

- if the interval belongs to the second and third quarters of the scale: an unknown bit **?** is sent to a waiting buffer for later definition and the interval transformed as shown in Figure 3.14
- if the interval belongs to more than two different quarters of the scale (the interval occupies parts of three or four different quarters), there is no need to transform the coding interval; the process exits from this loop and goes to the next step in the execution of the program; this means that if the probability in that interval is greater than 0.5, no bits are transmitted; this is the most important part of arithmetic coding which leads to a low bit rate.

A flow chart of the interval testing loop is shown in Figure 3.15 and a detailed example of coding a message is given in Figure 3.16. For every symbol to be coded we invoke the flow chart of Figure 3.15, except where the symbol is the end of file symbol, where the program stops. After each testing loop, the model can be updated (frequencies modified if the adaptive option has been chosen).

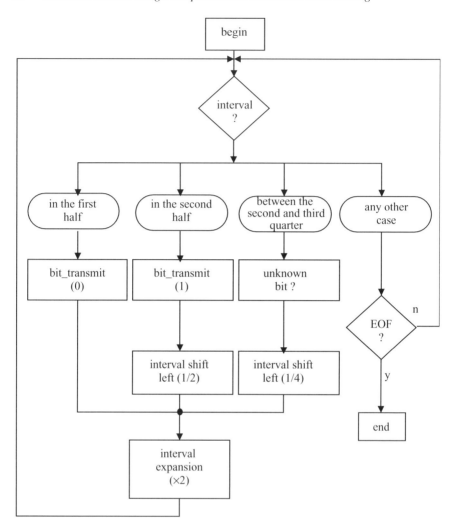

Figure 3.15  A flow chart for binary arithmetic coding

A C program of binary arithmetic coding is given in Figure 3.16.

The values of low and high are initialised to 0 and top, respectively. PSC_FIFO is first-in-first-out (FIFO) for buffering the output bits from the arithmetic encoder. The model is specified through cumul_freq [ ], and the symbol is specified using its index in the model.

### 3.4.2.3  An example of binary arithmetic coding

Because of the complexity of the arithmetic algorithm, a simple example is given below. To simplify the operations further, we use a fixed model with four symbols **a**,

```
#define q1 16384
#define q2 32768
#define q3 49152
#define top 65535

static long low, high, opposite_bits, length;
void encode_a_symbol (int index, int cumul_freq[])
{
        length = high - low +1;
        high = low -1 + (length * cumul_freq[index]) / cumul_freq[0];
        low += (length * cumul_freq[index+1])/ cumul_freq[0];
        for( ; ; ){
          if(high <q2){
            send out a bit "0" to PSC_FIFO;
            while (opposite_bits > 0) {
              send out a bit "1" to PSC_FIFO;
              opposite_bits--;
            }
          }
          else if (low >= q2) {
            send out a bit "1" to PSC_FIFO;
            while (opposite_bits > 0) {
            send out a bit "0" to PSC_FIFO;
            opposite_bits--;
          }
          low -= q2;
                high -= q2;
    }
    else if (low >= q1 && high < q3) {
      opposite_bits += 1;
      low -= q1;
      high -= q1;
    }
    else break;
    low *= 2;
    high = 2 * high + 1;
    }
}
```

*Figure 3.16   A C program of binary arithmetic coding*

**b**, **c** and **d**, where their fixed probabilities and cumulative probabilities are given in Table 3.7, and the message sequence of events to be encoded is **bbacd**.

Considering what we saw earlier, we start by initialising the whole range to [0, 1). In sending **b**, which has a range of [0.3, 0.8), since lower $= 0.3$ is in the second quarter and upper $= 0.8$ in the fourth quarter (occupying more than two quarters) nothing is sent, and no scaling is required, as shown in Table 3.8.

To send the next **b**, i.e. **bb**, since the previous width $= 0.8 - 0.3 = 0.5$, the new interval becomes $[0.3 + 0.3 \times 0.5, 0.3 + 0.8 \times 0.5) = [0.45, 0.7)$. This time lower $= 0.45$ is in the second quarter but upper $= 0.7$ is in the third quarter, so the unknown bit **?** is sent to the buffer, such that its value is to be determined later.

Note that since the range [0.45, 0.7) covers the second and third quarters, according to Figure 3.14, we have to shift both of them by a quarter (0.25) and then magnify by

Table 3.7  Probability and cumulative probability of four symbols as an example

| Symbol | pdf | cdf |
|---|---|---|
| a | 0.3 | [0.0, 0.3) |
| b | 0.5 | [0.3, 0.8) |
| c | 0.1 | [0.8, 0.9) |
| d | 0.1 | [0.9, 1.0) |

Table 3.8  Generated binary bits of coding **bbacd** message

| Encoding | Coding interval | Width | Bit |
|---|---|---|---|
| initialisation: | [0, 1) | 1.000 | |
| after b: | [0.3, 0.8) | 0.500 | |
| after bb: | [0.45, 0.7) | 0.250 | |
| after interval test | [0.4, 0.9) | 0.500 | ? |
| after bba: | [0.40, 0.55) | 0.150 | |
| after interval test | [0.3, 0.6) | 0.300 | ? |
| after interval test | [0.1, 0.7) | 0.600 | ? |
| after bbac: | [0.58, 0.64) | 0.060 | |
| after interval test | [0.16, 0.28) | 0.120 | 1000 |
| after interval test | [0.32, 0.56) | 0.240 | 0 |
| after interval test | [0.14, 0.62) | 0.480 | ? |
| after bbacd: | [0.572, 0.620) | 0.048 | |
| after interval test | [0.144, 0.240) | 0.096 | 10 |
| after interval test | [0.288, 0.480) | 0.192 | 0 |
| after interval test | [0.576, 0.960) | 0.384 | 0 |
| after interval test | [0.152, 0.920) | 0.768 | 1 |

a factor of 2, i.e. the new interval is $[(0.45 - 0.25) \times 2, (0.7 - 0.25) \times 2] = [0.4, 0.9)$ which has a width of 0.5. To code the next symbol **a**, the range becomes $[0.4 + 0 \times 0.5, 0.4 + 0.3 \times 0.5] = [0.4, 0.55)$. Again since lower and upper are at the second and third quarters, the unknown **?** is stored. According to Figure 3.14 both quarters are shifted by 0.25 and magnified by 2, $[(0.4 - 0.25) \times 2, (0.55 - 0.25) \times 2] = [0.3, 0.6)$.

Again [0.3, 0.6) is in the second and third interval so another **?** is stored. If we shift and magnify again $[(0.3 - 0.25) \times 2, (0.6 - 0.25) \times 2] = [0.1, 0.7)$, which now lies in the first and third quarters, so nothing is sent if there is no scaling. Now if we code **c**, i.e. **bbac**, the interval becomes $[0.1 + 0.8 \times 0.6, 0.1 + 0.9 \times 0.6] = [0.58, 0.64)$.

Now since [0.58, 0.64) is in the second half we send 1. We now go back and convert all **?** to 000 complementary to 1. Thus we have sent 1000 so far. Note that bits belonging to **?** are transmitted after finding a 1 or 0. Similarly, the subsequent symbols are coded and the final generated bit sequence becomes 1000010001. Table 3.8 shows this example in a tabular representation. A graphical representation is given in Figure 3.17.

#### 3.4.2.4 Adaptive arithmetic coding

In adaptive arithmetic coding, the assigned probability to the symbols changes as each symbol is coded [18]. For binary (integer) coding, this is accomplished by assigning a frequency of 1 to each symbol at the start of the coding (initialisation). As a symbol is coded, the frequency of that symbol is incremented by one. Hence the frequencies of symbols are adapted to their number of appearances so far. The decoder follows a similar procedure. At every stage of coding, a test is done to see if the cumulative frequency (sum of the frequencies of all symbols) exceeds *MAX_VAL*. If this is the case, all frequencies are halved (minimum 1), and encoding continues. For a better adaptation, the frequencies of the symbols may be initialised to a predefined distribution that matches the overall statistics of the symbols better. For better results, the frequencies are updated from only the $N$ most recently coded symbols. It has been shown that this method of adaptation with a limited past history can reduce the bit rate by more than 30 per cent below the first-order entropy of the symbols [19]. The reason for this is that, if some rare events that normally have high entropy could occur in clusters then, within only the $N$ most recent events, they now become the more frequent events, and hence require lower bit rates.

#### 3.4.2.5 Context-based arithmetic coding

A popular method for adaptive arithmetic coding is to adapt the assigned probability to a symbol, according to the context of its neighbours. This is called context-based arithmetic coding and forms an essential element of some image/video coding standards, such as JPEG2000 and MPEG-4. It is more efficient when it is applied to binary data, like the bit plane or the sign bits in JPEG2000 or binary shapes in MPEG-4. We explain this method with a simple example. Assume that binary symbols of $a$, $b$ and $c$, which may take values of 0 or 1, are the three immediate neighbours of a binary symbol $x$, as shown in Figure 3.18.

Due to high correlation between the symbols in the image data, if the neighbouring symbols of $a$, $b$ and $c$ are mainly 1 then it is logical to assign a high probability for coding symbol $x$, when its value is 1. Conversely, if the neighbouring symbols are mainly 0 the assigned probability of $x = 1$ should be reduced. Thus we can define the context for coding a 1 symbol as:

$$context = 2^2 c + 2^1 b + 2^0 a = 4c + 2b + a \tag{3.11}$$

For the binary values of $a$, $b$ and $c$, the context has a value between 0 and 7. Higher values of the context indicate that a higher probability should be assigned for coding of 1, and a complementary probability, when the value of $x$ is 0.

54  *Standard codecs: image compression to advanced video coding*

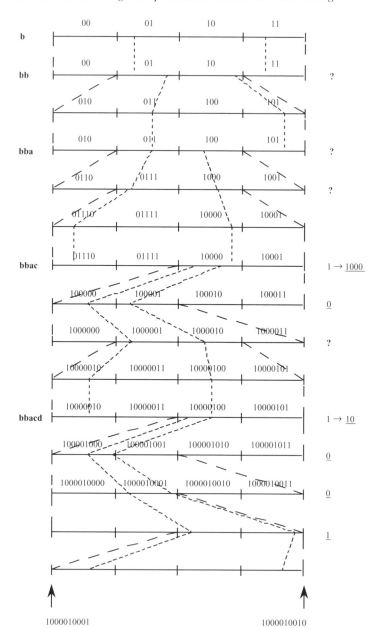

*Figure 3.17  Derivation of the binary bits for the given example*

*Figure 3.18    Three immediate neighbouring symbols to x*

## 3.5  A generic interframe video codec

Figure 3.19 shows a generic interframe encoder which is used in all the standard video codecs, such as H.261, H.263, MPEG-1, MPEG-2 and MPEG-4 [20,16,21,22,17]. In the following sections each element of this codec is described in a general sense. The specific aspects of these codecs will be addressed in more detail in the relevant chapters.

### 3.5.1  *Interframe loop*

In interframe predictive coding, the difference between pixels in the current frame and their prediction values from the previous frame is coded and transmitted. At the receiver, after decoding the error signal of each pixel, it is added to a similar prediction value to reconstruct the picture. The better the predictor, the smaller the error signal, and hence the transmission bit rate. If the scene is still, a good prediction for the current pixel is the same pixel in the previous frame. However, when there is motion, assuming that movement in the picture is only a shift of object position, then a pixel in the previous frame, displaced by a motion vector, is used.

### 3.5.2  *Motion estimator*

Assigning a motion vector to each pixel is very costly. Instead, a group of pixels are motion compensated, such that the motion vector overhead per pixel can be very small. In standard codecs a block of $16 \times 16$ pixels, known as a macroblock (MB) (to be differentiated from $8 \times 8$ DCT blocks), is motion estimated and compensated. It should be noted that motion estimation is only carried out on the luminance parts of the pictures. A scaled version of the same motion vector is used for compensation of chrominance blocks, depending on the picture format.

### 3.5.3  *Inter/intra switch*

Every MB is either interframe or intraframe coded, called inter/intra MBs. The decision on the type of MB depends on the coding technique, which will be explained in greater detail in the relevant chapters. For example, in JPEG, all MBs are intraframe coded, as JPEG is mainly used for coding of still pictures.

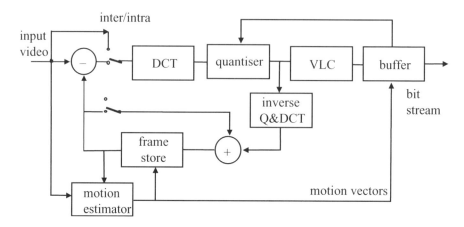

Figure 3.19  *A generic interframe predictive coder*

### 3.5.4  DCT

Every MB is divided into 8 × 8 luminance and chrominance pixel blocks. Each block is then transformed *via* the DCT. There are four luminance blocks in each MB, but the number of chrominance blocks depends on the colour resolutions (image format).

### 3.5.5  Quantiser

As mentioned in section 3.2, there are two types of quantiser. One with a dead zone for the AC coefficients and the DC coefficient of inter MB, the other without the dead zone is used for the DC coefficient of intra MB. The range of quantised coefficients can be from −2047 to +2047. With a dead zone quantiser, if the modulus (absolute value) of a coefficient is less than the quantiser step size $q$ it is set to zero, otherwise it is quantised according to eqn 3.6, to generate quantiser indices.

### 3.5.6  Variable length coding

The quantiser indices are variable length coded, according to the type of VLC used. Motion vectors, as well as the address of coded macroblocks, are also VLC coded.

### 3.5.7  IQ and IDCT

To generate a prediction for interframe coding, the quantised DCT coefficients are first inverse quantised and inverse DCT coded. These are added to their previous picture values (after a frame delay by the frame store), to generate a replica of the decoded picture. The picture is then used as a prediction for coding of the next picture in the sequence.

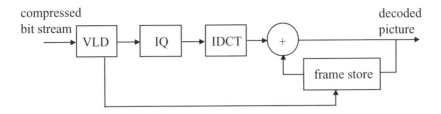

*Figure 3.20    Block diagram of a decoder*

## 3.5.8   Buffer

The bit rate generated by an interframe coder is variable. This is because the bit rate is primarily a function of picture activity (motion of objects and their details). Therefore, to transmit coded video into fixed rate channels (e.g. 2 Mbit/s links), the bit rate has to be regulated. Storing the coded data in a buffer and then emptying the buffer at the channel rate does this. However, if the picture activity is such that the buffer may overflow (violent motion) then a feedback from the buffer to the quantiser can regulate the bit rate. Here, as the buffer occupancy increases, the feedback forces the quantiser step size to be increased to reduce the bit rate. Similarly, if the picture activity is less (coding mainly slow motion parts of frames), then the quantiser step size is reduced to improve the picture quality.

## 3.5.9   Decoder

The compressed bit stream, after demultiplexing and variable length decoding (VLD), separates the motion vectors and the DCT coefficients. Motion vectors are used by motion compensation and the DCT coefficients after the inverse quatisation and inverse DCT are converted to error data. They are then added to the motion compensated previous frame, to reconstruct the decoded picture, as shown in Figure 3.20.

## 3.6   Constant and variable bit rates

The bit rate generated by the encoder of Figure 3.19 is called a constant bit rate (CBR), requiring a fixed channel bandwidth. An alternative solution is to use a transmission system which can adapt to the variable bit rate (VBR). For VBR coding, the feedback and the smoothing buffer are no longer needed. The quantiser step size in this case is fixed.

In an asynchronous transfer mode (ATM) system [23], the information is transmitted in the form of fixed length packets or cells and when a cell is full, it is transmitted. This can happen at any time, so that the transmission system has a bit rate capability that matches the encoder. The advantage is realised only when several video channels are multiplexed together. When one channel is generating cells in rapid succession, corresponding to high picture activity, it is probable that the other channels will generate cells at a lower rate. Only rarely will the total bit rate of the multiplex be exceeded and freeze out occur.

## 3.7 Problems

1. In the linear quantiser of Figure 3.4, derive the quantisation characteristics in terms of inputs and outputs, for each of the following conditions:

   a $th = q = 16$
   b $th = 0, q = 16$.

2. The following eight-bit resolution luminance samples are DPCM encoded with the prediction of previous sample:

   $$10, \quad 14, \quad 25, \quad 240, \quad 195, \quad 32$$

   If the quantisation is uniform with $th = 0$ and $q = 8$,

   a find the reconstructed samples (assume predictor is initialised to zero)
   b calculate the PSNR of the decoded samples.

3. A three-bit nonuniform quantiser is defined as:

   | input | output |
   |---|---|
   | if $\|x\| \leq 4$ | $y = \pm 2$ |
   | $4 < \|x\| \leq 10$ | $y = \pm 6$ |
   | $10 < \|x\| \leq 25$ | $y = \pm 15$ |
   | else | $y = \pm 50$ |

   If the DPCM data of problem 2 are quantised with this quantiser, find the reconstructed samples and the resulting PSNR value.

4. A step function signal with the following eight-bit digitised values:

   $$20;\ 20;\ 20;\ 20;\ 20;\ 231;\ 231;\ 231;\ 231;\ 231;\ 231;\ 231;\ 231;\ 231$$

   is DPCM coded with the nonuniform quantiser of problem 3. Plot the reconstructed samples and identify the positions where slope overload and granular noise occur.

5. Determine the elements of the $8 \times 8$ orthonormal forward and inverse DCT transformation matrices.

6. Use the DCT matrix of problem 5 to code the following eight pixels:

   $$35;\ 81;\ 190;\ 250;\ 200;\ 150;\ 100;\ 21$$

   a find the transform coefficients
   b why is only one AC coefficient significant?
   c reconstruct the pixels without any quantisation; comment on the reconstructed pixel values.

| 18 | 24 | 31 | 15 | 16 | 18 |
|----|----|----|----|----|----|
| 11 | 20 | 23 | 41 | 11 | 9  |
| 23 | 21 | 18 | 17 | 4  | 18 |
| 15 | 31 | 24 | 21 | 13 | 9  |

| 25 | 32 |
|----|----|
| 11 | 19 |

previous frame     current frame

*Figure 3.21  A block of 2 × 2 pixels in the current frame and its corresponding block in the previous frame shown in the shaded area*

7  The DCT coefficients of problem 6 are linearly quantised with a linear and dead zone quantiser with a step size of $th = q = 16$. Find the PSNR of the reconstructed pixels.

8  Find the PSNR of the reconstructed pixels, if in problem 7 the following coefficients are retained for quantisation and the remaining coefficients are set to zero:

   *a* DC coefficient
   *b* DC and the second AC coefficient.

9  A 2 × 2 block of pixels in the current frame is matched against a similar size block in the previous frame, as shown in Figure 3.21, within a search window of ±2 pixels horizontally and ±1 pixel vertically. Find the best matched motion vector of the block, if the distortion criterion is based on:

   *a* mean-squared error
   *b* mean absolute error.

10  For a maximum motion speed of six pixels/frame:

   *a* calculate the number of operations required by the full search method
   *b* if each block contains 16 × 16 pixels, calculate the number of multiplications and additions, if the cost function was:

      (i) mean-squared error (MSE)
      (ii) mean absolute error (MAE).

11  Repeat problem 10 for the following fast search methods:

   *a* TDL
   *b* TSS
   *c* CSA
   *d* OSA

12  Four symbols of **a**, **b**, **c** and **d** with probabilities $p(\mathbf{a}) = 0.2$, $p(\mathbf{b}) = 0.45$, $p(\mathbf{c}) = 0.3$ and $p(\mathbf{d}) = 0.05$ are Huffman coded. Derive the Huffman codes for these symbols and compare the average bit rate with that of the entropy.

13  In problem 12 a message comprising five symbols **cbdad** is Huffman coded
   a  write down the generated bit stream for this message
   b  if there is a single error in the:

   (i) first bit
   (ii) third bit
   (iii) fifth bit

   What is the decoded message in each case?
14  If the intervals of [0.0, 0.2), [0.2, 0.7), [0.7, 0.95) and [0.95, 1) are assigned for arithmetic coding of strings of **a**, **b**, **c** and **d** respectively, find the lower and upper values of the arithmetic coded string of **cbcab**.
15  With the interval of strings of **a**, **b**, **c** defined in problem 14, suppose the arithmetic decoder receives 0.83955:
   a  decode the first three symbols of the message
   b  decode the first five symbols of the message.
16  In arithmetic coding symbols can be decoded using the equation:

$$R_{n+1} = \frac{R_n - L_n}{U_n - L_n}$$

   where $R_0$ is the received number and $[L_n, U_n)$ is the interval of the $n$th symbol in the stream. Use this equation to decode the symbols in problem 15.
17  Find the binary arithmetic coding of string **cbcab** of problem 14.
18  Decimal numbers can be represented in binary form by their expansions in powers of $2^{-1}$. Derive the first 11 binary digits of the decimal number 0.83955. Compare your results with that of problem 17.
19  The binary digits of the arithmetic coded string **cbcab** are corrupted at the:
   a  first bit
   b  third bit
   c  fifth bit

   Decode the first five symbols of the string in each case.

## 3.8  References

1  JAIN, A.K.: 'Fundamentals of digital image processing' (Prentice Hall, 1989)
2  CHEN, W., SMITH, C., and FRALICK, S.: 'A fast computational algorithm for the discrete cosine transform', *IEEE Trans. Commun.*, 1979, **COM-25**, pp.1004–1009
3  ISHIGURO, T., and IINUMA, K.: 'Television bandwidth compression transmission by motion-compensated interframe coding', *IEEE Commun. Mag.*, 1982, **10**, pp.24–30
4  KAPPAGANTULA, S., and RAO, K.R.: 'Motion compensated predictive coding'. Proceedings of international technical symposium, *SPIE*, San Diego, CA, August 1983

5 BERGMANN, H.C.: 'Displacement estimation based on the correlation of image segments'. IRE conference on the *Electronic image processing*, York, U.K., July 1982
6 JAIN, J.R., and JAIN, A.K.: 'Displacement measurement and its application in interframe image coding', *IEEE Trans. Commun.*, 1981, **COM-29**, pp.1799–1808
7 SHANABLEH, T., and GHANBARI, M.: 'Heterogeneous video transcoding to lower spatio-temporal resolutions and different encoding formats', *IEEE Trans. Multimedia*, 2000, **2:2**, pp.101–110
8 KOGA, T., IINUMA, K., HIRANO, A., IIJIMA, Y., and ISHIGURO, T.: 'Motion compensated interframe coding for video conferencing'. Proceedings of national *Telecommunications* conference, New Orleans, LA, November 29–December 3, pp.G5.3.1–G5.3.5
9 CCITT Working Party XV/4: 'Description of reference model 8 (RM8)'. Specialists Group on Coding for Visual Telephony, doc. 525, June 1989
10 SRINIVASAN, R., and RAO, K.R.: 'Predictive coding based on efficient motion estimation'. IEEE International conference on *Communications*, Amsterdam, May 14–17, 1984, pp.521–526
11 GHANBARI, M.: 'The cross search algorithm for motion estimation', *IEEE Trans. Commun.*, 1990, **38:7**, pp.950–953
12 PURI, A., HANG, H.M., and SCHILLING, D.L.: 'An efficient block-matching algorithm for motion compensated coding'. Proceedings of IEEE *ICASSP'87*, 1987, pp.25.4.1–25.4.4
13 BIERLING, M.: 'Displacement estimation by hierarchical block matching', *Proc. SPIE, Visual Communications and Image Processing*, 1988, **1001**, pp.942–951
14 LANGDON, G.G.: 'An introduction to arithmetic coding', *IBM J. Res. Dev.*, 1984, **28:2**, pp.135–149
15 PENNEBAKER, W.B., and MITCHELL, J.L.: 'JPEG: still image compression standard' (Van Nostrand Reinhold, New York, 1993)
16 H.263: 'Draft ITU-T Recommendation H.263, video coding for low bit rate communication'. September 1997
17 MPEG-4: 'Testing and evaluation procedures document'. ISO/IEC JTC1/SC29/WG11, N999, July 1995
18 WITTEN, I.H., NEAL, R.M., and CLEARY, J.G.: 'Arithmetic coding for data compression', *Commun. ACM*, 1987, **30:6**, pp.520–540
19 GHANBARI, M.: 'Arithmetic coding with limited past history', *Electron. Lett.*, 1991, **27:13**, pp.1157–1159
20 H.261: 'ITU-T Recommendation H.261, video codec for audiovisual services at $p \times 64$ kbit/s'. Geneva, 1990
21 MPEG-1: 'Coding of moving pictures and associated audio for digital storage media at up to about 1.5 Mbit/s'. ISO/IEC 1117–2: video, November 1991
22 MPEG-2: 'Generic coding of moving pictures and associated audio information'. ISO/IEC 13818-2 video, draft international standard, November 1994
23 CUTHBURT, A.C., and SAPANEL, J.C.: 'ATM: the broadband telecommunications solution' (IEE publishing, 1993)

*Chapter 4*

# Subband and wavelet

Coding of still images under MPEG-4 [1] and the recent decision by the JPEG committee to recommend a new standard under JPEG2000 [2] has brought up a new image compression technique. The committee has decided to recommend a new way of coding still images based on the wavelet transform, in sharp contrast to the discrete cosine transform (DCT) used in the other standard codecs, as well as the original JPEG. In this Chapter we introduce this wavelet transform and show how it can be used for image compression.

## 4.1 Why wavelet transform?

Before describing the wavelet transform and its usage in image compression it is essential to answer two fundamental questions:

(i) What is wrong with the DCT and why should we use wavelet?
(ii) If wavelet is superior to DCT, why did the original JPEG not use it?

The answer to the first part is as follows. The DCT and the other block-based transforms partition an image into nonoverlapping blocks and process each block separately. At very low bit rates, the transform coefficients need to be coarsely quantised and so there will be a significant reconstruction error after the decoding. This error is more visible at the block boundaries, by causing a discontinuity in the image, and is best known as the blocking artefact. One way of reducing this artefact is to allow the basis functions to decay towards zero at these points, or to overlap over the adjacent blocks. The latter technique is called the lapped orthogonal transform [3]. The wavelet transform is a special type of such transform and hence it is expected to eliminate blocking artefacts.

The answer to the second part relates to the state of the art in image coding in the mid 1980s, the time when the original JPEG was under development. At that time although the wavelet transform and its predecessor, subband coding, were known, there was no efficient method of coding the wavelet transform coefficients, to be comparable with the DCT. In fact the proposals submitted to the JPEG committee were all DCT-based codecs, none on the wavelet. Also almost at the same time, out of the 15 proposals to the H.261 committee, there were 14 DCT-based codecs and one vector quartisation method and none on the wavelet transform. Thus in the mid 1980s, the compression performance, coupled with the considerable momentum already behind the DCT, led the JPEG committee to adopt DCT-based coding as the foundation of JPEG.

However, the state of the wavelet-transform-based image compression techniques have significantly improved since the introduction of the original JPEG. Much of the credit should go to Jussef Shapiro, who by the introduction of the embedded zero tree wavelet (EZW) made a significant breakthrough in coding of the wavelet coefficients [4]. At the end of this Chapter this method, and the other similar methods that exploit the multiresolution properties of the wavelet transform to give a computationally simple but efficient compression algorithm, will be introduced.

## 4.2 Subband coding

Before describing the wavelet transform, let us look at its predecessor, subband coding, which sometimes is called the early wavelet transform [5]. As we will see later, in terms of image coding they are similar. However, subband coding is a product designed by engineers [6], and the wavelet transform was introduced by mathematicians [7]. Therefore before proceeding to mathematics, which is sometimes cumbersome to follow, an engineering view of multiresolution signal processing may help us to understand it better.

Subband coding was first introduced by Crochiere *et al.* in 1976 [6], and has since proved to be a simple and powerful technique for speech and image compression. The basic principle is the partitioning of the signal spectrum into several frequency bands, then coding and transmitting each band separately. This is particularly suited to image coding. First, natural images tend to have a nonuniform frequency spectrum, with most of the energy being concentrated in the lower frequency band. Secondly, human perception of noise tends to fall off at both high and low frequencies and this enables the designer to adjust the compression distortion according to perceptual criteria. Thirdly, since images are processed in their entirety, and not in artificial blocks, there is no block structure distortion in the coded picture, as occurs in the block-transform-based image encoders, such as DCT.

Thus subband, like the Fourier transform, is based on frequency domain analysis of the image but its filter banks have a better decorrelation property that suits natural images better. This can be explained as follows. Fourier basis functions are very exact in frequency, but are spatially not precise. In other words, the signal energy of Fourier basis functions is not concentrated at one frequency, but is spread over all space. This would not be a problem if image pixels were always correlated. However, in reality, pixels in images of interest generally have low correlation, especially across the image discontinuities such as edges. In contrast to Fourier basis functions, the subband bases not only have fairly good frequency concentration but are also spatially compact.

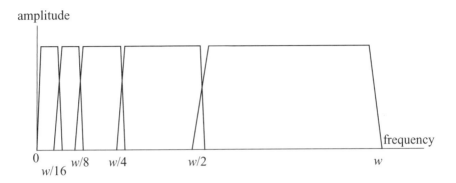

*Figure 4.1  A bank of bandpass filters*

If image edges are not too closely packed, most of the subband basis elements will not intersect with them, thus performing a better decorrelation on average.

In subband coding the band splitting is done by passing the image data through a bank of bandpass analysis filters, as shown in Figure 4.1. In order to adapt the frequency response of the decomposed pictures to the characteristics of the human visual system, filters are arranged into octave bands.

Since the bandwidth of each filtered version of the image is reduced, they can now in theory be downsampled at a lower rate, according to the Nyquist criteria, giving a series of reduced size subimages. The subimages are then quantised, coded and transmitted. The received subimages are restored to their original sizes and passed through a bank of synthesis filters, where they are interpolated and added to reconstruct the image.

In the absence of quantisation error, it is required that the reconstructed picture should be an exact replica of the input picture. This can only be achieved if the spatial frequency response of the analysis filters tiles the spectrum (i.e. the individual bands are set one after the other) without overlapping, which requires infinitely sharp transition regions and cannot be realised practically. Instead, the analysis filter responses have finite transition regions and do overlap, as shown in Figure 4.1, which means that the downsampling/upsampling processes introduce aliasing distortion into the reconstructed picture.

In order to eliminate the aliasing distortion, the synthesis and analysis filters have to have certain relationships such that the aliased components in the transition regions cancel each other out. To see how such a relation can make alias-free subband coding possible, consider a two-band subband, as shown in Figure 4.2.

The corresponding two-band subband encoder/decoder is shown in Figure 4.3. In this diagram, filters $H_0(z)$ and $H_1(z)$ represent the $z$ transform transfer functions of the respective lowpass and highpass analysis filters. Filters $G_0(z)$ and $G_1(z)$ are the corresponding synthesis filters. The downsampling and upsampling factors are 2.

At the encoder, downsampling by 2 is carried out by discarding alternate samples, the remainder being compressed into half the distance occupied by the original

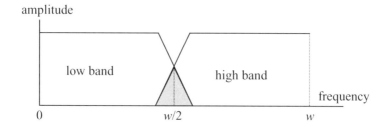

*Figure 4.2   A two-band analysis filter*

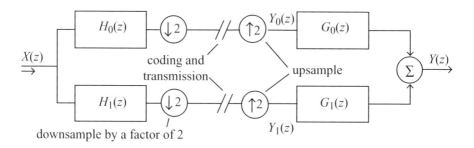

*Figure 4.3   A two-band subband encoder/decoder*

sequence. This is equivalent to compressing the source image by a factor of 2, which doubles all the frequency components present. The frequency domain effect of this downsampling/compression is thus to double the width of all components in the sampled spectrum.

At the decoder, the upsampling is a complementary procedure: it is achieved by inserting a zero-valued sample between each input sample, and is equivalent to a spatial expansion of the input sequence. In the frequency domain, the effect is as usual the reverse and all components are compressed towards zero frequency.

The problem with these operations is the impossibility of constructing ideal, sharp cut analysis filters. This is illustrated in Figure 4.4a. Spectrum A shows the original sampled signal which has been lowpass filtered so that some energy remains above $F_s/4$, the cutoff of the ideal filter for the task. Downsampling compresses the signal and expands to give B, and C is the picture after expansion or upsampling. As well as those at multiples of $F_s$, this process generates additional spectrum components at odd multiples of $F_s/2$. These cause aliasing when the final subband recovery takes place as at D.

In the highpass case, Figure 4.4b, the same phenomena occur, so that on recovery there is aliased energy in the region of $F_s/4$. The final output image is generated by adding the lowpass and highpass subbands regenerated by the upsamplers and associated filters. The aliased energy would normally be expected to cause interference.

*Subband and wavelet* 67

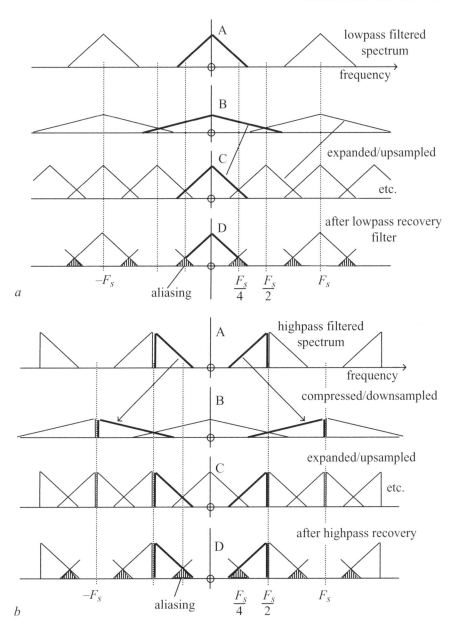

*Figure 4.4*   *a*   lowpass subband generation and recovery
               *b*   highpass subband generation and recovery

However, if the phases of the aliased components from the high and lowpass subbands can be made to differ by $\pi$, then cancellation occurs and the recovered signal is alias free.

How this can be arranged is best analysed by reference to $z$ transforms. Referring to Figure 4.3, after the synthesis filters, the reconstructed output in $z$ transform notation can be written as:

$$Y(z) = G_0(z) \cdot Y_0(z) + G_1(z) \cdot Y_1(z) \tag{4.1}$$

where $Y_0(z)$ and $Y_1(z)$ are inputs to the synthesis filters after upsampling. Assuming there are no quantisation and transmission errors, the reconstructed samples are given by:

$$Y_0(z) = \tfrac{1}{2}[H_0(z) \cdot X(z) + H_0(-z) \cdot X(-z)]$$
$$Y_1(z) = \tfrac{1}{2}[H_1(z) \cdot X(z) + H_1(-z) \cdot X(-z)] \tag{4.2}$$

where the aliasing components from the downsampling of the lower and higher bands are given by $H_0(-z)X(-z)$ and $H_1(-z)X(-z)$, respectively. By substituting these two equations in the previous one, we get:

$$Y(z) = \tfrac{1}{2}[H_0(z) \cdot G_0(z) + H_1(z) \cdot G_1(z)]X(z)$$
$$+ \tfrac{1}{2}[H_0(-z) \cdot G_0(z) + H_1(-z) \cdot G_1(z)]X(-z) \tag{4.3}$$

The first term is the desired reconstructed signal, and the second term is aliased components. The aliased components can be eliminated regardless of the amount of overlap in the analysis filters by defining the synthesis filters as:

$$G_0(z) = H_1(-z) \quad \text{and} \quad G_1(z) = -H_0(-z) \tag{4.4}$$

With such a relation between the synthesis and analysis filters, the reconstructed signal now becomes:

$$Y(z) = \tfrac{1}{2}[H_0(z) \cdot H_1(-z) - H_0(-z) \cdot H_1(z)]X(z) \tag{4.5}$$

If we define $P(z) = H_0(z)H_1(-z)$, then the reconstructed signal can be written as:

$$Y(z) = \tfrac{1}{2}[P(z) - P(-z)]X(z) \tag{4.6}$$

Now the reconstructed signal can be a perfect, but an $m$-sample delayed, replica of the input signal if:

$$P(z) - P(-z) = 2z^{-m} \tag{4.7}$$

Thus the $z$ transform input/output signals are given by:

$$Y(z) = z^{-m}X(z) \tag{4.8}$$

This relation in the pixel domain implies that the reconstructed pixel sequence $\{y(n)\}$ is an exact replica of the input sequence but delayed by $m$ pixels, $\{$i.e. $x(n-m)\}$.

In these equations $P(z)$ is called the product filter and $m$ is the delay introduced by the filter banks. The design of analysis/synthesis filters is based on factorisation of the product filter $P(z)$ into linear phase components $H_0(z)$ and $H_1(-z)$, with the constraint that the difference between the product filter and its image should be a simple delay. Then the product filter must have an odd number of coefficients. LeGall and Tabatabai [8] have used a product filter $P(z)$ of the kind:

$$P(z) = \tfrac{1}{16}(-1 + 9z^{-2} + 16z^{-3} + 9z^{-4} - z^{-6}) \qquad (4.9)$$

and by factorising have obtained several solutions for each pair of the analysis and synthesis filters:

$$H_0(z) = \tfrac{1}{4}(-1 + 3z^{-1} + 3z^{-2} - z^{-3}),$$
$$H_1(-z) = \tfrac{1}{4}(1 + 3z^{-1} + 3z^{-2} + z^{-3})$$

or

$$H_0(z) = \tfrac{1}{4}(1 + 3z^{-1} + 3z^{-2} + z^{-3}),$$
$$H_1(-z) = \tfrac{1}{4}(-1 + 3z^{-1} + 3z^{-2} - z^{-3})$$

or

$$H_0(z) = \tfrac{1}{8}(-1 + 2z^{-1} + 6z^{-2} + 2z^{-3} - z^{-4}),$$
$$H_1(-z) = \tfrac{1}{2}(1 + 2z^{-1} + z^{-2}) \qquad (4.10)$$

The synthesis filters $G_0(z)$ and $G_1(z)$ are then derived using their relations with the analysis filters, according to eqn. 4.4. Each of the above equation pairs gives the results $P(z) - P(-z) = 2z^{-3}$, which implies that the reconstruction is perfect with a delay of three samples.

## 4.3 Wavelet transform

The wavelet transform is a special case of subband coding and is becoming very popular for image and video coding. Although subband coding of images is based on frequency analysis, the wavelet transform is based on approximation theory. However, for natural images that are locally smooth and can be modelled as piecewise polynomials, a properly chosen polynomial function can lead to frequency domain analysis, like that of subband. In fact, wavelets provide an efficient means for approximating such functions with a small number of basis elements. Mathematically, a wavelet transform of a square-integrable function $x(t)$ is its decomposition into a set of basis functions, such as:

$$X_w(a,b) = \int_{-\infty}^{\infty} x(t)\Psi_{a,b}(t)\,dt \qquad (4.11)$$

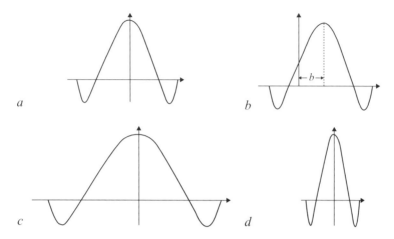

Figure 4.5  Effect of time dilation and translation on the mother wavelet
  a   mother wavelet $\Psi(t) = \Psi_{1,0}(t)$, $a = 1$, $b = 0$
  b   wavelet $\Psi_{1,b}(t)$, $a = 1$, $b \neq 0$
  c   wavelet $\Psi_{2,0}(t)$ at scale $a = 2$, $b = 0$
  d   wavelet $\Psi_{0.5,0}(t)$ at scale $a = 1/2$, $b = 0$

where $\Psi_{a,b}(t)$ is known as the basis function, which is a time dilation and translation version of a bandpass signal $\Psi(t)$, called the mother wavelet and is defined as:

$$\Psi_{a,b} = \frac{1}{\sqrt{a}} \Psi\left(\frac{t-b}{a}\right) \qquad (4.12)$$

where $a$ and $b$ are time dilation and translation parameters, respectively. The effects of these parameters are shown in Figure 4.5.

The width of the basis function varies with the dilation factor (or scale) $a$. The larger is $a$, the wider becomes the basis function in the time domain and hence narrower in the frequency domain. Thus it allows varying time and frequency resolutions (with trade-off between both) in the wavelet transform. It is this property of the wavelet transform that makes it suitable for analysing signals having features of different sizes as present in natural images. For each feature size, there is a basis function $\Psi_{a,b}(t)$ in which it will be best analysed. For example: in a picture of a house with a person looking through a window, the basis function with larger $a$ will analyse conveniently the house as a whole. The person by the window will be best analysed at a smaller scale and the eyes of the person at even smaller scale. So the wavelet transform is analogous to the analysis of a signal in the frequency domain using bandpass filters of variable central frequency (which depends on parameter $a$) but with a constant quality factor. Note that in filters, the quality factor is the ratio of the centre frequency to the bandwidth of the filter.

### 4.3.1 Discrete wavelet transform (DWT)

As the wavelet transform defined in eqn. 4.11 maps a one-dimensional signal $x(t)$ into a two-dimensional function $X_w(a, b)$, this increase in dimensionality makes it extremely redundant, and the original signal can be recovered from the wavelet transform computed on the discrete values of $a$ and $b$ [9]. The $a$ can be made discrete by choosing $a = a_0^m$, with $a_0 > 1$ and $m$ an integer. As $a$ increases, the bandwidth of the basis function (or frequency resolution) decreases and hence more resolution cells are needed to cover the region. Similarly, making $b$ discrete corresponds to sampling in time (sampling frequency depends on the bandwidth of the signal to be sampled which in turn is inversely proportional to $a$), it can be chosen as $b = nb_0 a_0^m$. For $a_0 = 2$ and $b_0 = 1$ there are choices of $\Psi(t)$ such that the function $\Psi_{m,n}(t)$ forms an orthonormal basis of space of square integrable functions. This implies that any square integrable function $x(t)$ can be represented as a linear combination of basis functions as:

$$x(t) = \sum_{m=-\infty}^{\infty} \sum_{n=-\infty}^{\infty} \alpha_{m,n} \Psi_{m,n}(t) \tag{4.13}$$

where $\alpha_{m,n}$ are known as the wavelet transform coefficients of $x(t)$ and are obtained from eqn. 4.11 by:

$$\alpha_{m,n} = \int_{-\infty}^{\infty} x(t) \Psi_{m,n}(t) \tag{4.14}$$

It is interesting to note that for every increment in $m$, the value of $a$ doubles. This implies doubling the width in the time domain and halving the width in the frequency domain. This is equivalent to signal analysis with octave band decomposition and corresponds to dyadic wavelet transform similar to that shown in Figure 4.1, used to describe the basic principles of subband coding, and hence the wavelet transform is a type of subband coding.

### 4.3.2 Multiresolution representation

Application of the wavelet transform to image coding can be better understood with the notion of multiresolution signal analysis. Suppose there is a function $\Phi(t)$ such that the set $\Phi(t - n), n \in Z$ is orthonormal. Also suppose $\Phi(t)$ is the solution of a two-scale difference equation:

$$\Phi(t) = \sum_{n=-\infty}^{\infty} c_n \sqrt{2} \Phi(2t - n) \tag{4.15}$$

where

$$c_n = \int_{-\infty}^{\infty} \Phi(t) \sqrt{2} \Phi(2t - n) \, dt \tag{4.16}$$

Let $x(t)$ be a square integrable function, which can be represented as a linear combination of $\Phi(t-n)$ as:

$$x(t) = \sum_{n=-\infty}^{\infty} c_n \Phi(t-n) \tag{4.17}$$

where $c_n$ is the expansion coefficient and is the projection of $x(t)$ onto $\Phi(t-n)$. Since dilation of a function varies its resolution, it is possible to represent $x(t)$ at various resolutions by dilating and contracting the function $\Phi(t)$. Thus $x(t)$ at any resolution $m$ can be represented as:

$$x_m(t) = 2^{-m/2} \sum_n c_n^m \Phi(2^{-m}t - n) \tag{4.18}$$

If $V_m$ is the space generated by $2^{-m/2}\Phi(2^{-m}t - n)$, then from eqn. 4.15, $\Phi(t)$ is such that for any $i > j$, the function that generates space $V_i$ is also in $V_j$, i.e. $V_i \subset V_j$ $(i > j)$. Thus spaces at successive scales can be nested such that $V_m$ for increasing $m$ can be viewed as the space of decreasing resolution. Therefore, a space at a coarser resolution $V_{j-1}$ can be decomposed into two subspaces: a space at a finer resolution $V_j$ and an orthogonal complement of $V_j$, represented by $W_j$ such that $V_j + W_j = V_{j-1}$ where $W_j \perp V_j$. The space $W_j$ is the space of differences between the coarser and the finer scale resolutions, which can be seen as the amount of detail added when going from a smaller resolution $V_j$ to a larger resolution $V_{j-1}$. The hierarchy of spaces is depicted in Figure 4.6.

Mallat [10] has shown that in general the basis for $W_j$ consists in translations and dilations of a single prototype function $\Psi(t)$, called a wavelet. Thus, $W_m$ is the space generated by $\Psi_{m,n}(t) = 2^{-m/2}\Psi(2^{-m}t - n)$. The wavelet $\Psi(t) \in V_{-1}$ can be obtained from $\Phi(t)$ as:

$$\Psi(t) = \sum_{n=-\infty}^{\infty} (-1)^n c_{1-n} \sqrt{2} \Phi(2t - n) \tag{4.19}$$

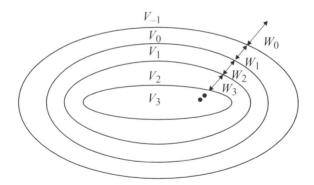

Figure 4.6  Multiresolution spaces

The function $\Phi(t)$ is called the scaling function of the multiresolution representation. Thus the wavelet transform coefficients of eqn. 4.14 correspond to the projection of $x(t)$ onto a detail space of resolution $m$, $W_m$. Hence, a wavelet transform basically decomposes a signal into spaces of different resolutions. In the literature, these kinds of decomposition are in general referred to as multiresolution decomposition. Here is an example of calculating the Haar wavelet through this technique.

**Example: Haar wavelet**

The scaling function of the Haar wavelet is the well known rectangular (rect) function:

$$\Phi(t) = \begin{cases} 1 & 0 \leq t \leq 1 \\ 0 & \text{otherwise} \end{cases}$$

The rect function satisfies eqn. 4.15 with $c_n$ calculated from eqn. 4.16 as:

$$c_n = \begin{cases} \frac{1}{\sqrt{2}} & n = 0, 1 \\ 0 & \text{otherwise} \end{cases}$$

Thus using eqn. 4.19, the Haar wavelet can be found as:

$$\Psi(t) = \Phi(2t) - \Phi(2t - 1)$$

$$\Psi(t) = \begin{cases} 1 & 0 \leq t < \frac{1}{2} \\ -1 & \frac{1}{2} \leq t < 1 \end{cases}$$

The scaling function $\Phi(t)$ (a rect function) and the corresponding Haar wavelet $\Psi(t)$ for this example are shown in Figure 4.7a and b, respectively. In terms of the approximation perspective, the multiresolution decomposition can be explained as follows. Let $x(t)$ be approximated at resolution $j$ by function $A_j x(t)$ through the expansion series of the orthogonal basis functions. Since $W_j$ represents the space of difference between a coarser scale $V_{j-1}$ and a finer scale $V_j$, then $D_j x(t) \in W_j$ represents the difference of approximation of $x(t)$ at the $j-1$th and $j$th resolution (i.e. $D_j x(t) = A_{j-1} x(t) - A_j x(t)$).

Thus the signal $x(t)$ can be split as $x(t) = A_{-1} x(t) = A_0 x(t) + D_0 x(t)$.

Figure 4.7c and d show the approximations of a continuous function at the two successive resolutions using a rectangular scaling function. The coarser approximation $A_0 x(t)$ is shown in Figure 4.7c and at higher resolution approximation $A_1 x(t)$ in Figure 4.7d where the scaling function is a dilated version of the rectangular function. For a smooth function $x(t)$, most of the variation (signal energy) is contained in $A_0 x(t)$, and $D_0 x(t)$ is nearly zero. By repeating this splitting procedure and partitioning $A_0 x(t) = A_1 x(t) + D_1 x(t)$, the wavelet transform of signal $x(t)$ can

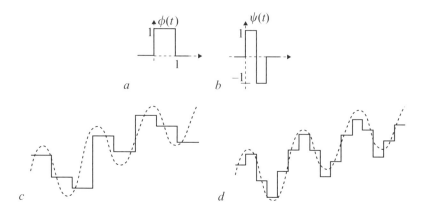

Figure 4.7  
  a  Haar scaling function  
  b  Haar wavelet  
  c  approximation of a continuous function, $x(t)$, at coarser resolution $A_0 x(t)$  
  d  higher resolution approximation $A_1 x(t)$

be obtained and hence the original function $x(t)$ can be represented in terms of its wavelets as:

$$x(t) = D_0 x(t) + D_1 x(t) + D_2 x(t) + \cdots + D_n x(t) + A_n x(t) \tag{4.20}$$

where $n$ represents the number of decompositions. Since the dynamic ranges of the detail signals $D_j x(t)$ are much smaller than the original function $x(t)$, they are easier to code than the coefficients in the series expansion of eqn. 4.13.

### 4.3.3 Wavelet transform and filter banks

For the repeated splitting procedure described above to be practical, there should be an efficient algorithm for obtaining $D_j x(t)$ from the original expansion coefficient of $x(t)$. One of the consequences of multiresolution space partitioning is that the scaling function $\Phi(t)$ possesses a self similarity property. If $\Phi(t)$ and $\bar{\Phi}(t)$ are the analysis and synthesis scaling functions, and $\Psi(t)$ and $\bar{\Psi}(t)$ are analysis and synthesis wavelets, then since $V_j \subset V_{j-1}$, these functions can be recursively defined as:

$$\Phi(t) = \sum_{n=-\infty}^{\infty} c_n \sqrt{2} \Phi(2t - n)$$

$$\bar{\Phi}(t) = \sum_{n=-\infty}^{\infty} \bar{c}_n \sqrt{2} \bar{\Phi}(2t - n) \tag{4.21}$$

$$\Psi(t) = \sum_{n=-\infty}^{\infty} d_n \sqrt{2} \Phi(2t - n) \quad \text{and} \quad \bar{\Psi}(t) = \sum_{n=-\infty}^{\infty} \bar{d}_n \sqrt{2} \bar{\Phi}(2t - n)$$

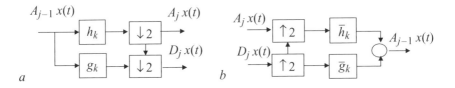

*Figure 4.8* One stage wavelet transform
    *a*   analysis
    *b*   synthesis

These recurrence relations provide ways of calculating the coefficients of the approximation of $x(t)$ at resolution $j$, $A_j x(t)$, and the coefficients of the detail signal $D_j x(t)$ from the coefficients of the approximation of $x(t)$ at a higher resolution $A_{j-1} x(t)$. In fact, simple mathematical manipulations can reveal that both the coefficients of approximation at a finer resolution and detail coefficients can be obtained by convolving the coefficient of approximation at a coarser resolution with a filter and downsampling it by a factor of 2. For a lower resolution approximation coefficient, the filter is a lowpass filter with taps $h_k = c_{-k}$, and for the details the filter is a highpass filter with taps $g_k = d_{-k}$. Inversely, the signal at a higher resolution can be recovered from its approximation at a lower resolution and coefficients of the corresponding detail signal. It can be accomplished by upsampling the coefficients of approximation at a lower resolution and detail coefficients by a factor of 2, convolving them with the synthesis filters of taps $\bar{h}_k = \bar{c}_k$ and $\bar{g}_k = \bar{d}_k$, respectively, and adding them together. One step of splitting and inverse process is shown in Figure 4.8, which is in fact the same as Figure 4.3 for subband. Thus the filtering process splits the signals into lowpass and highpass frequency components and hence increases frequency resolution by a factor of 2, but downsampling reduces the temporal resolution by the same factor. Hence, at each successive step, better frequency resolution at the expense of temporal resolution is achieved.

### 4.3.4 Higher order systems

The multidimensional wavelet transform can be obtained by extending the concept of the two-band filter structure of Figure 4.8 in each dimension. For example, decomposition of a two-dimensional image can be performed by carrying out one-dimensional decomposition in the horizontal and then in the vertical directions.

A seven-band wavelet transform coding of this type is illustrated in Figure 4.9, where band splitting is carried out alternately in the horizontal and vertical directions. In the Figure, L and H represent the lowpass and highpass analysis filters with a 2 : 1 downsampling, respectively. At the first stage of dyadic decomposition, three subimages with high frequency contents are generated. The subimage LH1 has mainly low horizontal, but high vertical frequency image details. This is reversed in the HL1 subimage. The HH1 subimage has high horizontal and high vertical image details. These image details at a lower frequency are represented by the LH2, HL2 and HH2

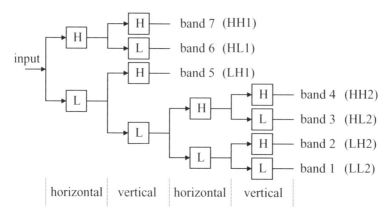

Figure 4.9 Multiband wavelet transform coding using repeated two-band splits

bands, respectively. The LL2 band is a lowpass subsampled image, which is a replica of the original image, but at a smaller size.

### 4.3.5 Wavelet filter design

As we saw in section 4.3.3, in practice the wavelet transform can be realised by a set of filter banks, similar to those of the subband. Relations between the analysis and synthesis of the scaling and wavelet functions also follow those of the synthesis and analysis filters of the subband. Hence we can use the concept of the product filter, defined in eqn. 4.7, to design wavelet filters. However, if we use the product filter as was used for the subband, we do not achieve anything new. But, if we add some constraints on the product filter, such that the property of the wavelet transform is maintained, then a set of wavelet filters can be designed.

One of the constraints required to be imposed on the product filter $P(z)$ is that the resultant filters $H_0(z)$ and $H_1(z)$ be continuous, as required by the wavelet definition. Moreover, it is sometimes desirable to have wavelets with the largest possible number of continuous derivatives. This property in terms of the $z$ transform means that the wavelet filters and consequently the product filter should have zeros at $z = -1$. A measure of the number of derivatives or number of zeros at $z = -1$ is given by the regularity of the wavelets and also called the number of vanishing moments [11]. This means that in order to have regular filters, filters must have a sufficient number of zeros at $z = -1$, the larger the number of zeros, the more regular the filter is.

Also since, in images, phase carries important information it is necessary that filters must have linear phase responses. On the other hand, although the orthogonal filters have an energy preserving property, most of the orthogonal filters do not have phase linearity. A particular class of filters that have linear phase in both their analysis and synthesis and are very close to orthogonal are known as biorthogonal filters. In biorthogonal filters, the lowpass analysis and the highpass synthesis filters are orthogonal to each other and similarly the highpass analysis and the lowpass synthesis are orthogonal to each other, hence the name biorthogonal. Note that, in the biorthogonal filters, since the lowpass and the highpass of either analysis or synthesis filters can be of different lengths, they are not themselves orthogonal to each other.

Thus for a wavelet filter to have at least $n$ zeros at $z = -1$, we chose the product filter to be [12]:

$$P(z) = (1 + z^{-1})^{2n} Q(z) \qquad (4.22)$$

where $Q(z)$ has $n$ unknown coefficients. Depending on the choice of $Q(z)$ and the regularity of the desired wavelets, one can design a set of wavelets as desired. For example with:

$$n = 2 \quad \text{and} \quad Q(z) = -1 + 4z^{-1} - z^{-2}$$

the product filter becomes:

$$P(z) = (1 + z^{-1})^4 (-1 + 4z^{-1} - z^{-2}) \qquad (4.23)$$

and with a proper weighting for orthonormality and then factorisation, it leads to two sets of $(5, 3)$ and $(4, 4)$ filter banks of eqn. 4.10. The weighting factor is determined from eqn. 4.7. These filters were originally derived for subbands, but as we see they can also be used for wavelets. Filter pair $(5, 3)$ is the recommended filter for the lossless image coding in JPEG2000 [13]. The coefficients of its analysis filters are given in Table 4.1.

As another example with:

$$n = 3 \quad \text{and} \quad Q(z) = 1 - 6z^{-1} + \frac{38}{3}z^{-2} - 6z^{-3} + z^{-4}$$

$$P(z) = (1 + z^{-1})^6 \left(1 - 6z^{-1} + \frac{38}{3}z^{-2} - 6z^{-3} + z^{-4}\right) \qquad (4.24)$$

the (9,3) pair of Daubechies filters [9] can be derived, which are given in Table 4.2. These filter banks are recommended for still image coding in MPEG-4 [14].

Table 4.1  Lowpass and highpass analysis filters of integer (5, 3) biorthogonal filter

| $n$ | lowpass | highpass |
|---|---|---|
| 0 | 6/8 | +1 |
| ±1 | 2/8 | −1/2 |
| ±2 | −1/8 | |

Table 4.2  Lowpass and highpass analysis filters of Daubechies (9, 3) biorthogonal filter

| n | lowpass | highpass |
|---|---|---|
| 0 | 0.99436891104360 | 0.70710678118655 |
| ±1 | 0.41984465132952 | −0.35355339059327 |
| ±2 | −0.17677669529665 | |
| ±3 | −0.06629126073624 | |
| ±4 | 0.03314563036812 | |

Table 4.3  Lowpass and highpass analysis filters of Daubechies (9, 7) biorthogonal filter

| n | lowpass | highpass |
|---|---|---|
| 0 | 0.85269865321930 | −0.7884848720618 |
| ±1 | 0.37740268810913 | 0.41809244072573 |
| ±2 | −0.11062402748951 | 0.04068975261660 |
| ±3 | −0.02384929751586 | −0.06453905013246 |
| ±4 | 0.03782879857992 | |

Another popular biorthogonal filter is the Duabechies (9,7) filter bank, recommended for lossy image coding in the JPEG2000 standard [13]. The coefficients of its lowpass and highpass analysis filters are tabulated in Table 4.3. These filters are known to have the highest compression efficiency.

The corresponding lowpass and highpass synthesis filters can be derived from the above analysis filters, using the relationship between the synthesis and analysis filters given by eqn. 4.4. That is $G_0(z) = H_1(-z)$ and $G_1(z) = -H_0(-z)$.

**Example**

As an example of the wavelet transform, Figure 4.10 shows all the seven subimages generated by the encoder of Figure 4.9 for a single frame of the flower garden test sequence, with a nine-tap low and three-tap highpass analysis Daubechies filter pair, (9, 3), given in Table 4.2. These filters have been recommended for coding of still images in MPEG-4 [14], which has been shown to achieve good compression efficiency.

The original image (not shown) dimensions were 352 pixels by 240 lines. Bands 1–4, at two levels of subdivision, are 88 by 60, and bands 5–7 are 176 by 120 pixels. All bands but band 1 (LL2) have been amplified by a factor of four and an offset of +128 to enhance visibility of the low level details they contain. The scope for

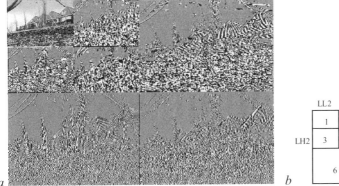
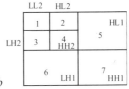

*Figure 4.10*  a  the seven subimages generated by the encoder of Figure 4.9
           b  layout of individual bands

bandwidth compression arises mainly from the low energy levels that appear in the highpass subimages.

Since at image borders all the input pixels are not available, a symmetric extension of the input texture is performed before applying the wavelet transform at each level [14]. The type of symmetric extension can vary. For example in MPEG-4, to satisfy the perfect reconstruction conditions with the Daubechies (9, 3) tap analysis filter pairs, two types of symmetric extension are used.

Type A is only used at the synthesis stage. It is used at the trailing edge of lowpass filtering and the leading edge of highpass filtering stages. If the pixels at the boundary of the objects are represented by **abcde**, then the type A extension becomes edcba|**abcde**, where the letters in bold type are the original pixels and those in plain are the extended pixels. Note that for a (9,3) analysis filter pair of Table 4.2, the synthesis filter pair will be (3, 9) with $G_0(z) = H_1(-z)$ and $G_1(z) = -H_0(-z)$, as was shown in eqn. 4.4.

Type B extension is used for both leading and trailing edges of the low and highpass analysis filters. For the synthesis filters, it is used at the leading edge of the lowpass, but at the trailing edge of the highpass. With this type of extension, the extended pixels at the leading and trailing edges become edcb|**abcde** and **abcde**|dcba, respectively.

## 4.4 Coding of the wavelet subimages

The lowest band of the wavelet subimages is a replica of the original image, but at a much reduced size, as can be seen from Figure 4.10. Efficient coding of this band depends on the number of wavelet decomposition levels. For example, if the number of wavelet decomposition levels is too high, then there is not much correlation between the pixels of the lowest band. In this case, pixel-by-pixel coding, as used in the JPEG2000 standard, is good enough. On the other hand, for MPEG-4, where not as many decomposition levels as JPEG2000 are used, there are some residual correlations between them. These can be reduced by DPCM coding. Also, depending whether wavelet transform is applied to still images or video, this band can be coded accordingly. However, in the relevant chapters coding of this band for appropriate application will be described.

For efficient compression of higher bands, as well as for a wide range of scalability, the higher order wavelet coefficients are coded with a zero tree structure like the embedded zero tree wavelet (EZW) algorithm first introduced by Shapiro [4]. This method and its variants are based on two concepts of quantisation by successive approximation, and exploitation of the similarities of the bands of the same orientation.

### 4.4.1 Quantisation by successive approximation

Quantisation by successive approximation is the representation of a wavelet coefficient value in terms of progressively smaller quantisation step sizes. The number of passes of the approximation depends on the desired quantisation distortions. To see how successive approximation can lead to quantisation, consider Figure 4.11, where a coefficient of length $L$ is successively refined to its final quantised value of $\hat{L}$.

The process begins by choosing an initial yardstick length $l$. The value of $l$ is set to half the largest coefficient in the image. If the coefficient is larger than the yardstick, it is represented with the yardstick value, otherwise its value is set to zero. After each pass the yardstick length is halved and the error magnitude, which is the difference between the original value of the coefficient and its reconstructed value, is compared with the new yardstick. The process is continued, such that the final error is acceptable. Hence, by increasing the number of passes the error in the representation of $L$ by $\hat{L}$ can be made arbitrarily small.

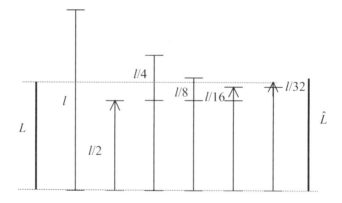

Figure 4.11   Principles of successive approximation

With regard to Figure 4.11, the quantised length $L$ can be expressed as:

$$\hat{L} = 0 \times l + 1 \times \frac{l}{2} + 0 \times \frac{l}{4} + 0 \times \frac{l}{8} + 1 \times \frac{l}{16} + 1 \times \frac{l}{32} \cdots = \frac{l}{2} + \frac{l}{16} + \frac{l}{32} \quad (4.25)$$

where only yardstick lengths smaller than the quantisation error are considered. Therefore, given an initial yardstick $l$, a length $L$ can be represented as a string of 1 and 0 symbols. As each symbol 1 or 0 is added, the precision in the representation of $L$ increases, and thus the distortion level decreases. This process is in fact equivalent to the binary representation of real numbers, called bit plane representation, where each number is represented by a string of 0s and 1s. By increasing the number of digits, the error in the representation can be made arbitrarily small.

Bit plane quantisation is another form of successive approximation that has been used in some standard codecs such as the JPEG2000 standard. Here, the wavelet coefficients are first represented by their maximum possible precision. This depends on the input pixel resolution (e.g. eight bits) and the dynamic range of the wavelet filter's coefficients. The symbols that represent the quantised coefficients are encoded one bit at a time, starting with the most significant bit (MSB) and preceding to the least significant bit (LSB). Thus for an $M$-bit plane quantisation with the finest quantiser step size of $\Delta$, the yardstick is $\Delta 2^{M-1}$. $\Delta$ is called the basic quantiser step size.

### 4.4.2 Similarities among the bands

A two-stage wavelet transform (seven bands) of the flower garden image sequence with the position of the bands was shown in Figure 4.10. It can be seen that the vertical bands look like scaled versions of each other, as do the horizontal and diagonal bands. Of particular interest in these subimages is the fact that the nonsignificant coefficients from bands of the same orientation tend to be in the same corresponding locations. Also, the edges are approximately at the same corresponding positions. Considering that subimages of lower bands (higher stages of decomposition) have half the dimensions of their higher bands, then one can make a quad tree representation of the bands of the same orientation, as shown in Figure 4.12 for a ten-band (three-stage wavelet transform).

In this Figure a coefficient in the lowest vertical band, $LH_3$, corresponds to four coefficients of its immediately higher band $LH_2$, which relates to 16 coefficients in $LH_1$. Thus, if a coefficient in $LH_3$ is zero, it is likely that its children in the higher bands of $LH_2$ and $LH_1$ are zero. The same is true for the other horizontal and diagonal bands. This tree of zeros, called zero tree, is an efficient way of representing a large group of zeros of the wavelet coefficients. Here, the root of the zero tree is required to be identified and then the descendant children in the higher bands can be ignored.

## 4.5 Embedded zero tree wavelet (EZW) algorithm

The combination of the zero tree roots with successive approximation has opened up a very interesting coding tool for not only efficient compression of wavelet coefficients, but also as a means for spatial and SNR scalability [4].

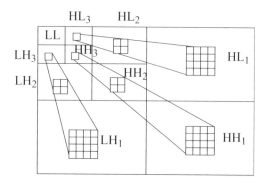

*Figure 4.12  Quad tree representation of the bands of the same orientation*

The encoding algorithm with slight modification on the successive approximation, for efficient coding, according to EZW [4] is described as follows:

1. The image mean is computed and extracted from the image. This depends on how the lowest band LL is coded. If it is coded independently of other bands, such as with DPCM in MPEG-4, then this stage can be ignored.
2. An $R$ stage ($3R + 1$ band) wavelet transform is applied to the (zero mean) image.
3. The initial yardstick length $l$ is set to half of the maximum absolute value of the wavelet coefficients.
4. A list of the coordinates of the coefficients, called the dominant list, is generated. This list determines the order in which the coefficients are scanned. It must be such that coefficients from a lower frequency band (higher scale) are always scanned before the ones from a higher frequency band. Two empty lists of coefficient coordinates, called the subordinate list and the temporary list, are also created.
5. The wavelet transform of the image is scanned and, if a wavelet coefficient is smaller than the current yardstick length $l$, it is reconstructed to zero. Otherwise, it is reconstructed as $\pm 3l/2$, according to its sign.
6. *Dominant pass*: the reconstructed coefficients are scanned again, according to the order in the dominant list, generating a string of symbols as follows. If a reconstructed coefficient is positive or negative, a + or a − is added to the string, and the coordinates of this coefficient are appended to the subordinate list. If a reconstructed coefficient is zero, its coordinates are appended to the temporary list. In the case of a zero-valued reconstructed coefficient, two different symbols can be appended to the string; if all its corresponding coefficients in bands of the same orientation and higher frequencies are zero, a zero tree root (ZT) is added to the string, and its corresponding coefficients are removed from the dominant list and added to the temporary list (since they are already known to be zero, they do not need to be scanned again). Otherwise, an isolated zero (Z) is added to the string. The strings generated from the four-symbol

alphabet of +, −, ZT and Z are encoded with an adaptive arithmetic encoder [15], whose model is updated to four symbols at the beginning of this pass. However, during the scanning of the highest horizontal, vertical and diagonal frequency bands ($HL_1$, $LH_1$ and $HH_1$ of Figure 4.12), no zero tree roots can be generated. Therefore, just before the scanning of the first coefficient of these bands, the model of the arithmetic coder is updated to three symbols of +, − and Z.

7   The yardstick length $l$ is halved.
8   *Subordinate pass*: the coefficients which previously have not been reconstructed as zero are scanned again according to their order in the subordinate list, and each one has added to it either $+l/2$ or $-l/2$ in order to minimise the magnitude of its reconstruction error. If $l/2$ is added, a + is appended to the string, and if $l/2$ is subtracted, a − is appended. At the end of the subordinate pass the subordinate list is reordered so that the coefficients whose reconstructed values have higher magnitudes come first. The + and − symbols of this pass are encoded with the arithmetic coder, which had its model updated to two symbols (+ and −) at the beginning of this pass.
9   The dominant list is replaced by the temporary list, and the temporary list is emptied.
10  The whole process is repeated from step 5. It stops at any point when the size of the bit stream exceeds the desired bit rate budget.

An observation has to be made on the dominant pass (step 6). In this pass, only the reconstructed values of the coefficients that are still in the dominant list can be affected. Therefore, in order to increase the number of zero tree roots, the coefficients not in the dominant list can be considered zero for determining if a zero-valued coefficient is either a zero tree root or an isolated zero.

The bit stream includes a header giving extra information to the decoder. The header contains the number of wavelet transform stages, the image dimensions, the initial value of the yardstick length and the image mean. Both the encoder and decoder initially have identical dominant lists. As the bit stream is decoded, the decoder updates the reconstructed image, as well as its subordinate and temporary lists. In this way, it can exactly track the stages of the encoder, and can therefore properly decode the bit stream. It is important to observe that the ordering of the subordinate list in step 8 is carried out based only on the reconstructed coefficient values, which are available to the decoder. If it was not so, the decoder would not be able to track the encoder, and thus the bit stream would not be properly decoded.

### 4.5.1 Analysis of the algorithm

The above algorithm has many interesting features, which make it especially significant to note. Among them one can say:

*a*   The use of zero trees exploits similarities among the bands of the same orientation and reduces the number of symbols to be coded.

84  *Standard codecs: image compression to advanced video coding*

b   The use of a very small alphabet to represent an image (maximum number of four symbols) makes adaptive arithmetic coding very efficient, because it adapts itself very quickly to any changes in the statistics of the symbols.
c   Since the maximum distortion level of a coefficient at any stage is bounded by the current yardstick length, the average distortion level in each pass is also given by the current yardstick, being the same for all bands.
d   At any given pass, only the coefficients with magnitudes larger than the current yardstick length are encoded nonzero. Therefore, the coefficients with higher magnitudes tend to be encoded before the ones with smaller magnitudes. This implies that the EZW algorithm tends to give priority to the most important information in the encoding process. This is aided by the ordering of the subordinate in step 8. Thus for the given bit rate, the bits are spent where they are needed most.
e   Since the EZW algorithm employs a successive approximation process, the addition of a new symbol (+, −, ZT and Z) to the string just further refines the reconstructed image. Furthermore, while each symbol is being added to the string it is encoded into the bit stream; hence the encoding and decoding can stop at any point, and an image with a level of refinement corresponding to the symbols encoded/decoded so far can be recovered. Therefore, the encoding and decoding of an image can stop when the bit rate budget is exhausted, which makes possible an extremely precise bit rate control. In addition, due to the prioritisation of the more important information mentioned in item *d*, no matter where in the bit stream the decoding is stopped, the best possible image quality for that bit rate is achieved.
f   *Spatial/SNR scalability*: in order to achieve spatial or SNR scalability, two different scanning methods are employed in this scheme. For spatial scalability, the wavelet coefficients are scanned in the subband-by-subband fashion, from the lowest to the highest frequency subbands. For SNR scalability, the wavelet coefficients are scanned in each tree from the top to the bottom. The scanning method is defined in the bit stream.

## 4.6 Set partitioning in hierarchical trees (SPIHT)

The compression efficiency of EZW is to some extent due to the use of arithmetic coding. Said and Pearlman [16] have introduced a variant of coding of wavelet coefficients by successive approximation, that even without arithmetic coding outperforms EZW (see Figure 4.20). They call it set partitioning in hierarchical trees (SPIHT). Both EZW and SPIHT are spatial tree-based encoding techniques that exploit magnitude correlation across bands of the decomposition. Each generates a fidelity progressive bit stream by encoding, in turn, each bit plane of a quantised dyadic subband decomposition. Both use a significance test on sets of coefficients to efficiently isolate and encode high magnitude coefficients. However, the crucial parts in the SPIHT coding process is the way the subsets of the wavelet coefficients are partitioned and the significant information is conveyed.

One of the main features of this scheme in transmitting the ordering data is that it is based on the fact that the execution path of an algorithm is defined by the results of the comparisons of its branching points. So, if the encoder and decoder have the same sorting algorithm, then the decoder can duplicate the encoder's execution path if it receives the results of the magnitude comparisons. The ordering information can be recovered from the execution path.

The sorting algorithm divides the set of wavelet coefficients, $\{C_{i,j}\}$, into partitioning subsets $T_m$ and performs the magnitude test:

$$\max_{\substack{(i,j) \\ (i,j) \in T_m}} \{|C_{i,j}|\} \geq 2^n ? \tag{4.26}$$

If the decoder receives a no to that answer (the subset is insignificant), then it knows that all coefficients in $T_m$ are insignificant. If the answer is yes (the subset is significant), then a certain rule shared by the encoder and decoder is used to partition $T_m$ into new subset $T_{m,l}$ and the significant test is then applied to the new subsets. This set division process continues until the magnitude test is done to all single coordinate significant subsets in order to identify each significant coefficient.

To reduce the number of magnitude comparisons (message bits) a set partitioning rule that uses an expected ordering in the hierarchy defined by the subband pyramid is defined (similar to Figure 4.12, used in zero tree coding). In section 4.4.2 we saw how the similarities among the subimages of the same orientation can be exploited to create a spatial orientation tree (SOT). The objective is to create new partitions such that subsets expected to be insignificant contain a huge number of elements and subsets expected to be significant contain only one element.

To make clear the relationship between magnitude comparisons and message bits, the following function is used:

$$S_n(T) = 1, \text{ if } \max_{\substack{(i,j) \\ (i,j) \in T}} \{|C_{i,j}|\} \geq 2^n$$

$$= 0, \text{ otherwise} \tag{4.27}$$

to indicate the significance of a set of coordinates $T$. To simplify the notation of single pixel sets, $S_n(\{(i, j)\})$ is represented by $S_n(i, j)$.

To see how SPHIT can be implemented, let us assume $O(i, j)$ to represent a set of coordinates of all offsprings of node $(i, j)$. For instance, except for the highest and lowest pyramid levels, $O(i, j)$ is defined in terms of its offsprings as:

$$O(i, j) = \{(2i, 2j), (2i, 2j + 1), (2i + 1, 2j), (2i + 1, 2j + 1)\} \tag{4.28}$$

We also define $D(i, j)$ as a set of coordinates of all descendants of the node $(i, j)$, and $H$, a set of coordinates of all spatial orientation tree roots (nodes in the highest pyramid level). Finally, $L(i, j)$ is defined as:

$$L(i, j) = D(i, j) - O(i, j) \tag{4.29}$$

The $O(i, j)$, $D(i, j)$ and $L(i, j)$ in a spatial orientation tree are shown in Figure 4.13.

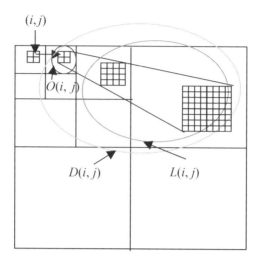

*Figure 4.13    Spatial orientation tree and the set partitioning in SPIHT*

With the use of parts of the spatial orientation trees as the partitioning subsets in the sorting algorithm, the set partitioning rules are defined as follows:

1.  The initial partition is formed with the sets $\{(i, j)\}$ and $D(i, j)$, for all $(i, j) \in H$.
2.  If $D(i, j)$ is significant, then it is partitioned into $L(i, j)$ plus the four single-element sets with $(k, l) \in O(i, j)$.
3.  If $L(i, j)$ is significant, then it is partitioned into the four sets $D(k, l)$, with $(k, l) \in O(i, j)$.
4.  Each of the four sets now has the format of the original set and the same partitioning can be used recursively.

### 4.6.1    Coding algorithm

Since the order in which the subsets are tested for significance is important, in a practical implementation the significance information is stored in three ordered lists, called the list of insignificant sets (LIS), list of insignificant pixels (LIP) and list of significant pixels (LSP). In all lists each entry is identified by a coordinate $(i, j)$, which in the LIP and LSP represents individual pixels, and in the LIS represents either the set $D(i, j)$ or $L(i, j)$. To differentiate between them, it is said that an LIS entry is of type A if it represents $D(i, j)$, and of type B if it represents $L(i, j)$ [16].

During the sorting pass, the pixels in the LIP, which were insignificant in the previous pass, are tested, and those that become significant are moved to the LSP. Similarly, sets are sequentially evaluated following the LIS order, and when a set is found to be significant it is removed from the list and partitioned. The new subsets with more than one element are added back to the LIS and the single coordinate sets are added to the end of the LIP or the LSP depending whether they are insignificant or significant, respectively. The LSP contains the coordinates of the pixels that are visited in the refinement pass.

Thus the algorithm can be summarised as:

1. Initialisation: let the initial yardstick, $n$, be $n = \lfloor \log_2(\max_{(i,j)}\{|C_{i,j}|\}) \rfloor$. Set the LSP as an empty set list, and add the coordinates $(i, j) \in H$ to the LIP, and only those with descendants also to the LIS, as the type A entries.
2. Sorting pass:
    2.1 for each entry $(i, j)$ in the LIP do:
        2.1.1 output $S_n(i, j)$;
        2.1.2 if $S_n(i, j) = 1$ then move $(i, j)$ to the LSP and output the sign of $C_{i,j}$;
    2.2 for each entry $(i, j)$ in the LIS do:
        2.2.1 if the entry is of type A then
            2.2.1.1 output $S_n(D(i, j))$;
            2.2.1.2 if $S_n(D(i, j)) = 1$ then
                2.2.1.2.1 for each $(k, l) \in O(i, j)$ do:
                    2.2.1.2.1.1 output $S_n(k, l)$;
                    2.2.1.2.1.2 if $S_n(k, l) = 1$, then add $(k, l)$ to the LSP and output the sign of $C_{k,l}$;
                    2.2.1.2.1.3 if $S_n(k, l) = 0$ then add $(k, l)$ to the end of the LIP;
                2.2.1.2.2 if $L(i, j) \neq \Phi$ then move $(i, j)$ to the end of the LIS, as an entry of type B, and go to step 2.2.2; otherwise, remove entry $(i, j)$ from the LIS;
        2.2.2 if the entry is of type B then
            2.2.2.1 output $S_n(L(i, j))$;
            2.2.2.2 if $S_n(L(i, j)) = 1$ then
                2.2.2.2.1 add each $(k, l) \in O((i, j)$ to the end of the LIS as an entry of type A;
                2.2.2.2.2 remove $(i, j)$ from the LIS.
3. Refinement pass: for each entry $(i, j)$ in the LSP, except those included in the last sorting pass (i.e. with the same $n$), output the $n$th most significant bit of $|C_{i,j}|$.
4. Quantisation step update: decrement $n$ by 1 and go to step 2.

One important characteristic of the algorithm is that the entries added to the end of the LIS above are evaluated before the same sorting pass ends. So, when it is said for each entry in the LIS, it means those that are being added to its end. Also similar to EZW, the rate can be precisely controlled because the transmitted information is formed of single bits. The encoder can estimate the progressive distortion reduction and stop at a desired distortion value.

Note that, in this algorithm, the encoder outputs all branching conditions based on the outcome of the wavelet coefficients. Thus, to obtain the desired decoder's algorithm, which duplicates the encoder's execution path as it sorts the significant coefficients, we simply replace the word output with input. The ordering information is recovered when the coordinate of the significant coefficients is added to the end of the LSP. But note that whenever the decoder inputs data, its three control lists (LIS, LIP and LSP) are identical to the ones used by the encoder at the moment it outputs that data, which means that the decoder indeed recovers the ordering from the execution path.

An additional task done by the decoder is to update the reconstructed image. For the value of $n$ when a coordinate is moved to the LSP, it is known that $2^n \leq |C_{i,j}| < 2^{n+1}$. So, the decoder uses that information, plus the sign bit that is input just after the insertion in the LSP, to set the reconstructed coefficients $\hat{C}_{i,j} = \pm 1.5 \times 2^n$. Similarly, during the refinement pass, the decoder adds or subtracts $2^{n-1}$ to $\hat{C}_{i,j}$ when it inputs the bits of the binary representation of $|C_{i,j}|$. In this manner, the distortion gradually decreases during both the sorting and refinement passes.

Finally it is worth mentioning some of the differences between EZW and SPIHT. The first difference is that they use a slightly different spatial orientation tree (SOT). In EZW, each root node in the top LL band has three offspring, one in each high frequency subband at the same decomposition level and all other coefficients have four children in the lower decomposition subband of the same orientation. However, in SPIHT, in a group of $2 \times 2$ root nodes in the top LL band, the top left node has no descendant and the other three have four offspring each in the high frequency band of the corresponding orientation. Thus SPIHT uses fewer trees with more elements per tree than in EZW. Another important difference is in their set partitioning rules. SPIHT has an additional partitioning step in which a descendant (type A) set is split into four individual child coefficients and a grand descendant (type B) set. EZW explicitly performs a breadth first search of the hierarchical trees, moving from coarser to finer subbands. Although it is not explicit, SPIHT does a roughly breadth first search as well. After partitioning a grand descendant set, SPIHT places the four new descendant sets at the end of the LIS. It is the appending to the LIS that results in the approximate breadth first traversal.

## 4.7 Embedded block coding with optimised truncation (EBCOT)

Embedded block coding with optimised truncation EBCOT [17] is another wavelet-based coding algorithm that has the capability of embedding many advanced features in a single bit stream while exhibiting state-of-the-art compression performance. Due to its rich set of features, modest implementation complexity and excellent compression performance, the EBCOT algorithm has been adopted in the evolving new still image coding standard, under the name of JPEG2000. Thus, because of the important role of EBCOT in the JPEG2000 standard, we describe this coding technique in some detail. Before describing this new method of wavelet coding, let us investigate the problem with EZW and SPIHT that caused their rejection for JPEG2000.

As we will see in Chapter 5, spatial and SNR scalability of images are among the many requirements from the JPEG2000 standard. Spatial scalability means, from the compressed bit stream, to be able to decode pictures of various spatial resolutions. This is in fact an inherent property of the wavelet transform, irrespective of how the wavelet coefficients are coded, since at any level of the decomposition the lowest band of that level gives a smaller replica of the original picture. The SNR scalability means, from the compressed bit stream, to be able to decode pictures of various qualities. In both EZW and SPIHT, this is achieved by successive approximation or bit plane encoding of the wavelet coefficients. Thus it appears that EZW and SPIHT coding of the wavelet coded images can meet these two scalability requirements. However, the bit streams of both EZW and SPIHT inherently offer only SNR scalability. If spatial scalability is required, then the bit stream should be modified accordingly, which is then not SNR scalable. This is because the zero tree-based structure used in these methods involves downward dependencies between the subbands produced by the successive wavelet decompositions. These dependencies interfere with the resolution scalability. Moreover, the interband dependency by the use of zero tree structure causes the error to propagate through the bands. This again does not meet the error resilience requirement of JPEG2000.

These shortfalls can be overcome if each subband is coded independently. Even to make coding more flexible, subband samples can be partitioned into small blocks with each block coded independently. The dependencies may exist within a block but not between different blocks. The size of the blocks determines the degree to which one is prepared to sacrifice coding efficiency in exchange for flexibility in the ordering of information within the final compressed bit stream. The independent block coding paradigm is the heart of EBCOT. This independent coding allows local processing of the samples in each block, which is advantageous for hardware implementation. It also makes highly parallel implementation possible, where multiple blocks are encoded and decoded simultaneously. More importantly, due to flexibility in the rearrangement of bits in the EBCOT, simultaneous SNR and spatial scalability is possible. Obviously, the error encountered in any block's bit stream will clearly have no influence on other blocks and hence improves the robustness. Finally, since unlike EZW and SPHIT similarities between the bands are not exploited, there is small deficiency in the compression performance. This is compensated for by the use of more efficient context-based arithmetic coding and the post compression rate distortion (PCRD) optimisation.

In EBCOT, each subband is partitioned into relatively small block of samples, called code blocks. Typical code block size is either $32 \times 32$ or $64 \times 64$ and each code block is coded independently. The actual coding algorithm in EBCOT can be

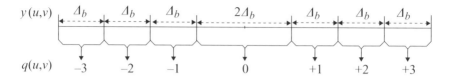

*Figure 4.14   Uniform dead zone quantiser with step size $\Delta_b$*

divided into three stages:

(i) bit plane quantisation
(ii) binary arithmetic coding (tier 1 coding)
(iii) bit stream organisation (tier 2 coding).

Each of these steps is described briefly in the following subsections.

### 4.7.1   Bit plane quantisation

All code blocks of a subband use the same quantiser. The basic quantiser step size $\Delta_b$ of a subband $b$ is selected based on either perceptual importance of the subband or rate control. The quantiser maps the magnitude of a wavelet coefficient to a quantised index, as shown in Figure 4.14, keeping its sign. It has uniform characteristics (equal step size) with a dead zone of twice the step size.

In bit plane coding, the quantiser index is encoded one bit at a time, starting from the most significant bit (MSB) and preceding to the least significant bit (LSB). If $K$ is the sufficient number of bits to represent any quantisation index in a code block, then the maximum coefficient magnitude will be $\Delta_b(2^K - 1)$. An embedded bit stream for each code block is formed by first coding the most significant bit, i.e. $(K-1)$th bit together with the sign of any significant sample for all the samples in the code block. Then the next most significant bit, i.e. $(K-2)$th bit, is coded until all the bit planes are encoded. If the bit stream is truncated then some or all of the samples in the block may be missing one or more least significant bits. This is equivalent to having used a coarser dead zone quantiser with step size $\Delta_b 2^p$, where $p$ is the index of the last available bit plane for the relevant sample or $p$ least significant bits of quantiser index still remain to be encoded.

### 4.7.2   Conditional arithmetic coding of bit planes (tier 1 coding)

During progressive bit plane coding, substantial redundancies exist between the successive bit planes. The EBCOT algorithm has exploited these redundancies in two ways. The first is to identify whether a coefficient should be coded, and the second how best the entropy coding can be adapted to the statistics of the neighbouring coefficients. Both of these goals are achieved through the introduction of the binary significance state $\sigma$, which is defined to signify the importance of each coefficient in a code block. At the beginning of the coding process, the significant states of all

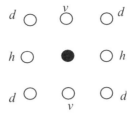

*Figure 4.15  Eight immediate neighbouring symbols*

the samples in the code block are initialised to 0 and then changed to 1 immediately after coding the first nonzero magnitude bit for the sample. Since the neighbouring coefficients generally have similar significance resulting in clusters of similar binary symbols in a plane, then the significance state, $\sigma$, is a good indicator for a bit plane of a coefficient to be a candidate for coding.

Also, in EBCOT, adaptive binary arithmetic is used for entropy coding of each symbol. Again, the clustering of significance of neighbours can be exploited to adapt the probability model, based on the significance states of its immediate neighbours. This is called context-based binary arithmetic coding, where the probability assigned to code a binary symbol of 0 or 1 or the positive/negative sign of a coefficient, is derived from the context of its eight immediate neighbours, shown in Figure 4.15.

In general, the eight immediate neighbours can have 256 different contextual states, or as many contextual states for the positive and negative signs. In EBCOT, the probability models used by the arithmetic coder are limited within 18 different contexts: nine for the significance propagation, one for run length, five for sign coding and three for refinement. These contexts are explained in the relevant parts.

Since the status of the eight immediate neighbours affects the formation of the probability model, the way coefficients in a code block are scanned should be defined. In the JPEG2000 standard, in every bit plane of a code block, the coefficients are visited in a stripe scan order with a height of four pixels, as shown in Figure 4.16.

Starting from the top left, the first four bits of the first column are scanned. Then the four bits of the second column, until the width of the code block is covered. Then the second four bits of the first column of the next stripe are scanned and so on. The stripe height of four has been chosen to facilitate hardware and software implementations.

### 4.7.3  Fractional bit plane coding

The quantised coefficients in a code block are bit plane encoded independent of other code blocks in the subbands. Instead of encoding the entire bit plane in one pass, each bit plane is encoded in three subbit plane passes, called fractional bit plane coding. The reason for this is to be able to truncate the bit stream at the end of each pass to create a near optimum bit stream. This is also known as post compression

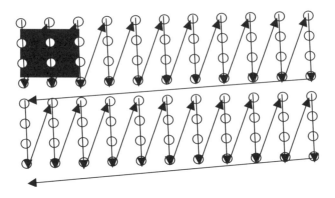

Figure 4.16  Stripe scanned order in a code block

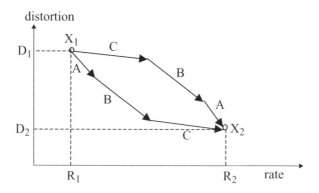

Figure 4.17  The impact of order of fractional bit plane coding in distortion reduction

rate distortion (PCRD) optimisation [18]. Here, the pass that results in the largest reduction in distortion for the smallest increase in bit rate is encoded first.

The goal and benefits of fractional bit plane coding can be understood with the help of Figure 4.17. Suppose $(R_1, D_1)$ and $(R_2, D_2)$ are the rate distortion pairs corresponding to two adjacent bit planes $p_1$ and $p_2$. Also, assume during the encoding of each pass that the increase in bit rate and reduction in distortion follow the characteristics identified by labels A, B and C. As we see, in coding the whole bit plane, that is going from point $X_1$ to point $X_2$, no matter which route is followed, the ultimate bit rate is increased from $R_1$ to $R_2$, and the distortion is reduced from $D_1$ to $D_2$. But if due to the limit in the bit rate budget, the bit rate has to be truncated between $R_1$ and $R_2$, $R_1 \leq R \leq R_2$, then it is better to follow the passes in the sequence ABC rather than CBA.

This sort of fractional bit plane coding is in fact a method of optimising the rate distortion curve, with the aim of generating a finely embedded bit stream, which

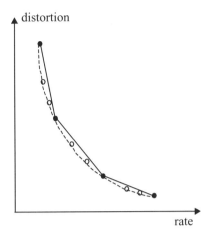

*Figure 4.18  Rate distortion with optimum trunctaion*

is known as post compression rate distortion (PCRD) optimisation in EBCOT [18]. Figure 4.18 compares the optimised rate distortion associated with the fractional bit plane encoding *versus* that of the conventional bit plane encoding. The solid dots represent the rate distortion pairs at the end of each biplane, and the solid line is the rate distortion curve that one could expect to obtain by truncating the bit stream produced by this method to an arbitrary bit rate. On the other hand, the end points associated with each pass of the fractional bit plane coding are shown with blank circles and the broken line illustrates its rate distortion curve. Since initial coding passes generally have steeper rate distortion slopes, the end point for each coding pass lies below the convex interpolation of the bit plane termination point. Thus fractional bit plane encoding results in a near optimum coding performance compared with simple bit plane coding.

Generally, in coding a coefficient the largest reduction in distortion occurs when the coefficient is insignificant, but it is more likely to become significant during the coding. Moderate reduction in distortion is when the coefficient is already significant and the coding refines it. Finally, the least reduction in distortion is when the insignificant coefficient after the encoding is likely to remain insignificant. These are in fact the remaining coefficients not coded in the two previous cases. Thus it is reasonable to divide bit plane encoding into three passes, and encode each pass in the above encoding order. In JPEG2000, the fractional bit plane is carried out in three passes. The roles played by each encoding pass, and their order of significance in generating an optimum bit stream, are given below.

### 4.7.3.1  Significance propagation pass

This is the first pass of the fractional bit plane encoding that gives the largest reduction in the encoding distortion. In this pass, the bit of a coefficient in a given bit plane is encoded, if and only if, prior to this pass, the state of the coefficient was insignificant but at least one of its eight immediate neighbours had significant states. If the coefficient is to be coded, the magnitude of its bit, 0 or 1, is arithmetically coded with a derived probability model from the context of its eight immediate neighbours, shown in Figure 4.15. The probability assigned to bit 0 is complementary to the probability assigned to bit 1. The context selection is based upon the significance of the sample's eight immediate neighbours, which are grouped in three categories.

$$\begin{aligned}\text{horizontal: } & h_i(u, v) = \sigma_i(u-1, v) + \sigma_i(u+1, v) \\ \text{vertical: } & v_i(u, v) = \sigma_i(u, v-1) + \sigma_i(u, v+1) \\ \text{diagonal: } & d_i(u, v) = \sum_{m=\pm 1}\sum_{n=\pm 1} \sigma_i(u+m, v+n)\end{aligned} \quad (4.30)$$

where $h_i(u, v)$, $v_i(u, v)$ and $d_i(u, v)$ are the horizontal, vertical and diagonal neighbours, for the $i$th coefficient at coordinates $(u, v)$, and $\sigma_i(u, v)$ is the significance state of a coefficient at those coordinates. Neighbours that lie outside the code block are interpreted as insignificant for the purpose of constructing these three quantities. To optimise both the model adaptations of cost and implementation complexity, 256 possible neighbourhood configurations are mapped to nine distinct coding contexts based on eqn. 4.30, as shown in Table 4.4.

To make identical context assignment for LH and HL bands, the code blocks of HL subbands are transposed before the coding. The LH subband responds most strongly to horizontal edges in the original image, so the context mapping gives more emphasis on the horizontal neighbours.

Note that the significance propagation pass includes only those bits of coefficients that were insignificant before this pass and have a nonzero context. If the bit of the coefficient is 1 (the coefficient becomes significant for the first time), then its state

Table 4.4  Assignment of the nine contexts based on neighbourhood significance

| LL, LH and HL bands | | | | HH band | | |
|---|---|---|---|---|---|---|
| $h_i(u, v)$ | $v_i(u, v)$ | $d_i(u, v)$ | context | $d_i(u, v)$ | $h_i(u, v) + v_i(u, v)$ | context |
| 0 | 0 | 0 | 0 | 0 | 0 | 0 |
| 0 | 0 | 1 | 1 | 0 | 1 | 1 |
| 0 | 0 | >1 | 2 | 0 | >1 | 2 |
| 0 | 1 | X | 3 | 1 | 0 | 3 |
| 0 | 2 | X | 4 | 1 | 1 | 4 |
| 1 | 0 | 0 | 5 | 1 | >1 | 5 |
| 1 | 0 | >0 | 6 | 2 | 0 | 6 |
| 1 | >0 | X | 7 | 2 | >0 | 7 |
| 2 | X | X | 8 | >2 | X | 8 |

of significance, $\sigma$, is changed to 1, to affect the context of its following neighbours. Thus the significance of states of coefficients propagates throughout the coding, and hence the name given to this pass is the significance propagation pass. Note also that if a sample is located at the boundary of a block, then only the available immediate neighbours are considered and the significance state of the missing neighbours are assumed to be zero.

Finally, if a coefficient is found to be significant, its sign is also arithmetically coded. Since the sign bits from the adjacent samples exhibit substantial statistical dependencies, they can be effectively exploited to improve the arithmetic coding efficiency. For example, the wavelet coefficients of horizontal and vertical edges are likely to be of the same polarity. Those after and before the edge are of mainly opposite polarity. In the EBCOT algorithm, the arithmetic coding of a sign bit employs five contexts. Context design is based upon the relevant sample's immediate horizontal and vertical neighbours, each of which may be in one of the three states: significant and positive, significant and negative, insignificant. There are thus $3^4 = 81$ unique neighbourhood configurations. The details of the symmetry configurations and approximations to map these 81 configurations to one of the five context levels can be found in [18].

### 4.7.3.2 Magnitude refinement pass

The magnitude refinement pass is the second most efficient encoding pass. During this pass, the magnitude bit of a coefficient that has already become significant in a previous bit plane is arithmetically coded. The magnitude refinement pass includes the bits from the coefficients that are already significant, except those that have just become significant in the immediately preceding significance propagation pass. There are three contexts for the arithmetic coder, which are derived from the summation of the significance states of the horizontal, vertical and diagonal neighbours. These are the states as currently known to the decoder and not the states used before the significance decoding pass. Further, it is dependent on whether this is the first refinement bit (the bit immediately after the significance and sign bits) or not.

In general, the refinement bits have an even distribution, unless the coefficient has just become significant in the previous bit plane (i.e. the magnitude bit to be encoded is the first refinement bit). This condition is first tested and, if it is satisfied, the magnitude bit is encoded using two contexts, based on the significance of the eight immediate neighbours (see Figure 4.15). Otherwise, it is coded with a single context regardless of the neighbouring values.

### 4.7.3.3 Clean up pass

All the bits not encoded during the significance propagation and refinement passes are encoded in the clean up pass. That is the coefficients that are insignificant and had the context value of zero (none of the eight immediate neighbours was significant) during the significance propagation pass. Generally, the coefficients coded in this pass have a very small probability of being significant and hence are expected to remain insignificant. Therefore, a special mode, called the run mode, is used to aggregate the coefficients of remaining insignificance. A run mode is entered if all the four samples in a vertical column of the stripe of Figure 4.16 have insignificant neighbours. Specifically, run mode is executed when each of the following conditions holds:

- four consecutive samples must all be insignificant, i.e. $\sigma_i(u+m,v) = 0$, for $0 \leq m \leq 3$
- the samples must have insignificant neighbours, i.e. $h_i(u+m,v) = v_i(u+m,v) = d_i(u+m,v) = 0$, for $0 \leq m \leq 3$
- samples must reside within the same subblock
- the horizontal index of the first sample, $u$, must be even.

In the run mode a binary symbol is arithmetically coded with a single context to specify whether all four samples in the vertical column remain insignificant. Symbol 0 implies that all the four samples are insignificant and symbol 1 implies that at least one of four samples becomes significant in the current bit plane. If the symbol is 1, then two additional arithmetically coded bits are used to specify the location of the first nonzero coefficient in the vertical column.

Since it is equally likely that any one of the four samples in the column is the first nonzero sample, then the arithmetic coding uses a uniform context. Thus run mode has a negligible role in coding efficiency. It is primarily used to improve the throughput of the arithmetic encoder through symbol aggregation.

After specifying the position of the first nonzero symbol in the run, the remaining samples in the vertical column are coded in the same manner as in the significance propagation pass and use the same nine contexts. Similarly, if at least one of the four coefficients in the vertical column has a significant neighbour, the run mode is disabled and all the coefficients in that column are again coded with the procedure used for the significance propagation pass.

For each code block, the number of MSB planes that are entirely zero is signalled in the bit stream. Since the significance state of all the coefficients in the first nonzero MSB is zero, this plane only uses the clean up pass and the other two passes are not used.

**Example**

In order to show how the wavelet coefficients in a fractional bit plane are coded, Figure 4.19 illustrates a graphical demonstration of step-by-step encoding from bit plane to bit plane and pass to pass.

The Barbara image of size $256 \times 256$ pixels with two levels of wavelet decomposition generates seven subimages, as shown in Figure 4.19. Except for the lowest band, the magnitudes of all the other bands are magnified by a factor of four, for better illustration of image details. The code block size is assumed to be a square array of $64 \times 64$ coefficients. Hence, every high frequency band of LH1, HL1 and HH1 is coded in four code blocks, and the remaining bands of LL2, LH2, HL2 and HH2 in one code block each. That is, the whole image is coded in 16 code blocks. The bit

Subband and wavelet 97

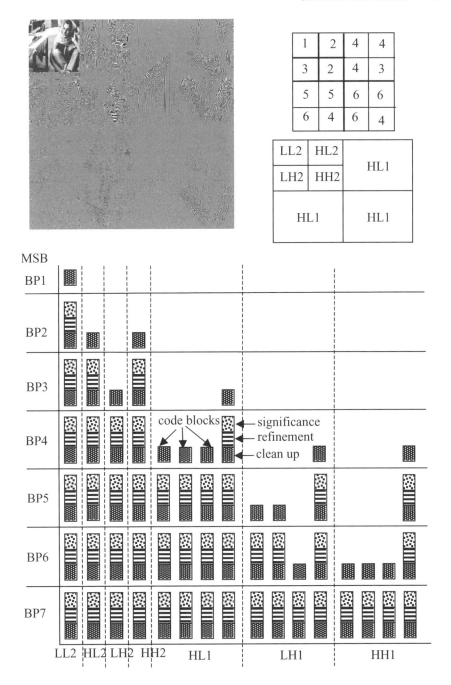

Figure 4.19 An illustration of fractional bit plane encoding

streams generated by each pass in every bit plane are also shown in square boxes with different textures.

Before the coding starts, the significance states of all the code blocks are initialised to zero. For each code block the encoding starts from the most significant bit plane. Since the LL2 band has a higher energy (larger wavelet coefficients) than the other bands, in this example only some of the MSB of the code block of this band are significant at the first scanned bit plane, PB1. In this bit plane the MSB of none of the other code blocks is significant (they are entirely zero), and are not coded at all.

Since the code block of LL2 band in BP1 is coded for the first time (the significant states of all the coefficients are initialised to zero), then the MSB bit of every coefficient has a nonsignificant neighbour and hence cannot be coded at the significance propagation pass. Also, none of the bits are coded at the refinement pass, because these coefficients had not been coded in the previous bit plane. Therefore, all the bits are left to be coded in the clean up pass, and they constitute the bit stream of this pass.

At the second bit plane, BP2, some of the coefficients in the HL2 and HH2 code blocks become significant for the first time. Hence, as was explained earlier, they are coded only at the clean up pass, as shown in the Figure. For the code block of LL2, since the coefficients with significant magnitudes of this code block have already been coded at the clean up pass of BP1, the block uses all the three passes. Those coefficients of the bit plane BP2 with insignificant states that have at least one state significant immediate neighbour are coded with the significance propagation pass. The state significant coefficients are now refined in the refinement pass. The remaining coefficients are coded at the clean up pass. Other code blocks in this bit plane are not coded at all.

At the third bit plane, BP3, some of the coefficients of the code block of subband LH2 and those of the code block of subband HL1 become significant for the first time, hence they are coded only in the clean up pass. The code blocks of LL2, HL2 and HH2 are now coded at the three passes. The remaining code blocks are not coded at all. The bit plane numbers of those code blocks that are coded for the first time are also shown in the Figure with numbers from 1 to 6. As we see, after bit plane 7 all the code blocks of the bands are coded in all three passes.

### 4.7.4 Layer formation and bit stream organisation (tier 2 coding)

The arithmetic coding of the bit plane data is referred as tier 1 coding. The tier 1 coding generates a collection of bit streams with one independent embedded bit stream for each code block. The purpose of tier 2 coding is to multiplex the bit streams for transmission and to signal the ordering of the resulting coded bit plane pass in an efficient manner. The second tier coding process can be best viewed as a somewhat elaborated parser for recovering pointers to code block segments in the bit stream. It is this coding step which enables the bit stream to have SNR, spatial and arbitrary progression and scalability.

The compressed bit stream from each code block is distributed across one or more layers in the final compressed bit stream. Each layer represents a quality increment. The number of passes included in a specific layer may vary from one code block to another and is typically determined by the encoder as a result of PCRD optimisation. The quality layers provide the feature of SNR scalability of the final bit stream such that truncating the bit stream to any whole number of layers yields approximately an optimal rate distortion representation of the image. However, employing a large number of quality layers can minimise the approximation. On the other hand, more quality layers implies a large overhead as auxiliary information to identify the contribution made by each code block to each layer. When the number of layers is large, only a subset of the code blocks will contribute to any given layer, introducing substantial redundancy in this auxiliary information. This redundancy is exploited in tier 2 coding to efficiently code the auxiliary information for each quality layer.

*4.7.5 Rate control*

Rate control refers to the process of generating an optimal image for a bit rate and is strictly an encoder issue. In section 4.7.3 we introduced fractional bit plane coding as one of these methods. In [17], Taubman proposes an efficient rate control method for the EBCOT compression algorithm that achieves a desired rate in a single iteration with minimum distortion. This method is known as post compression rate distortion (PCRD) optimisation. A JPEG2000 encoder with several possible variations can also use this method.

In another form of PCRD, each subband is first quantised using a very fine step size, and the bit planes of the resulting code blocks are entropy coded (tier 1 coding). This typically generates more coding passes for each code block than will be eventually included in the final bit stream. Next, a Lagrangian R–D optimisation is performed to

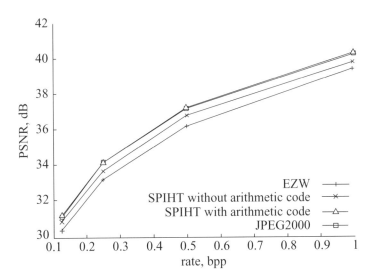

*Figure 4.20    Compression performance of various wavelet coding algorithms*

determine the number of coding passes from each code block that should be included in the final compressed bit stream to achieve the desired bit rate. If more than a single quality layer is desired, this process can be repeated at the end of each layer to determine the additional number of coding passes from each code block that need to be included in the next layer. The details of PCRD optimisation can be found in [18].

At the end of this Chapter, it is worth comparing the compression efficiency of the three methods of wavelet coding, namely: EZW, SPIHT and the EBCOT. Figure 4.20 shows the quality of the Lena image coded with these methods at various bit rates. As the Figure shows, EZW has the poorest performance of all. The SPIHT method, even without arithmetic coding, outperforms EZW by about 0.3–0.4 dB. Adding arithmetic coding into SPIHT improves the coding efficiency by another 0.3 dB. The EBCOT algorithm, adopted in the JPEG2000 standard, is as good as the best of SPIHT.

## 4.8 Problems

1 Derive the analysis and synthesis subband filters of the product filter $P(z)$ defined by its $z$ transform as:

$$P(z) = \tfrac{1}{16}(-1 + 9z^{-2} + 16z^{-3} + 9z^{-4} - z^{-6})$$

2 The $z$ transform of a pair of lowpass and highpass analysis filters is given by:

$$H_0(z) = \tfrac{1}{\sqrt{2}}(1 + z^{-1}) \quad \text{and} \quad H_1(z) = \tfrac{1}{\sqrt{2}}(1 - z^{-1})$$

  a Calculate the product filter, and deduce the amount of end to end encoding/decoding delay
  b Derive the corresponding pairs of the synthesis filters.

3 The product filter for wavelets with $n$ zeros at $z = -1$ is given by:

$$P(z) = (1 + z^{-1})^{2n} Q(z)$$

Use eqn. 4.7 to calculate the weighting factor of the product filter $P(z)$ and the corresponding end-to-end coding delay, for:

  a $n = 2$ and $Q(z) = -1 + 4z^{-1} - z^{-2}$
  b $n = 3$ and $Q(z) = 1 - 6z^{-1} + \tfrac{38}{3}z^{-2} - 6z^{-3} + z^{-4}$

4 Show that the low and highpass analysis filters of $a$ in problem 3 are in fact the $(5, 3)$ and $(4, 4)$ subband filters of LeGall and Tabatabai.

5 Show that the filters for $b$ in problem 3 are the 9/3 Daubechies filter pairs.

6 Derive the corresponding pairs of the synthesis filters of problems 4 and 5.

7 List major similarities and differences between EZW and SPIHT.

|  |  |  |  |  |  |  |  |
|---|---|---|---|---|---|---|---|
| 63 | −34 | 49 | 10 | 7 | 13 | −12 | 7 |
| −31 | 23 | 14 | −13 | 3 | 4 | 6 | −1 |
| 15 | 14 | 3 | −12 | 5 | −7 | 3 | 9 |
| −9 | −7 | −14 | 8 | 4 | −2 | 3 | 2 |
| −5 | 9 | −1 | 47 | 4 | 6 | −2 | 2 |
| 3 | 0 | −3 | 2 | 3 | −2 | 0 | 4 |
| 2 | −3 | 6 | −4 | 3 | 6 | 3 | 6 |
| 2 | 11 | 5 | 6 | 0 | 3 | −4 | 4 |

*Figure 4.21*

8  Consider a simple wavelet transform of 8 × 8 image shown in Figure 4.21. Assume it has been generated by a three-level wavelet transform:

   a  Show different steps in the first dominant pass (most significant bit plane) of the EZW algorithm
   b  Assuming that EZW uses two bits to code symbols in the alphabet {POS, NEG, ZT, Z} and one bit to code the sign bit, calculate total number of bits outputted in this pass.

9  a  In Figure 4.21, calculate the number of bits outputted in the first significant pass of the SPIHT algorithm
   b  Comment on the results of 8b and 9a.

## 4.9  References

1  MPEG-4: 'Video shape coding'. ISO/IEC JTC1/SC29/WG11, N1584, March 1997
2  SKODRAS, A., CHRISTOPOULOS, C., and EBRAHIMI, T.: 'The JPEG 2000 still image compression standard', *IEEE Signal Process. Mag.*, September 2001, pp.36–58
3  MALAVER, H.S., and STAELIN, D.H.: 'The LOT: transform coding without blocking effects', *IEEE Trans. Acoustic, Speech Signal Process.*, 1989, **37:4**, pp.553–559

4 SHAPIRO, J.M.: 'Embedded image coding using zero-trees of wavelet coefficients', *IEEE Trans. Signal Process.*, 1993, **4:12**, pp.3445–3462
5 USEVITCH, B.E.: 'A tutorial on modern lossy wavelet image compression: foundation of JPEG 2000', *IEEE Signal Process. Mag.*, September 2001, pp.22–35
6 CROCHIERE, R.E., WEBER, S.A., and FLANAGAN, J.L.: 'Digital coding of speech in sub-bands', *Bell Syst. Tech. J.*, 1967, **55**, pp.1069–1085
7 DAUBECHIES, I.: 'Orthonormal bases of compactly supported wavelets', *Communications on Pure and Applied Mathematics*, 1988, XLI, pp.909–996
8 LE GALL, D., and TABATABAI, A.: 'Subband coding of images using symmetric short kernel filters and arithmetic coding techniques'. IEEE international conference on *Acoustics, speech and signal processing*, ICASSP'98, 1988, pp.761–764
9 DAUBECHIES, I.: 'The wavelet transform, time frequency localization and signal analysis', *IEEE Trans. Inf. Theory*, 1990, **36:5**, pp.961–1005
10 MALLAT, S.: 'A theory of multi-resolution signal decomposition: the wavelet representation', *IEEE Trans. Pattern Anal. Mach. Intell.*, 1989, **11:7**, pp.674–693
11 DAUBECHIES, I.: 'Orthogonal bases of compactly supported wavelets II, variations on a theme', *SIAM J. Math. Anal.*, 1993, **24:2**, pp.499–519
12 DA SILVA, E.A.B.: 'Wavelet transforms in image coding'. PhD thesis, 1995, University of Essex, UK
13 RABBANI, M., and Josh, R.: 'An overview of the JPEG2000 still image compression standard', *Signal Process., Image Commun.*, 2002, **17:1**, pp.15–46
14 MPEG-4: 'Video verification model version-11'. ISO/IEC JTC1/SC29/WG11, N2171, Tokyo, March 1998
15 WITTEN, I.H., NEAL, R.M., and CLEARY, J.G.: 'Arithmetic coding for data compression', *Commun. ACM*, 1987, **30:6**, pp.520–540
16 SAID, A., and PEARLMAN, W.A.: 'A new, fast and efficient image codec based on set partitioning in hierarchical trees', *IEEE Trans. Circuits Syst. Video Technol.*, 1996, **6:3**, pp.243–250
17 TAUBMAN, D.: 'High performance scalable image compression with EBCOT', *IEEE Trans. Image Process.*, 2000, **9:7**, pp.1158–1170
18 TAUBMAN, D., Ordenlich, E., WEINBERGER, M., and SEROUSSI, G.: 'Embedded block coding in JPEG2000', *Signal Process., Image Commun.*, 2002, **17:1**, pp.1–24

*Chapter 5*
# Coding of still pictures (JPEG and JPEG2000)

In the mid 1980s joint work by the members of the ITU-T (International Telecommunication Union) and the ISO (International Standards Organisation) led to standardisation for the compression of grey scale and colour still images [1]. This effort was then known as JPEG: the Joint Photographic Experts Group. As is apparent, the word joint refers to the collaboration between the ITU-T and ISO. The JPEG encoder is capable of coding full colour images at an average compression ratio of 15 : 1 for subjectively transparent quality [2]. Its design meets special constraints, which make the standard very flexible. For example, the JPEG encoder is parametrisable, so that the desired compression/quality trade-offs can be determined based on the application or the wishes of the user [3].

JPEG can also be used in coding of video, on the basis that video is a succession of still images. In this case the process is called motion JPEG. Currently, motion JPEG has found numerous applications, the most notable one being video coding for transmission over packet networks with unspecified bandwidth or bit rates (UBR). A good example of UBR networks is the Internet where, due to unpredictability of the network load, congestion may last for a significant amount of time. Since in motion JPEG each frame is independently coded, it is an ideal encoder of video for such a hostile environment.

Another application of motion JPEG is video compression for recording on magnetic tapes, where again the independent coding of pictures increases the flexibility of the encoder for recording requirements, such as editing, pause, fast forward, fast rewind etc. Also, such an encoder can be very resilient to loss of information, since the channel error will not propagate through the image sequence. However, since the coding of I-pictures in the MPEG-2 standard is similar to motion JPEG, normally video compression for recording purposes is carried out with the I-picture part of the MPEG-2 encoder. The I-pictures are those which are encoded without reference to previous or subsequent pictures. This will be explained in Chapter 7.

At the turn of the millennium, the JPEG committee decided to develop another standard for compression of still images, named the JPEG2000 standard [4]. This was in response to growing demands for multimedia, Internet and a variety of digital imagery applications. However, in terms of compression methodology these two standards are very different. Hence in order to discriminate between them, throughout the book, the original JPEG is called JPEG and the new one JPEG2000.

## 5.1 Lossless compression

The JPEG standard specifies two classes of encoding and decoding, namely lossless and lossy compression. Lossless compression is based on a simple predictive DPCM method using neighbouring pixel values, and DCT is employed for the lossy mode.

Figure 5.1 shows the main elements of a lossless JPEG image encoder.

The digitised source image data in the form of either RGB or $YC_bC_r$ is fed to the predictor. The image can take any format from 4:4:4 down to 4:1:0, with any size and amplitude precision (e.g. 8 bits/pixel). The predictor is of the simple DPCM type (see Figure 3.1), where every individual pixel of each colour component is differentially encoded. The prediction for an input pixel $x$ is made from combinations of up to three neighbouring pixels at positions $a$, $b$ and $c$ from the same picture of the same colour component, as shown in Figure 5.2.

The prediction is then subtracted from the actual value of the pixel at position $x$, and the difference is losslessly entropy coded by either Huffman or arithmetic coding. The entropy table specifications unit determines the characteristics of the variable length codes of either entropy coding method.

The encoding process might be slightly modified by reducing the precision of input image samples by one or more bits prior to lossless coding. For lossless processes, sample precision is specified to be between 2 and 16 bits. This achieves higher

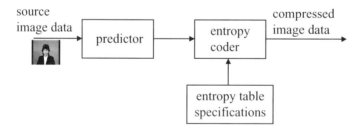

*Figure 5.1    Lossless encoder*

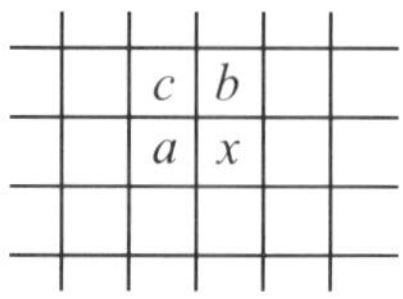

*Figure 5.2    Three-sample prediction neighbourhood*

compression than normal lossless coding, but has lower compression than DCT-based lossy coding for the same bit rate and image quality. Note that this is in fact a type of lossy compression, since reduction in the precision of input pixels by $b$ bits is equivalent to the quantisation of the difference samples by a quantiser step size of $2^b$.

## 5.2 Lossy compression

In addition to lossless compression, the JPEG standard defines three lossy compression modes. These are called the baseline sequential mode, the progressive mode and the hierarchical mode. These modes are all based on the discrete cosine transform (DCT) to achieve a substantial compression while producing a reconstructed image with high visual fidelity. The main difference between these modes is the way in which the DCT coefficients are transmitted.

The simplest DCT-based coding is referred to as the baseline sequential process and it provides capability that is sufficient for many applications. The other DCT-based processes which extend the baseline sequential process to a broader range of applications are referred to as extended DCT-based processes. In any extended DCT-based decoding processes, baseline decoding is required to be present in order to provide a default decoding capability.

### 5.2.1 Baseline sequential mode compression

Baseline sequential mode compression is usually called baseline coding for short. In this mode, an image is partitioned into $8 \times 8$ nonoverlapping pixel blocks from left to right and top to bottom. Each block is DCT coded, and all the 64 transform coefficients are quantised to the desired quality. The quantised coefficients are immediately entropy coded and output as part of the compressed image data, thereby minimising coefficient storage requirements.

Figure 5.3 illustrates the JPEG's baseline compression algorithm. Each eight-bit sample is level shifted by subtracting $2^{8-1=7} = 128$ before being DCT coded. This is known as DC level shifting. The 64 DCT coefficients are then uniformly quantised according to the step size given in the application-specific quantisation matrix. The use of a quantisation matrix allows different weighting to be applied according to the sensitivity of the human visual system to a coefficient of the frequency.

Two examples of quantisation tables are given in Tables 5.1 and 5.2. [5]. These tables are based on psychovisual thresholding and are derived empirically using luminance and chrominance with a 2 : 1 horizontal subsampling. These tables may not be suitable for any particular application, but they give good results for most images with eight-bit precision.

If the elements of the quantisation tables of luminance and chrominance are represented by $Q(u, v)$, then a quantised DCT coefficient with horizontal and vertical spatial frequencies of $u$ and $v$, $F^q(u, v)$, is given by:

$$F^q(u, v) = \left\lfloor \frac{F(u, v)}{Q(u, v)} \right\rfloor \tag{5.1}$$

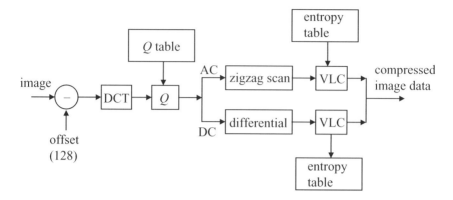

*Figure 5.3   Block diagram of a baseline JPEG encoder*

*Table 5.1   Luminance Q table*

| 16 | 11 | 10 | 16 | 24  | 40  | 51  | 61  |
|----|----|----|----|-----|-----|-----|-----|
| 12 | 12 | 14 | 19 | 26  | 58  | 60  | 55  |
| 14 | 13 | 16 | 24 | 40  | 57  | 69  | 56  |
| 14 | 17 | 22 | 29 | 51  | 87  | 80  | 62  |
| 18 | 22 | 37 | 56 | 68  | 109 | 103 | 77  |
| 24 | 35 | 55 | 64 | 81  | 104 | 113 | 92  |
| 49 | 64 | 78 | 87 | 103 | 121 | 120 | 101 |
| 72 | 92 | 95 | 98 | 112 | 100 | 103 | 99  |

*Table 5.2   Chrominance Q table*

| 17 | 18 | 24 | 47 | 99 | 99 | 99 | 99 |
|----|----|----|----|----|----|----|----|
| 18 | 21 | 26 | 66 | 99 | 99 | 99 | 99 |
| 24 | 26 | 56 | 99 | 99 | 99 | 99 | 99 |
| 47 | 66 | 99 | 99 | 99 | 99 | 99 | 99 |
| 99 | 99 | 99 | 99 | 99 | 99 | 99 | 99 |
| 99 | 99 | 99 | 99 | 99 | 99 | 99 | 99 |
| 99 | 99 | 99 | 99 | 99 | 99 | 99 | 99 |
| 99 | 99 | 99 | 99 | 99 | 99 | 99 | 99 |

where $F(u, v)$ is the transform coefficient value prior to quantisation, and $\lfloor . \rfloor$ means rounding the division to the nearest integer. At the decoder the quantised coefficients are inverse quantised by:

$$F^Q(u, v) = F^q(u, v) \times Q(u, v) \tag{5.2}$$

to reconstruct the quantised coefficients.

A quality factor $q\_JPEG$ is normally used to control the elements of the quantisation matrix $Q(u, v)$ [6]. The range of $q\_JPEG$ percentage values is between 1 and 100 per cent. The JPEG quantisation matrices of Tables 5.1 and 5.2 are used for $q\_JPEG = 50$, for the luminance and chrominance, respectively. For other quality factors, the elements of the quantisation matrix, $Q(u, v)$, are multiplied by the compression factor $\alpha$, defined as [6]:

$$\alpha = \frac{50}{q\_JPEG} \quad \text{if } 1 \leq q\_JPEG \leq 50$$
$$\alpha = 2 - \frac{2 \times q\_JPEG}{100} \quad \text{if } 50 \leq q\_JPEG \leq 99 \tag{5.3}$$

subject to the condition that the minimum value of the modified quantisation matrix elements, $\alpha Q(u, v)$, is 1. For 100 per cent quality, $q\_JPEG = 100$, that is lossless compression; all the elements of $\alpha Q(u, v)$ are set to 1.

After quantisation, the DC (commonly referred to as (0,0)) coefficient and the 63 AC coefficients are coded separately as shown in Figure 5.3. The DC coefficients are DPCM coded with prediction of the DC coefficient from the previous block, as shown in Figure 5.4, i.e. DIFF $= DC_i - DC_{i-1}$. This separate treatment from the AC coefficients is to exploit the correlation between the DC values of adjacent blocks and to code them more efficiently as they typically contain the largest portion of the image energy. The 63 AC coefficients starting from coefficient AC(1,0) are run length coded following a zigzag scan as shown in Figure 5.4.

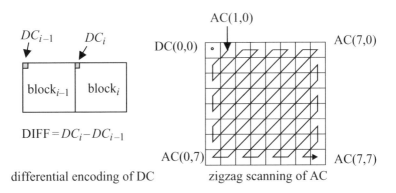

Figure 5.4    Preparing of the DCT coefficients for entropy coding

The adoption of a zigzag scanning pattern is to facilitate entropy coding by encountering the most likely nonzero coefficients first. This is due to the fact that, for most natural scenes, the image energy is mainly concentrated in a few low frequency transform coefficients.

## 5.2.2 Run length coding

Entropy coding of the baseline encoder is accomplished in two stages. The first stage is the translation of the quantised DCT coefficients into an intermediate set of symbols. In the second stage, variable length codes are assigned to each symbol. For the JPEG standard a symbol is structured in two parts: a variable length code (VLC) for the first part, normally referred to as symbol-1, followed by a binary representation of the amplitude for the second part, symbol-2.

### 5.2.2.1 Coding of DC coefficients

Instead of assigning individual variable length codewords (e.g. Huffman code) to each DIFF, the DIFF values are categorised based on the magnitude range called CAT. The CAT is then variable length coded. Table 5.3 shows the categories for the range of amplitudes in the baseline JPEG. Since the DCT coefficient values are in the range $-2047$ to $2047$, there are 11 categories for nonzero coefficients. Category zero is not used for symbols; it is used for defining the end of block (EOB) code.

The CAT after being VLC coded is appended with additional bits to specify the actual DIFF values (amplitude) within the category. Here CAT is symbol-1 and the appended bits represent symbol-2.

When the DIFF is positive the appended bits are just the lower order bits of the DIFF. When it is negative, the appended bits become the lower order bits of DIFF-1. The lower order bits start from the point where the most significant bit of the appended

Table 5.3  The category (CAT) of the baseline encoder

| CAT | Range |
|---|---|
| 0 | — |
| 1 | $-1, 1$ |
| 2 | $-3, -2, 2, 3$ |
| 3 | $-7, \ldots -4, 4, \ldots 7$ |
| 4 | $-15, \ldots -8, 8, \ldots 15$ |
| 5 | $-31, \ldots -16, 16, \ldots 31$ |
| 6 | $-63, \ldots -32, 32, \ldots 63$ |
| 7 | $-127, \ldots -64, 64, \ldots 127$ |
| 8 | $-255, \ldots -128, 128, \ldots 255$ |
| 9 | $-511, \ldots -256, 256, \ldots 511$ |
| 10 | $-1023, \ldots -512, 512, \ldots 1023$ |
| 11 | $-2047, \ldots -1024, 1024, \ldots 2047$ |

bit sequence is one for positive differences and zero for negative differences. For example: for DIFF = 6 = 0000...00110, the appended bits start from 1, hence it would be 110. This is because DIFF is positive and the most significant bit of the appended bits should be 1. Also, since 6 is in the range of 4 to 7 (Table 5.3) the value of CAT is 3. From Table B.1 of Appendix B, the codeword for CAT = 3 is 100. Thus the overall codeword for DIFF = 6 is **100110**, where 100 is the VLC code of CAT (symbol-1) and 110 is the appended codeword (symbol-2).

For a negative DIFF, such as DIFF = −3, first of all −3 is in the range of −3 to −2, thus from Table 5.3, CAT = 2, and its VLC code from Table B.1 of Appendix B is 011. However, to find the appended bits, DIFF − 1 = −4 = 1111...100, where the lower order bits are 00. Note the most significant bit of the appended bits is 0. Thus the codeword becomes **01100**.

### 5.2.2.2 Coding of AC coefficients

For each nonzero AC coefficient in zigzag scan order, symbol-1 is described as a two-dimensional event of (RUN, CAT), sometimes called (RUN, SIZE). For the baseline encoder, CAT is the category for the amplitude of a nonzero coefficient in the zigzag order, and RUN is the number of zeros preceding this nonzero coefficient. The maximum length of run is limited to 15. Encoding of runs greater than 15 is done by a special symbol (15, 0), which is a run length of 15 zero coefficients followed by a coefficient of zero amplitude. Hence it can be interpreted as the extension symbol with 16 zero coefficients. There can be up to three consecutive (15, 0) symbols before the terminating symbol-1 followed by a single symbol-2. For example a (RUN = 34, CAT = 5) pair would result in three symbols $a$, $b$, and $c$, with $a = (15, 0)$, $b = (15, 0)$ and $c = (2, 5)$.

An end of block (EOB) is designated to indicate that the rest of the coefficients of the block in the zigzag scanning order are quantised to zero. The EOB symbol is represented by (RUN = 0, CAT = 0).

The AC code table for symbol-1 consists of one Huffman codeword (maximum length 16 bits, not including additional bits) for each possible composite event. Table B.2 of Appendix B shows the codewords for all possible combinations of RUN and CAT of symbol-1 [4]. The format of the additional bits (symbol-2) is the same as in the coding of DIFF in DC coefficients. For the $k$th AC coefficient in the zigzag scan order, ZZ($k$), the additional bits are either the lower order bits of ZZ($k$) when ZZ($k$) is positive, or the lower order bits of ZZ($k$) − 1, when ZZ($k$) is negative. In order to clarify this, let us look at a simple example.

Example: the quantised DCT coefficients of a luminance block are shown in Figure 5.5. If the DC coefficient in the previous luminance block was 29, find the codewords for coding of the DC and AC coefficients.

*Codeword for the DC coefficient*
DIFF = 31 − 29 = 2. From Table 5.3, CAT = 2 and according to Table B.1 of Appendix B, the Huffman code for this value of CAT is 011. To find the appended

110  *Standard codecs: image compression to advanced video coding*

|  |  |  |  |  |  |  |  |
|---|---|---|---|---|---|---|---|
| 31 | 18 | 0 | 0 | 0 | 0 | 0 | 0 |
| −21 | −13 | 0 | 0 | 0 | 0 | 0 | 0 |
| 0 | 5 | 0 | 0 | 0 | 0 | 0 | 0 |
| 0 | 0 | 0 | 0 | 0 | 0 | 0 | 0 |
| 0 | 0 | 0 | 0 | 0 | 0 | 0 | 0 |
| 0 | 0 | 0 | 0 | 0 | 0 | 0 | 0 |
| 0 | 0 | 0 | 0 | 0 | 0 | 0 | 0 |
| 0 | 0 | 0 | 0 | 0 | 0 | 0 | 0 |

*Figure 5.5   Quantised DCT coefficients of a luminance block*

bits, since DIFF $= 2 > 0$, then $2 = 000\ldots 00\underline{10}$. Thus the appended bits are 10. Hence the overall codeword for coding the DC coefficient is 01110.

*Codewords for the AC coefficients*

Scanning starts from the first nonzero AC coefficient that has a value of 18. From Table 5.3, the CAT value for 18 is 5, and since there is no zero value AC coefficient before it, then RUN $= 0$. Hence, symbol-1 is (0, 5). From Table B.2 of Appendix B, the codeword for (0, 5) is 11010. The symbol-2 is the lower order bits of $ZZ(k) = 18 = 000\ldots 0\underline{10010}$, which is 10010. Thus the first AC codeword is 1101010010.

The next nonzero AC coefficient in the zigzag scan order is −21. Since it has no zero coefficient before it and it is in the range of −31 to −16, then it has a RUN $= 0$, and CAT $= 5$. Thus symbol-1 of this coefficient is (0, 5), which again from Table B.2 of Appendix B, has a codeword of 11010. For symbol-2, since $-21 < 0$, then $ZZ(k) - 1 = -21 - 1 = -22 = 111\ldots 1 1\underline{01010}$, and symbol-2 becomes 01010 (note the appended bits start from where the most significant bit is 0). Thus the overall codeword for the second nonzero AC coefficient is 1101001010.

The third nonzero AC coefficient in the scan is −13, which has one zero coefficient before it. Then RUN $= 1$ and, from its range, CAT is 4. From Table B.2 of Appendix B, the codeword for (RUN $= 1$, CAT $= 4$) is 111110110. To find symbol-2, $ZZ(k) - 1 = -13 - 1 = -14 = 111\ldots 11\underline{0010}$. Thus symbol-2 = 0010, and the whole codeword becomes 1111101100010.

The fourth and the final nonzero AC coefficient is 5 (CAT $= 3$), which is preceded by three zeros (RUN $= 3$). Thus symbol-1 is (3, 3), which, from Table B.2 of Appendix B, has a codeword of 111111110101. For symbol-2, since $ZZ(k) = 5 = 000\ldots 00\underline{101}$, then the lower order bits are 101, and the whole codeword becomes 111111110101101.

Since 5 is the last nonzero AC coefficient, then the encoding terminates here and the end of block (EOB) code is transmitted which is defined as the (0, 0) symbol with no appended bits. From Table B.2 of Appendix B, its codeword is 1010.

### 5.2.2.3 Entropy coding

For coding of the magnitude categories or run length events, the JPEG standard specifies two alternative entropy coding methods, namely Huffman coding and arithmetic coding. Huffman coding procedures use Huffman tables, and the type of table is determined by the entropy table specifications, shown in Figure 5.3. Arithmetic coding methods use arithmetic coding conditioning tables, which may also be determined by the entropy table specification. There can be up to four different Huffman and arithmetic coding tables for each DC and AC coefficient. No default values for Huffman tables are specified, so the applications may choose tables appropriate for their own environment. Default tables are defined for the arithmetic coding conditioning. Baseline sequential coding uses Huffman coding, and extended DCT-based and lossless processes may use either Huffman or arithmetic coding (see Table 5.4).

*Table 5.4    Summary: essential characteristics of coding process*

Baseline process (required for all DCT-based decoders)
- DCT-based process
- source image: 8-bit samples within each component
- sequential
- Huffman coding: 2 AC and 2 DC tables
- decoders shall process scans with 1, 2, 3 and 4 components
- interleaved and noninterleaved scans

Extended DCT-based processes
- DCT-based process
- source image: 8-bit or 12-bit samples
- sequential or progressive
- Huffman or arithmetic coding: 4 AC and 4 DC tables
- decoder shall process scans with 1, 2, 3 and 4 components
- interleaved and noninterleaved scans

Lossless process
- predictive process (not DCT-based)
- source image; $N$-bit samples ($2 \leq N \leq 16$)
- sequential
- Huffman or arithmetic coding: 4 tables
- decoders shall process scans with 1, 2, 3 and 4 components
- interleaved and noninterleaved scans

Hierarchical processes
- multiple layers (nondifferential and differential)
- uses extended DCT-based or lossless processes
- decoders shall process scans with 1, 2, 3 and 4 components
- interleaved and noninterleaved scans

In arithmetic coding of AC coefficients, the length of zero run is no longer limited to 15; it can go up to the end of the block (e.g. 62). Also, arithmetic coding may be made adaptive to increase the coding efficiency. Adaptive means that the probability estimates for each context are developed based on a prior coding decision for that context. The adaptive binary arithmetic coder may use a statistical model to improve encoding efficiency. The statistical model defines the contexts which are used to select the conditional probability estimates used in the encoding and decoding procedures.

## 5.2.3 Extended DCT-based process

The baseline encoder only supports basic coding tools, which are sufficient for most image compression applications. These include input image with eight-bit/pixel precision, Huffman coding of the run length, and sequential transmission. If other modes, or any input image precision, are required, and in particular if arithmetic coding is employed to achieve higher compression, then the term extended DCT-based process is applied to the encoder. Table 5.4 summarises all the JPEG supported coding modes.

Figure 5.6 illustrates the reconstruction of a decoded image in a sequential mode (baseline or extended). As mentioned, as soon as a block of pixels is coded, its 64 coefficients are quantised, coded and transmitted. The receiver after decoding the coefficients, inverse quantisation and inverse transformation, sequentially adds them to the reconstructed image. Depending on the channel rate, it might take some time to reconstruct the whole image. In Figure 5.6, reconstructed images at 25, 50, 75 and 100 per cent of the image are shown.

In the progressive mode, the quantised coefficients are stored in the local buffer and transmitted later. There are two procedures by which the quantised coefficients in the buffer may be partially encoded within a scan. Firstly, for the highest image quality (lowest quantisation step size) only a specified band of coefficients from the zigzag scanned sequence need to be coded. This procedure is called spectral selection, since each band typically contains coefficients which occupy a lower or higher part of the frequency spectrum for the $8 \times 8$ block. Secondly, the coefficients within the current band need not be encoded to their full accuracy within each scan (coarser quantisation). On a coefficient's first encoding, a specified number of the most significant bits are encoded first. In subsequent scans, the less significant bits are then encoded. This procedure is called successive approximation or bit plane encoding. Either procedure may be used separately, or they may be mixed in flexible combinations.

Figure 5.7 shows the reconstructed image quality with the first method. In this Figure, the first image is reconstructed from the DC coefficient only, with its full quantised precision. The second image is made up of DC (coefficient 0) plus the AC coefficients 1 and 8, according to the zigzag scan order. That is, after receiving the two new AC coefficients, a new image is reconstructed from these coefficients and the previously received DC coefficients. The third image is made up of coefficients 0, 1, 8, 16, 9, 2, 3, 10, 17 and 24. In the last image, all the significant coefficients (up to EOB) are included.

*Coding of still pictures (JPEG and JPEG2000)* 113

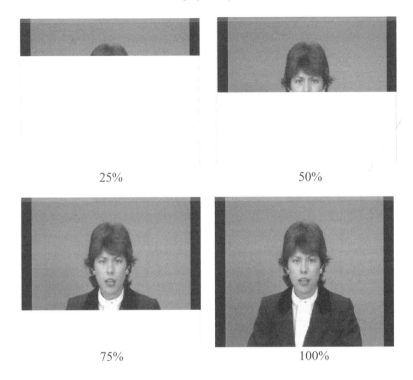

*Figure 5.6    Reconstructed images in sequential mode*

### 5.2.4   Hierarchical mode

In the hierarchical mode, an image is coded as a sequence of layers in a pyramid. Each lower size image provides prediction for the next upper layer. Except for the top level of the pyramid, for each luminance and colour component at the lower levels the difference between the source components and the reference reconstructed image is coded. The coding of the differences may be done using only DCT-based processes, only lossless processes, or DCT-based processes with a final lossless process for each component.

Downsampling and upsampling filters, similar to those of Figures 2.4 and 2.6, may be used to provide a pyramid of spatial resolution, as shown in Figure 5.8. The hierarchical coder including the downsampling and upsampling filters is shown in Figure 5.9.

In this Figure, the image is lowpass filtered and subsampled by 4 : 1, in both directions, to give a reduced image size 1/16. The baseline encoder then encodes the reduced image. The decoded image at the receiver may be interpolated by 1 : 4 to give the full size image for display. At the encoder, another baseline encoder encodes the difference between the subsampled input image by 2 : 1 and the 1 : 2 upsampled decoded image. By repeating this process, the image is progressively coded, and

114  *Standard codecs: image compression to advanced video coding*

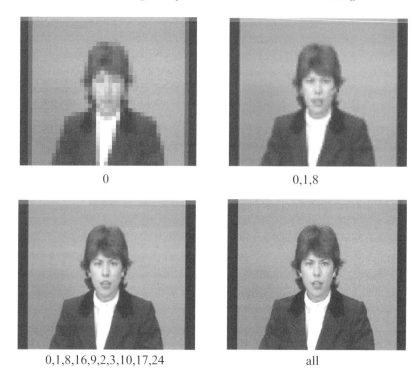

*Figure 5.7  Image reconstruction in progressive mode*

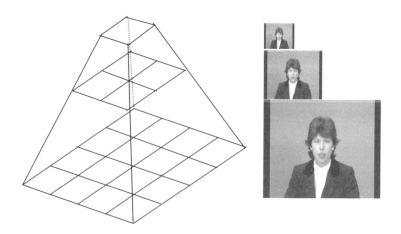

*Figure 5.8  Hierarchical multiresolution encoding*

*Coding of still pictures (JPEG and JPEG2000)* 115

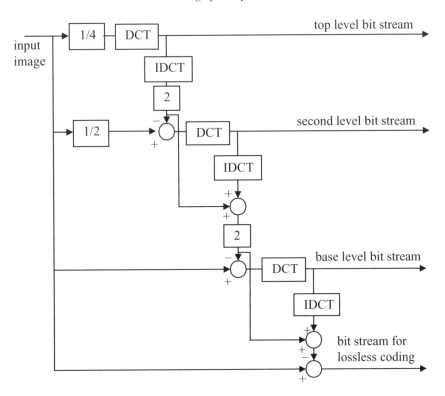

Figure 5.9  *A three-level hierarchical encoder*

at the decoder it is progressively built up. The bit rate at each level depends on the quantisation step size at that level. Finally, for lossless reversibility of the coded image, the difference between the input image and the latest decoded image is lossless entropy coded (no quantisation).

As we see, the hierarchical mode offers a progressive representation similar to the progressive DCT-based mode, but it is useful in environments which have multiresolution requirements. The hierarchical mode also offers the capability of progressive transmission to a final lossless stage, as shown in Figure 5.9.

### 5.2.5 Extra features

In coding of colour pictures, encoding is called noninterleaved if all blocks of a colour component are coded before beginning to code the next component. Encoding is interleaved if the encoder compresses a block of 8 × 8 pixels from each component in turn, considering the image format. For example, with the 4 : 4 : 4 format, one block from each luminance and two chrominance components are coded. In the 4 : 2 : 0

format, the encoder codes four luminance blocks before coding one block from each of $C_b$ and $C_r$.

The encoder is also flexible to allow different blocks within a single component to be compressed using different quantisation tables resulting in a variation in the reconstructed image quality, based on their perceived relative importance. This method of coding is called region of interest (ROI) coding. Also, the standard can allow different regions within a single image block to be compressed at different rates.

## 5.3 JPEG2000

Before describing the basic principles of the JPEG2000 standard, it might be useful to understand why we need another standard. Perhaps the most convincing explanation is that, since the introduction of JPEG in the 1980s, too much has changed in the digital image industry. For example, current demands for compressed still images range from web logos of sizes less than 10 Kbytes to high quality scanned images of the order of 5 Gbytes!! [8]. The existing JPEG surely is not optimised to efficiently code such a wide range of images. Moreover, scalability and interoperability requirements of digital imagery in a heterogeneous network of ATM, Internet, mobile etc., make the matter much more complicated.

The JPEG2000 standard is devised with the aim of providing the best quality or performance and capabilities to market evolution that the current JPEG standard fails to cater for. In the mean time it is assumed that Internet, colour facsimile, printing, scanning, digital photography, remote sensing, mobile, medical imagery, digital libraries/archives and e-commerce are among the most immediate demands. Each application area imposes a requirement that JPEG2000 should fulfil. Some of the most important features [9] that this standard aims to deliver are:

*Superior low bit rate performance*: this standard should offer performance superior to the current standards at low bit rates (e.g. below 0.25 bit per pixel for highly detailed greyscale images). This significantly improved low bit rate performance should be achieved without sacrificing performance on the rest of the rate distortion spectrum. Examples of applications that need this feature include image transmission over networks and remote sensing. This is the highest priority feature.

*Continuous tone and bilevel compression*: it is desired to have a standard coding system that is capable of compressing both continuous tone and bilevel images [7]. If feasible, the standard should strive to achieve this with similar system resources. The system should compress and decompress images with various dynamic ranges (e.g. 1 bit to 16 bit) for each colour component. Examples of applications that can use this feature include compound documents with images and text, medical images with annotation overlays, graphic and computer generated images with binary and near-to-binary regions, alpha and transparency planes, and facsimile.

*Lossless and lossy compression*: it is desired to provide lossless compression naturally in the course of progressive decoding (i.e. difference image encoding, or any other technique, which allows for the lossless reconstruction to be valid). Examples of applications that can use this feature include medical images where loss is not always tolerable, image archival pictures where the highest quality is vital for preservation but not necessary for display, network systems that supply devices with different capabilities and resources, and prepress imagery.

*Progressive transmission by pixel accuracy and resolution*: progressive transmission that allows images to be reconstructed with increasing pixel accuracy or spatial resolution is essential for many applications. This feature allows the reconstruction of images with different resolutions and pixel accuracy, as needed or desired, for different target devices. Examples of applications include web browsing, image archiving and printing.

*Region of interest coding*: often there are parts of an image that are more important than others. This feature allows a user-defined region of interest (ROI) in the image to be randomly accessed and/or decompressed with less distortion than the rest of the image.

*Robustness to bit errors:* it is desirable to consider robustness to bit errors while designing the code stream. One application where this is important is wireless communication channels. Portions of the code stream may be more important than others in determining decoded image quality. Proper design of the code stream can aid subsequent error correction systems in alleviating catastrophic decoding failures. Use of error confinement, error concealment, restart capabilities, or source channel coding schemes can help minimise the effects of bit errors.

*Open architecture:* it is desirable to allow open architecture to optimise the system for different image types and applications. With this feature, the decoder is only required to implement the core tool set and a parser that understands the code stream. If necessary, unknown tools are requested by the decoder and sent from the source.

*Protective image security*: protection of a digital image can be achieved by means of methods such as: watermarking, labelling, stamping, fingerprinting, encryption, scrambling etc. Watermarking and fingerprinting are invisible marks set inside the image content to pass a protection message to the user. Labelling is already implemented in some imaging formats such as SPIFF, and must be easy to transfer back and forth to the JPEG2000 image file. Stamping is a mark set on top of a displayed image that can only be removed by a specific process. Encryption and scrambling can be applied on the whole image file or limited to part of it (header, directory, image data) to avoid unauthorised use of the image.

## 5.4 JPEG2000 encoder

The JPEG2000 standard follows the generic structure of the intraframe still image coding introduced for the baseline JPEG. That is, decorrelating the pixels within a frame by means of transformation and then quantising and entropy coding of the quantised transform coefficients for further compression. However, in order to meet the design requirements set forth in section 5.3, in addition to the specific requirements

*Figure 5.10  A general block diagram of the JPEG2000 encoder*

from the transformation and coding, certain preprocessing on the pixels and post processing of the compressed data is necessary. Figure 5.10 shows a block diagram of a JPEG2000 encoder.

In presenting this coder, we only talk about the fundamentals behind this standard. More details can be found in the ISO standardisation documents and several key papers [9–11].

### 5.4.1  Preprocessor

Image pixels prior to compression are preprocessed to make certain goals easier to achieve. There are three elements in this preprocessor.

#### 5.4.1.1  Tiling

Partitioning the image into rectangular nonoverlapping pixel blocks, known as tiling, is the first stage in preprocessing. The tile size is arbitrary and can be as large as the whole image size down to a single pixel. A tile is the basic unit of coding, where all the encoding operations, from transformation down to bit stream formation, are applied to tiles independent of each other. Tiling is particularly important for reducing memory requirement and, since they are coded independently, any part of the image can be accessed and processed differently from the other parts of the image. However, due to tiling, the correlation between the pixels in adjacent tiles is not exploited, and hence as the tile size is reduced, the compression gain of the encoder is also reduced.

#### 5.4.1.2  DC level shifting

Similar to DC level shifting in the JPEG standard (Figure 5.3), values of the RGB colour components within the tiles are DC shifted by $2^{B-1}$, for $B$ bits per colour component. Such an offset makes certain processing, such as numerical overflow, arithmetic coding, context specification etc. simpler. In particular, this allows the lowest subband, which is a DC signal, to be encoded along with the rest of the AC wavelet coefficients. At the decoder, the offset is added back to the colour component values.

#### 5.4.1.3  Colour transformation

There are significant correlations between the RGB colour components. Hence, prior to compression by the core encoder, they are decorrelated by some form of transformation. In JPEG2000 two types of colour decorrelation transform are recommended.

In the first type, the decorrelated colour components $Y$, $C_b$ and $C_r$, are derived from the three colour primaries R, G and B according to:

$$\begin{bmatrix} Y \\ C_b \\ C_r \end{bmatrix} = \begin{bmatrix} 0.299 & 0.587 & 0.114 \\ -0.16875 & -0.33126 & 0.500 \\ 0.500 & -0.41869 & -0.08131 \end{bmatrix} \begin{bmatrix} R \\ G \\ B \end{bmatrix} \qquad (5.4)$$

Note that this transformation is slightly different from the one used for coding colour video (see section 2.2). Also note that since transformation matrix elements are approximated (not exact), then even if $YC_bC_r$ are losslessly coded, the decoded RGB colour components cannot be free from loss. Hence this type of colour transformation is irreversible, and it is called irreversible colour transformation (ICT). ICT is used only for lossy compression.

The JPEG2000 standard also defines a colour transformation for lossless compression. Therefore, the transformation matrix elements are required to be integer. In this mode, the transformed colour components are referred to as $Y$, $U$ and $V$, and are defined as:

$$\begin{bmatrix} Y \\ U \\ V \end{bmatrix} = \begin{bmatrix} 0.25 & 0.5 & 0.25 \\ 1 & -1 & 0 \\ 0 & -1 & 1 \end{bmatrix} \begin{bmatrix} R \\ G \\ B \end{bmatrix} \qquad (5.5)$$

Here the colour decorrelation is not as good as ICT, but it has the property that if $YUV$ are losslessly coded, then exact values of the original RGB can be retrieved. This type of transformation is called reversible colour transformation (RCT). RCT may also be used for lossy coding, but since ICT has a better decorrelation property than RCT, use of RCT can reduce the overall compression efficiency.

It is worth mentioning that in compression of colour images, colour fidelity may be traded for that of luminance. In the JPEG standard this is done by subsampling the chrominance components $C_b$ and $C_r$, or $U$ and $V$, like the 4:2:2 and 4:2:0 image formats. In JPEG2000, image format is always 4:4:4, and the colour subsampling is done by the wavelet transform of the core encoder. For example, in coding of a 4:4:4 image format, if the highest LH, HL and HH bands of $C_b$ and $C_r$ chrominance components are set to zero, it has the same effect as coding of a 4:2:0 image format.

### 5.4.2 Core encoder

Each transformed colour component of $YC_bC_r/YUV$ is coded by the core encoder. As in the JPEG encoder, the main elements of the core encoder are: transformation, quantisation and entropy coding. Thus a more detailed block diagram of JPEG2000 is given by Figure 5.11.

In the following sections these elements and their roles in image compression are presented.

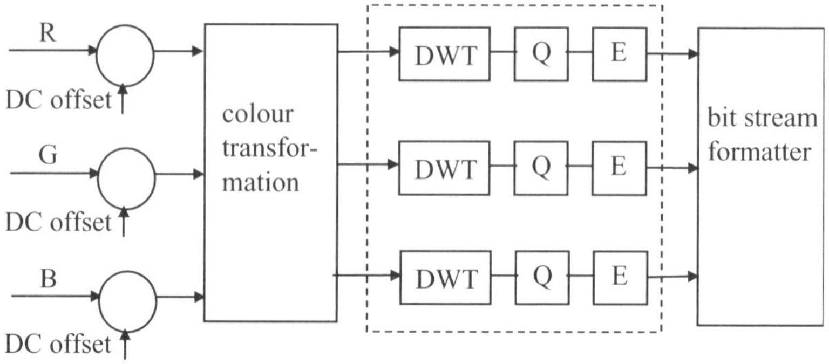

120  *Standard codecs: image compression to advanced video coding*

*Figure 5.11  The encoding elements of JPEG2000*

### 5.4.2.1  Discrete wavelet transform

In JPEG2000, transformation of pixels that in the JPEG standard used DCT has been replaced by the discrete wavelet transform (DWT). This has been chosen to fulfil some of the requirements set forth by the JPEG2000 committee. For example:

a   Multiresolution image representation is an inherent property of the wavelet transform. This also provides simple SNR and spatial scalability, without sacrificing compression efficiency.
b   Since the wavelet transform is a class of lapped orthogonal transforms then, even for small tile sizes, it does not create blocking artefacts.
c   For larger dimension images, the number of subband decomposition levels can be increased. Hence, by exploiting a larger area of pixel intercorrelation a higher compression gain can be achieved. Thus for images coded at low bit rates, DWT is expected to produce better compression gain than the DCT which only exploits correction with $8 \times 8$ pixels.
d   DWT with integer coefficients, such as the (5,3) tap wavelet filters, can be used for lossless coding. Note that in DCT, since the cosine elements of the transformation matrix are approximated, lossless coding is not then possible.

The JPEG2000 standard recommends two types of filter bank for lossy and lossless coding. The default irreversible transform used in the lossy compression is the Daubechies 9-tap/7-tap filter [12]. For the reversible transform, with a requirement for lossless compression, it is LeGall and Tabatabai's 5-tap/3-tap filters as they have integer coefficients [13]. Table 5.5 shows the normalised coefficients (rounded to six decimal points) of the lowpass and highpass analysis filters $H_0(z)/H_1(z)$ of the 9/7 and 5/3 filters. Those of the synthesis $G_0(z)$ and $G_1(z)$ filters can be derived from the usual method of $G_0(z) = H_1(-z)$ and $G_1(z) = -H_0(-z)$.

Note that to preserve image energy in the pixel and wavelet domains, the integer filter coefficients in lossless compression are normalised for unity gain. Since the lowpass and highpass filter lengths are not equal, these types of filter are called biorthogonal. The lossy 9/7 Daubechies filter pairs [12] are also biorthogonal.

*Table 5.5  Analysis lowpass and highpass filter banks*

| Coefficients | Lossy compression (9/7) | | Lossless compression (5/3) | |
|---|---|---|---|---|
| | lowpass $H_0(z)$ | highpass $H_1(z)$ | lowpass $H_0(z)$ | highpass $H_1(z)$ |
| 0 | +0.602949 | +1.115087 | 3/4 | 1 |
| ±1 | +0.266864 | −0.591272 | 1/4 | −1/2 |
| ±2 | −0.078223 | −0.057544 | −1/8 | |
| ±3 | −0.016864 | +0.091272 | | |
| ±4 | +0.026729 | | | |

### 5.4.2.2  Quantisation

After the wavelet transform, all the coefficients are quantised linearly with a dead band zone quantiser (Figure 3.5). The quantiser step size can vary from band to band and since image tiles are coded independently, it can also vary from tile to tile. However, one quantiser step size is allowed per subband of each tile. The choice of the quantiser step size can be driven by the perceptual importance of that band on the human visual system (HVS), similar to the quantisation weighting matrix used in JPEG (Table 5.1), or by other considerations, such as the bit rate budget.

As mentioned in Chapter 4, wavelet coefficients are most efficiently coded when they are quantised by successive approximation, which is the bit plane representation of the quantised coefficients. In this context the quantiser step size in each subband, called the basic quantiser step size, $\Delta$, is related to the dynamic range of that subband, such that the initial quantiser step size, after several passes, ends up with the basic quantiser step size $\Delta$. In JPEG2000, the basic quantiser step size for band $b$, $\Delta_b$, is represented with a total of two bytes, an 11-bit mantissa $\mu_b$ and a five-bit exponent $\varepsilon_b$ according to the relationship:

$$\Delta_b = 2^{R_b - \varepsilon_b}\left(1 + \frac{\mu_b}{2^{11}}\right) \tag{5.6}$$

where $R_b$ is the number of bits representing the nominal dynamic range of subband $b$. That is $2^{R_b}$ is greater than the magnitude of the largest coefficient in subband $b$. Values of $\mu_b$ and $\varepsilon_b$ for each subband are explicitly transmitted to the decoder. For lossless coding, used with reversible (5,3) filter banks, $\mu_b = 0$ and $\varepsilon_b = R_b$, which results in $\Delta_b = 1$. On the other hand, the maximum value of $\Delta_b$ is almost twice the dynamic range of the input sample when $\varepsilon_b = 0$ and $\mu_b$ has its maximum value, which is sufficient for all practical cases of interest.

### 5.4.2.3  Entropy coding

The indices of the quantised coefficients in each subband are entropy coded to create the compressed bit stream. In Chapter 4 we introduced three efficient methods of coding these indices, namely: EZW, SPIHT and EBCOT. As mentioned in section 4.7, the JPEG committee chose embedded block coding with optimised truncation (EBCOT), due to its many interesting features that fulfil the JPEG2000 objectives. Details of EBCOT were given in section 4.7, and here we only summarise its principles and show how it is used in the JPEG2000 standard.

In EBCOT, each subband of an image tile is partitioned into small rectangular blocks, called code blocks, and code blocks are encoded independently. The dimensions of the code blocks are specified by the encoder and, although they may be chosen freely, there are some constraints; they must be an integer power of two; the total number of coefficients in a code block cannot exceed 4096 and the height of the code block cannot be less than four. Thus the maximum length of the code block is 1024 coefficients.

The quantiser indices of the wavelet coefficients are bit plane encoded one bit at a time, starting from the most significant bit (MSB) and preceding to the least significant bit (LSB). During this progressive bit plane encoding, if the quantiser index is still zero, that coefficient is called insignificant. Once the first nonzero bit is encoded, the coefficient becomes significant and its sign is encoded. For significant coefficients, all subsequent bits are referred to as refinement bits. Since in the wavelet decomposition the main image energy is concentrated at lower frequency bands, many quantiser indices of the higher frequency bands will be insignificant at the earlier bit planes. Clustering of insignificant coefficients in bit planes creates strong redundancies among the neighbouring coefficients that are exploited by JPEG2000 through a context-based adaptive arithmetic coding.

In JPEG2000, instead of encoding the entire bit plane in one pass, each bit plane is encoded in three subbit plane passes. This is called fractional bit plane encoding, and the passes are known as: significance propagation pass, refinement pass and clean up pass. The reason for this is to be able to truncate the bit stream at the end of each pass to create the near optimum bit stream. Here, the pass that results in the largest reduction in distortion for the smallest increase in bit rate is encoded first.

In the significance propagation pass, the bit of a coefficient in a given bit plane is encoded if and only if, prior to this pass, the coefficient was insignificant and at least one of its eight immediate neighbours was significant. The bit of the coefficient in that bit plane, 0 or 1, is then arithmetically coded with a probability model derived from the context of its eight immediate neighbours. Since neighbouring coefficients are correlated, it is more likely that the coded coefficient becomes significant, resulting in a large reduction in the coding distortion. Hence this pass is the first to be executed in fractional bit plane coding.

In the refinement pass, a coefficient is coded if it was significant in the previous bit plane. Refining the magnitude of a coefficient reduces the distortion moderately. Finally, those coefficients that were not coded in the two previous passes are coded in the clean up pass. These are mainly insignificant coefficients (having eight insignificant immediate neighbours) and are likely to remain insignificant. Hence their contribution in reducing distortions is minimal and is used in the last pass. For more details of coding, refer to EBCOT in section 4.7.

## 5.4.3 Postprocessing

Once the entire image has been compressed, the bit stream generated by the individual code blocks is postprocessed to facilitate various functionalities of the JPEG2000 standard. This part is similar to the layer formation and bit stream organisation of EBCOT known as tier 2 (see section 4.7).

To form the final bit stream, the bits generated by the three spatially consistent coded blocks (one from each subband at each resolution level) comprise a packet partition location, called a precinct [10]. A collection of packets, one from each precinct, at each resolution level comprises the layer. Figure 5.12 shows the relationship between the packetised bit stream and the units of image, such as the code block, precinct, tile and the image itself.

Here, the smallest unit of compressed data is the coded bits from a code block. Data from three code blocks of a precinct makes a packet, with an appropriate header, addressing the precinct position in the image. Packets are then grouped into the layer and finally form the bit stream, all with their relevant headers to facilitate flexible decoding. Since precincts correspond to spatial locations, a packet could be interpreted as one quality increment for one resolution at one spatial location. Similarly, a layer could be viewed as one quality increment for the entire image.

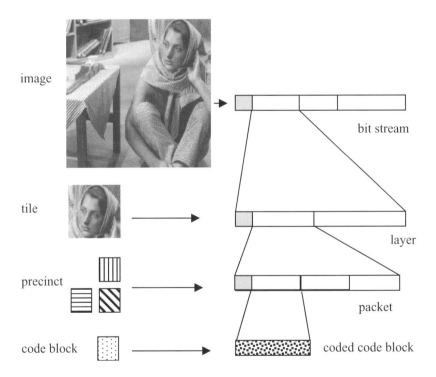

*Figure 5.12   Correspondence between the spatial data and bit stream*

Each layer successively and gradually improves the image quality and resolution, so that the decoder is able to decode the code block contributions contained in the layer, in sequence. Since ordering of packets into the layer and hence into the bit stream can be as desired, various forms of progressive image transmission can be realised.

## 5.5 Some interesting features of JPEG2000

Independent coding of code blocks and flexible multiplexing of quality packets into the bit stream exhibit some interesting phenomena. Some of the most remarkable features of JPEG2000 are outlined in the following.

### 5.5.1 Region of interest

In certain applications, it might be desired to code parts of a picture at a higher quality than the other parts. For example, in web browsing one might be interested in a logo of a complex web page image that needs to be seen first. This part needs to be given higher priority for transmission. Another example is in medical images, where the region of interest (ROI) might be an abnormality in the part of the whole image that requires special attention.

Figure 5.13 shows an example of a ROI, where the head and scarf of Barbara are coded at a higher quality than the rest of the picture, called the background. Loss of image quality outside the region of interest (outside the white box), in particular on the tablecloth, trousers and the books, is very clear.

Coding of the region of interest in the JPEG2000 standard is implemented through the so-called maxshift method [14]. The basic principle in this method is to scale (shift) up the coefficients, such that their bits are placed at a higher level than the bits associated with the background data, as shown in Figure 5.14. Depending on the scale value, $S$, some bits of the ROI coefficients might be encoded together with those of the background, like Figure 5.14a, or all the bits of the ROI are encoded before any background data are coded, as shown in Figure 5.14b. In any case, the ROI at the decoder is decoded or refined before the rest of the image.

It is interesting to note that, if the value of scaling $S$ is computed such that the minimum coefficient belonging to the ROI is larger than the maximum coefficient of the background, then it is possible to have arbitrary shaped ROIs, without the need for defining the ROI shape to the decoder. This is because every received coefficient that is smaller than $S$ belongs to the background, and can be easily identified.

### 5.5.2 Scalability

Scalability is one of the important parts of all the image/video coding standards. Scalable image coding means being able to decode more than one quality or resolution image from the bit stream. This allows the decoders of various capabilities to decode images according to their processing powers or needs. For example, although low

Coding of still pictures (JPEG and JPEG2000)    125

*Figure 5.13*   *Region of interest with better quality*

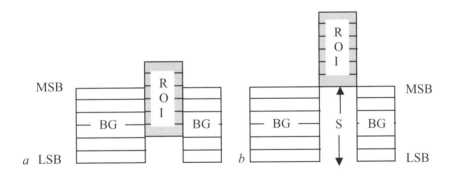

*Figure 5.14*   *Scaling of the ROI coefficients*

performance decoders may decode only a small portion of the bit stream, providing basic quality or resolution images, high performance decoders may decode a larger portion of the bit stream, proving higher quality images. The most well known types of scalability in JPEG2000 are the SNR and spatial scalabilities. Since in JPEG2000, code blocks are individually coded, bit stream scalability is easily realised. In order to have either SNR or spatial scalability, the compressed data from the code blocks should be inserted into the bit stream in the proper order.

126  *Standard codecs: image compression to advanced video coding*

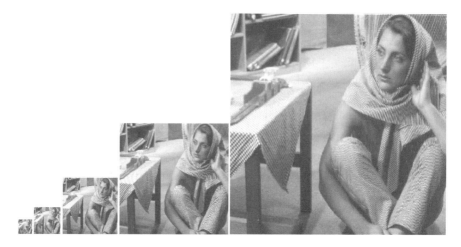

*Figure 5.15  Spatial scalable decoding*

#### 5.5.2.1 Spatial scalability

In spatial scalability, from a single bit stream, images of various spatial resolutions can be decoded. The smaller size picture with an acceptable resolution is the base layer and the parts of the bit stream added to the base layer to create higher resolution images comprise the next enhancement layer, as shown in Figure 5.15.

In JPEG2000, due to the octave band decomposition of the wavelet transform, spatial scalability is easily realised. In this mode, compressed data of the code blocks has to be packed into the bit stream such that all the bit planes of the lower level subbands precede those of the higher bands.

#### 5.5.2.2 SNR scalability

The signal-to-noise ratio (SNR) scalability involves producing at least two levels of image of the same spatial resolutions, but at different quality, from a single bit stream. The lowest quality image is called the base layer, and the parts of the bit stream that enhance the image quality are called enhancement layers. In JPEG2000, through bit plane encoding, the lowest significant bit plane that gives an acceptable image quality can be regarded as the base layer image. Added quality from the subsequent bit planes produces a set of enhanced images. Figure 5.16 shows a nine-layer SNR scalable image produced by bit plane coding from a single layer, where the compressed code block data from a bit plane of all the subbands is packed before the data from the next bit plane.

In Figure 5.16 the first picture is made up from coding the most significant bit of the lowest LL band. As bit plane coding progresses towards lower bits, more bands are coded, improving the image quality. Any of the images can be regarded as the base layer, but for an acceptable quality, picture number four or five may just meet

# Coding of still pictures (JPEG and JPEG2000) 127

*Figure 5.16   SNR scalable decoding*

the criterion. The remaining higher quality images become its enhanced versions at different quality levels.

## 5.5.3   Resilience

Compressed bit streams, especially those using variable length codes (e.g. arithmetic coding) are extremely sensitive to channel errors. Even a single bit error may destroy the structure of the following valid codewords, corrupting a very large part of the image. Resilience and robustness to channel errors are the most desired features expected from any image/video encoder. Fortunately in JPEG2000, since the individual quality packets can be independently decoded, the effect of channel errors can be confined to the area covered by these packets. This is not the case with the other wavelet transform encoders such as EZW and SPIHT. Figure 5.17 shows the impact of a single bit error on the reconstructed picture, encoded by the SPIHT and JPEG2000.

128  *Standard codecs: image compression to advanced video coding*

*a*     SPIHT          *b*     JPEG2000

Figure 5.17   *Effect of single bit error on the reconstructed image, encoded by SPIHT and JPEG2000*

As the Figure shows, although a single bit error destroys the whole picture encoded by SPIHT, its effect on JPEG2000 is only limited to a small area, around Barbara's elbow.

It is worth noting that the SPIHT encoder, even without arithmetic coding, is very sensitive to channel errors. For example, a single bit error in the early pass of LIS (refer to section 4.6) can corrupt the whole image, as can be seen from Figure 5.17a. In fact, this picture is not arithmetically coded at all, and the single bit error was introduced at the first pass of the LIS data. Of course it is possible to guard EZW and SPIHT compressed data against channel errors. For instance, if the bit stream generated by each tree of EZW/SPIHT can be marked, then propagation of errors into the trees can be prevented. This requires some extra bits to be inserted between the tree's bit streams as resynchronisation markers. This inevitably increases the bit rate. Since the resynchronisation marker bits are fixed in rate, irrespective of the encoded bit rate, the increase in bit rate is more significant at lower bit rates than at higher bit rates. For example, in coding Barbara with SPIHT at 0.1, 0.5 and 1 bit/pixel, the overhead in bits will be 2.7, 0.64 and 0.35 per cent, respectively.

## 5.6  Problems

1  The luminance quantisation Table 5.1 is used in the baseline JPEG for a quality factor of 50 per cent. Find the quantisation table for the following quality factors:

   *a*  25 %
   *b*  99 %
   *c*  100 %

2 In problem 1 find the corresponding tables for the chrominance.

3 The DCT transform coefficients (luminance) of an 8 × 8 pixel block prior to quantisation are given by:

| 1000 | −2  | 35  | 18  | 15 | −8  | 62  | 5   |
|------|-----|-----|-----|----|-----|-----|-----|
| −4   | 15  | −21 | −4  | 51 | 2   | −11 | 1   |
| 9    | −8  | 13  | −11 | 43 | −20 | 7   | −3  |
| −17  | 16  | −11 | 3   | −2 | 5   | −13 | 6   |
| −6   | 12  | 42  | −15 | 31 | −2  | 7   | −3  |
| 12   | −7  | 2   | −11 | 15 | −5  | 3   | 18  |
| −19  | 52  | 6   | 13  | 4  | −10 | 8   | 10  |
| 35   | −11 | −7  | 3   | 5  | 9   | 7   | 382 |

Find the quantisation indices for the baseline JPEG with the quality factors of:

*a* 50 %

*b* 25 %

4 In problem 3, if the quantised index of the DC coefficient in the previous block was 50, find the pairs of symbol-1 and symbol-2 for the given quality factors.

5 Derive the Huffman code for the 25 per cent quality factor of problem 4 and hence calculate the number of bits required to code this block.

6 A part of the stripe of the wavelet coefficients of a band is given as:

| 20  | 30 |
|-----|----|
| −16 | 65 |
| 31  | 11 |
| 50  | 24 |

Assume the highest bit plane is 6. Using EBCOT identify which coefficient is coded at bit plane 6 and which one at bit plane 5. In each case identify the type of fractional bit plane used.

## 5.7 References

1 ISO 10918-1 (JPEG): 'Digital compression and coding of continuous-tone still images'. 1991
2 FURHT, B.: 'A survey of multimedia compression techniques and standards. Part I: JPEG standard', *Real-Time Imaging*, 1995, pp.1–49
3 WALLACE, G.K.: 'The JPEG still picture compression standard', *Commun. ACM*, 1991, **34:4**, pp.30–44
4 JPEG2000: 'JPEG2000 part 2, final committee draft'. ISO/IEC JTC1/SC29/WG1 N2000, December 2000
5 PENNEBAKER, W.B., and MITCHELL, J.L.: 'JPEG: still image compression standard' (Van Nostrand Reinhold, New York, 1993)

6 The independent JPEG Group: 'The sixth public release of the Independent JPEG Group's free JPEG software'. C source code of JPEG encoder release 6b, March 1998, [ftp://ftp.uu.net/graphics/jpeg/jpegsrc_v6b_tar.gz]
7 WANG, Q., and GHANBARI, M.: 'Graphics segmentation based coding of multimedia images', *Electron. Lett.*, 1995, **31:6**, pp.542–544
8 SKODRAS, A., CHRISTOPOULOS, C., and EBRAHIMI, T.: 'The JPEG2000 still image compression standard', *IEEE Signal Process. Mag.*, September 2001, pp.36–58
9 'Visual evaluation of JPEG-2000 colour image compression performance'. ISO/IECJTC1/SC29/WG1 N1583, March 2000
10 RBBANI, M., and JOSHI, R.: 'An overview of the JPEG2000 image compression standard', *Signal Process. Image Commun.*, **17:1**, January 2002, pp.3–48
11 SANTA-CRUZ, D., GROSBOIS, R., and EBRAHIMI, T.: 'JPEG2000 performance evaluation and assessment', *Signal Process., Image Commun.* **17:1**, January 2002, pp.
12 DAUBECHIES, I.: 'The wavelet transform, time frequency localization and signal analysis, *IEEE Trans Inf. Theory*, 1990, **36:5**, pp.961–1005
13 LE GALL, D., and TABATABAI, A.: 'Subband coding of images using symmetric short kernel filters and arithmetic coding techniques'. IEEE international conference on *Acoustics, speech and signal processing*, ICASSP'88, 1988, pp.761–764
14 CHRISTOPOULOS, C.A., ASKELF, J., and LARSSON, M.: 'Efficient methods for encoding regions of interest in the up-coming JPEG2000 still image coding standard', *IEEE Signal Process. Lett.*, 2000, **7**, pp.247–249

## Chapter 6
# Coding for videoconferencing (H.261)

The H.261 standard defines the video coding and decoding methods for digital transmission over ISDN at rates of $p \times 64$ kbit/s, where $p$ is in the range of 1–30 [1]. The video bit rates will lie between approximately 64 kbit/s and 1920 kbit/s. The recommendation is aimed at meeting projected customer demand for videophone, videoconferencing and other audio-visual services. It was ratified in December 1990.

The coding structure of H.261 is very similar to that of the generic codec of Chapter 3 (Figure 3.18). That is, it is an interframe DCT-based coding technique. Interframe prediction is first carried out in the pixel domain. The prediction error is then transformed into the frequency domain, where the quantisation for bandwidth reduction takes place. Motion compensation can be included in the prediction stage, although it is optional. Thus the coding technique removes temporal redundancy by interframe prediction and spatial redundancy by transform coding. Techniques have been devised to make the codec more efficient, and at the same time suitable for telecommunications.

It should be noted that any recommendation only specifies what is expected for a decoder; it does not give information on how to design it. Even less information is given about the encoder. Therefore the design of the encoder and the decoder is at the discretion of the manufacturer, provided they comply with the syntax bit stream. Since the aim of this book is an introduction to the fundamentals of video coding standards, rather than giving instructions on the details of a specific codec, we concentrate on the reference model (RM) codec. The reference model is a software-based codec, which is devised to be used in laboratories to study the core elements as a basis for the design of flexible hardware specifications.

During the development of H.261, from May 1988 to May 1989, the reference model underwent eight refinement cycles. The last version, known as reference model eight (RM8) [2], is in fact the basis of the current H.261. However, the two may not be exactly identical (although very similar), and the manufacturers may decide on a different approach for better optimisation of their codecs. Herein we interchangeably use RM8 for H.261. Before describing this codec, we will first look at the picture format, and spatio–temporal resolutions of the images to be coded with H.261.

## 6.1 Video format and structure

Figure 6.1 shows a block diagram of an H.261-based audio-visual system, where a preprocessor converts the CCIR-601 video (video at the output of a camera) to a new format. The coding parameters of the compressed video signal are multiplexed and then combined with the audio, data and end-to-end signalling for transmission. The transmission buffer controls the bit rate, either by changing the quantiser step size at the encoder, or in more severe cases by requesting reduction in frame rate, to be carried out at the preprocessor.

The H.261 standard also allows up to three pictures to be interpolated between transmitted pictures, so reducing the frame rates to 15, 10 and 7.5, respectively. The use of quarter-CIF, or QCIF, resolution will reduce the sample rate even further to suit low bit rate channels.

In CIF and QCIF, DCT blocks are grouped into macroblocks of four luminance and two corresponding $C_b$ and $C_r$ chrominance blocks. The macroblocks (MB) are in turn grouped into layers termed groups of blocks (GOB). A CIF frame has 12 GOBs and QCIF has three, as illustrated in Figure 6.2.

The objectives of structuring an image into macroblocks and layers are as follows:

- similar inter/intra coding mode for luminance and chrominance blocks at the same area
- the use of one motion vector for both luminance and chrominance blocks
- efficient coding of the large number of 8 × 8 DCT blocks that will be expected to be without coded information in interframe coding; this is implemented *via* the inclusion of VLC codes for coded block pattern (CBP) and macroblock addressing [1]
- to allow synchronisation to be reestablished when bits are corrupted by the insertion of start codes in the GOB headers; note that since DCT coefficients are VLC

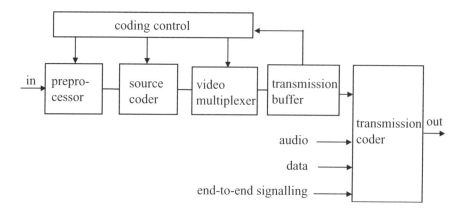

*Figure 6.1* A block diagram of an H.261 audio-visual encoder

## Coding for videoconferencing (H.261) 133

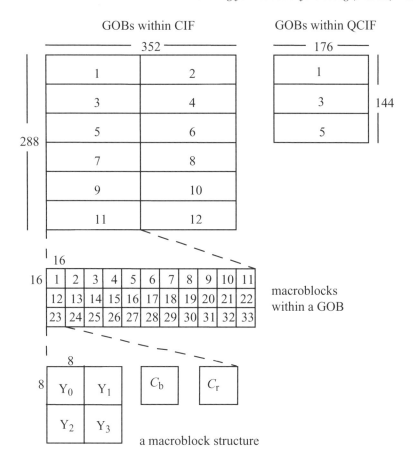

*Figure 6.2  Block, macroblock and GOB structure of CIF and QCIF formatted pictures*

coded, then any error during the transmission renders the remaining VLC coded data undecodable, hence, with a GOB structure, only a portion of the picture is degraded
• to carry side information appropriate for GOB, macroblock or higher layers; this includes picture format, temporal references, macroblock type, quantiser index etc.

## 6.2 Video source coding algorithm

The video coding algorithm is shown in Figure 6.3, and is similar to the generic interframe coder of Figure 3.18 in Chapter 3. The main elements are prediction including motion compensation, transform coding, quantisation, VLC and rate control.

134  *Standard codecs: image compression to advanced video coding*

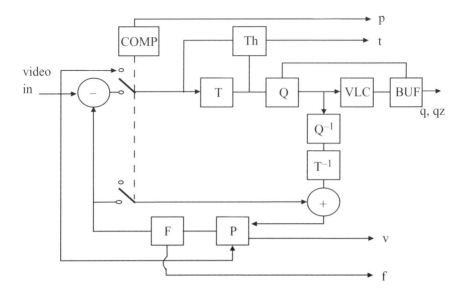

Figure 6.3  *A block diagram of H.261 video encoder*

The prediction error (inter mode) or the input picture (intra mode) is subdivided into 16 × 16 macroblock pixels, which may or may not be transmitted. Macroblocks that are to be transmitted are divided into 8 × 8 pixel blocks, which are transform coded (DCT), quantised and VLC coded for transmission. As we discussed in section 6.1, the atomic coding unit in all standard video codecs is a macroblock. Hence in describing the codec we will explain how each macroblock is coded.

In Figure 6.3, the function of each coding element and the messages carried by each flag are:

COMP   a comprator for deciding the inter/intra coding mode for an MB
Th   threshold, to extend the quantisation range
T   transform coding blocks of 8 × 8 pixels
$T^{-1}$   inverse transform
Q   quantisation of DCT coefficients
$Q^{-1}$   inverse quantisation
P   picture memory with motion compensated variable delay
F   loop filter
p   flag for inter/intra
t   flag for transmitted or not
q   quantisation index for transform coefficients
qz   quantiser indication
v   motion vector information
f   switching on/off of the loop filter

Details of the functions of each block are described in the following sections.

### 6.2.1 Prediction

The prediction is interpicture, which may include motion compensation, since motion compensation in H.261 is optional. The decoder accepts one motion vector per macroblock. Both horizontal and vertical components of these motion vectors have integer values not exceeding ±15 pixels/frame. Motion estimation is only based on the luminance pixels and the vector is used for motion compensation of all four luminance blocks in the macroblock. Halving the component values of the macroblock motion vector and truncating them towards zero derives the motion vector for each of the two chrominance blocks. Motion vectors are restricted such that all pixels referenced by them are within the coded picture area.

For the transmission of motion vectors, their differences are variable length coded. The differential technique is based on one-dimensional prediction, that is the difference between the successive motion vectors in a row of GOBs. For the first macroblock in the GOB, the initial vector is set to zero.

### 6.2.2 MC/NO_MC decision

Not all the macroblocks in a picture are motion compensated. The decision whether a macroblock should be motion compensated or not depends on whether motion compensated prediction can substantially reduce the prediction error. Figure 6.4 shows the region (shaded) where motion compensation is preferred. In this Figure the absolute values of frame difference, *fd*, and those of motion compensated frame difference, *mfd*, normalised to $16 \times 16 = 256$ pixels inside the macroblock are compared.

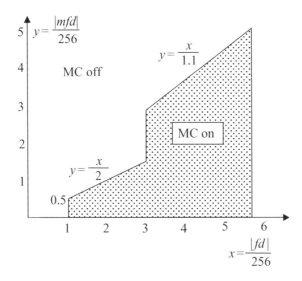

*Figure 6.4* Characteristics of MC/NO_MC

From the Figure we see that if motion compensated error is slightly, but not significantly, less than the nonmotion compensated error, we prefer to use nonmotion compensation. This is because motion compensation entails a motion vector overhead (even if it might be zero); hence, if the difference between MC and NO_MC error cannot justify the extra bits, there is no advantage in using motion compensation.

### 6.2.3 Inter/intra decision

Sometimes it might be advantageous to intraframe code a macroblock, rather than interframe coding it. There are at least two reasons for intraframe coding:

1. Scene cuts or, in the event of violent motion, interframe prediction error, may not be less than that of the intraframe. Hence intraframe pictures might be coded at lower bit rates.
2. Intraframe coded pictures have a better error resilience to channel errors. Note that, in interframe coding, at the decoder the received data is added to the previous frame to reconstruct the coded picture. In the event of channel error, the error propagates into the subsequent frames. If that part of the picture is not updated, the error can persist for a long time.

Similar to the MC/NO_MC decision, one can make a similar decision for coding a macroblock in inter or intra mode. In this case the variance of intraframe MB is compared with the variance of inter (motion compensated or not). The smallest is chosen. Figure 6.5 shows the characteristics of the function for inter/intra decision. Here for large variances no preference between the two modes is given, but for smaller

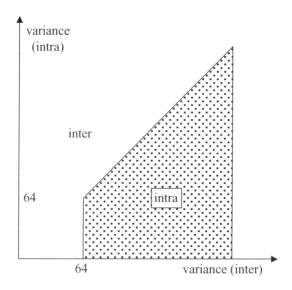

*Figure 6.5*  *Characteristics of inter/intra*

variances interframe is preferred. The reason is that, in intra mode, the DC coefficients of the blocks have to be quantised with a quantiser without a dead zone and with eight-bit resolutions. This increases the bit rate compared with that of the interframe mode, and hence interframe is preferred.

### 6.2.4 Forced updating

As mentioned, intraframe coded MB increases the resilience of the H.261 codec to channel errors. In case in the inter/intra macroblock decision no intra mode is chosen, some of the macroblocks in a frame are forced to be intra coded. The specification recommends that a macroblock should be updated at least once every 132 frames. This means that for common intermediate format (CIF) pictures with 396 macroblocks per frame, on average three MBs of every frame are intraframe coded. This has an important impact on the quality of pictures due to errors. For example, in CIF pictures at 10 Hz, the effect of channel errors may corrupt up to 132 frames, and be visible for almost 13 s.

## 6.3 Other types of macroblock

In H.261 there are as many as eight different types of macroblock:

1. *Inter coded*: interframe coded macroblocks with no motion vector or with a zero motion vector.
2. *MC coded*: motion compensated MB, where the MC error is significant and needs to be DCT coded.
3. *MC not coded*: these are motion compensated error MBs, where the motion compensated error is insignificant. Hence there is no need to be DCT coded.
4. *Intra*: intraframe coded macroblocks.
5. *Skipped*: if all the six blocks in a macroblock, without motion compensation, have an insignificant energy they are not coded. These MBs are sometimes called skipped, not coded or fixed MBs. These types of MB normally occur at the static parts of the image sequence. Fixed MBs are therefore not transmitted and at the decoder they are copied from the previous frame.

Since the quantiser step sizes are determined at the beginning of each GOB or row of GOBs, they have to be transmitted to the receiver. Hence the first MBs have to be identified with a new quantiser parameter. Therefore we can have some new macroblock types, which are:

6. *Inter coded + Q*
7. *MC coded + Q*
8. *Intra + Q*

To summarise the type of macroblock selection, we can draw a flow chart indicating how each one of the 396 MBs in a picture is coded. Decisions on the types of coding follow Figure 6.6 from left to right.

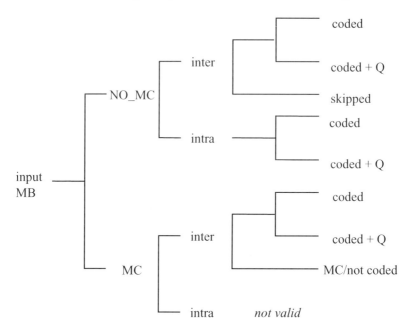

*Figure 6.6   Decision tree for macroblock type*

### 6.3.1  Addressing of macroblocks

If all the quantised components in one of the six blocks in an MB are zero, the block is declared as not coded. When all six blocks are not coded, the MB is declared not coded (fixed MB or skipped MB). In other cases the MBs are declared coded, and their type are variable length coded. The shortest code is assigned to inter code MB and the longest to intra + Q, as they are the most frequent and most rare MB types, respectively.

Once the type of a macroblock is identified and it is VLC coded, its position inside the GOB should also be determined. Considering that H.261 is a videoconferencing codec, normally used for coding head-and-shoulders pictures, it is more likely that coded macroblocks are in the foreground of the picture. Hence they are normally clustered in regions. Therefore the overhead information for addressing the positions of the coded MB is minimised if they are relatively addressed to each other. The relative addresses are represented by run lengths, which are the number of fixed MBs to the next coded MB. Figure 6.7 shows an example of addressing the coded MBs within a GOB. Numbers represent the relative addressing value of the number of fixed macroblocks preceding a nonfixed MB. The GOB start code indicates the beginning of the GOB. These relative addressing numbers are finally VLC coded.

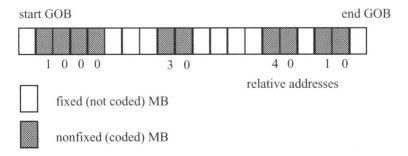

*Figure 6.7   Relative addressing of coded MB*

### 6.3.2  Addressing of blocks

Since an MB has six blocks, four luminance and two chrominance, there will be $2^6 = 64$ different combinations of the coded/noncoded blocks. Except the one with all six blocks not coded (fixed MB), the remaining 63 are identified within 63 different patterns. The pattern information consists of a set of 63 coded block patterns (CBP) indicating coded/noncoded blocks within a macroblock. With a coding order of $Y_0$, $Y_1$, $Y_2$, $Y_3$, $C_b$ and $C_r$, the block pattern information or pattern number is defined as:

$$pattern\_number = 32Y_0 + 16Y_1 + 8Y_2 + 4Y_3 + 2C_b + C_r \qquad (6.1)$$

where in the equation the coded and noncoded blocks are assigned 1 and 0, respectively. Each pattern number is then VLC coded. It should be noted that if a macroblock is intracoded (or intra+Q), its pattern information is not transmitted. This is because, in intraframe coded MB, all blocks have significant energy and will be definitely coded. In other words, there will not be any noncoded blocks in an intra coded macroblock. Figure 6.8 illustrates two examples of the coded block pattern, where some of the luminance or chrominance blocks are not coded.

### 6.3.3  Addressing of motion vectors

The motion vectors of the motion compensated macroblocks are differentially coded with a prediction from their immediate preceding motion vector. The prediction vector is set to zero if either: the macroblock is the first macroblock of each row of GOB, the previous macroblock was not coded, coded with zero motion vector or it was intra coded.

The differential vector is then variable length coded (VLC) and is known as the motion vector data (MVD). The MVD consists of a pair of VLC, the first component for the differential horizontal value and the second component for the differential vertical displacement. Since most head-and-shoulders type scenes normally move in a rigid fashion, differential encoding of motion vectors can significantly reduce

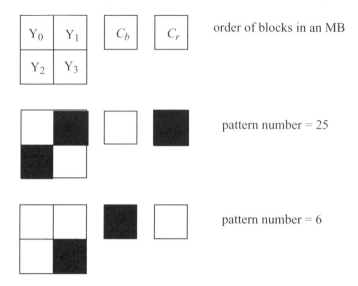

*Figure 6.8* Examples of bit pattern for indicating the coded/not coded blocks in an MB (black coded, white not coded)

the motion vector overhead. However, this makes motion vectors very sensitive to channel errors, but the extent of the error propagation is limited within the GOB.

## 6.4 Quantisation and coding

Every one of the six blocks of a selected MB is transform coded with a two-dimensional DCT. The DCT coefficients of each block are then quantised and coded. In section 3.2 we described two types of quantiser: the one without a dead zone which is used for quantising the DC coefficient of intra MB, for the H.261 standard, this quantiser uses a fixed step size of eight; the second type is with a dead zone for coding AC coefficients and the DC coefficient of interframe coded MB (MC or NO_MC).

For the latter case, a threshold, *th*, may be added to the quantiser scale, such that the dead zone is increased, causing more zero coefficients for efficient compression. Figure 6.9 shows this quantiser, where a threshold, *th*, is added to every step size. The value of the threshold is sent to the receiver as side information (see Figure 6.3). Ratios of the quantised coefficients to the quantiser step size, called indices, are to be coded.

### 6.4.1 Two-dimensional variable length coding

For transmission of the quantisation parameters, a special order is defined which increases the efficiency of capturing the nonzero components. Starting from the DC

## Coding for videoconferencing (H.261)  141

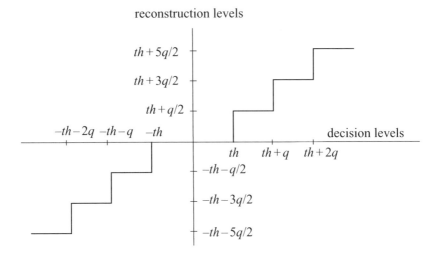

Figure 6.9    A uniform quantiser with threshold

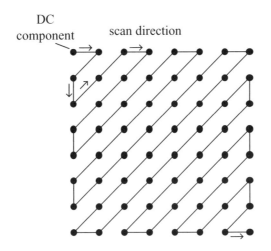

Figure 6.10    Zigzag scanning of 8 × 8 transform coefficients

coefficient on the top left corner of an 8 × 8 coefficient matrix, the values are scanned in a zigzag sequence as shown in Figure 6.10.

The justification for this is that in natural images the main energy of the transform coefficients is concentrated in the lower frequencies (top left corner). Hence the coefficients which normally have the larger values are scanned first. Scanning of the indices terminates when the last nonzero coefficient has been reached.

142  *Standard codecs: image compression to advanced video coding*

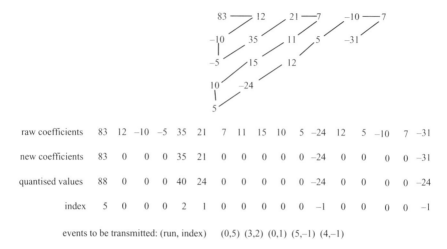

Figure 6.11   *Zigzag scanning and run–index generation*

To increase the coding efficiency a two-dimensional variable length code (2D-VLC) has been adopted. The 2D-VLC is performed in two stages. In the first stage, an event is produced for every nonzero index. The event is a combination of the index magnitude (index) and the number of zeros preceding that index (run).

To see how a two-dimensional index and run generation makes 2D-VLC coding very efficient, the following example is based on a coder having a quantiser step size of $q = 16$ with equal threshold levels $th = q$. Let us assume a pixel block is DCT coded with coefficient values as shown partly in Figure 6.11.

After zigzag scanning, coefficients are quantised. For a dead zone of $th = 16$, coefficient values less than this threshold are set to zero. Larger values are quantised according to the quantisation characteristics (see Figure 6.9). Here we see that rather than 1D-VLC coding of 17 individual coefficients we need to code only five two-dimensional events, which requires substantially fewer bits.

In this 2D-VLC, since the range of index (possible values of indices) can vary from $-127$ to $+127$, and the range of run (number of zeros preceding an index) may vary from 0 to 63, there will be $2 \times 128 \times 64 = 16\,384$ possible events. Design of a Huffman code for this large number of symbols is impractical. Some codewords might be as long as 200 bits! Here we use what might be called a modified Huffman code. In this code, all the symbols with small probabilities are grouped together and are identified with an *ESCAPE* symbol. The *ESCAPE* symbol has a probability equal to the sum of all it represents. Now the most commonly occurring events and the *ESCAPE* symbol are encoded with variable length codes (Huffman code) in the usual way. Events with low probabilities are identified with a fixed length run and index, appended to the *ESCAPE* code. The end of block code (*EOB*) is also one of the symbols to be variable length coded.

# Coding for videoconferencing (H.261)   143

In H.261, *ESCAPE* is six bits long (i.e. 000001), thus rare events with six bits run (0–63) and eight bits index (−127 to +127) require 20 bits [1]. The EOB code is represented with a two-bit word. The DC/intra index is linearly quantised with a step size of eight and no dead zone. The resulting value is coded with an eight-bit resolution.

Figure 6.12 shows an example of a 2D-VLC table for positive values of indices, derived from statistics of coding the Claire test image sequence. As we see, most frequent events are registered at low index and low run values. The sum of the rare events, which represents the frequency of the *ESCAPE*, is even less than some frequent events. The corresponding 2D-VLC table is also shown next to the frequency table. Also in this example the sum has the same frequency as the event (run = 4, index = 1).

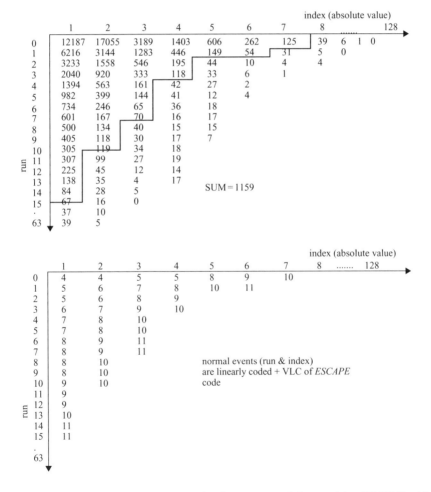

Figure 6.12   An example of run and index frequency and the resulting 2D-VLC table

144  *Standard codecs: image compression to advanced video coding*

They are expected to have the same word length. Other events can be defined as *ESCAPE* + normal run + normal index, with:

*ESCAPE* code = 6 bits,
normal run = 6 bits (1 out of 64 possible values)
normal index = 8 bits (1 out of 128 values) plus the sign bit
total bits for the modified Huffman coded events = 6 + 6 + 8 = 20

## 6.5 Loop filter

At low bit rates the quantiser step size is normally large. Larger step sizes can force many DCT coefficients to zero. If only the DC and a few AC coefficients remain, then the reconstructed picture appears blocky. When the positions of blocky areas vary from one frame to another, it appears as a high frequency noise, commonly referred to as mosquito noise. The blockiness degradations at the slant edges of the image appear as staircase noise. Figure 6.13 illustrates single shots of a CIF size Claire test image sequence and its coded version at 256 kbit/s. The sequence is in colour, with 352 pixels by 288 lines at 30 Hz, but only the luminance is shown. The colour components have a quarter resolution of luminance (176 pixels by 144 lines). As can be seen at this bit rate the coded image quality is very good with no visible distortions.

At lower bit rates, artefacts begin to appear. This is shown in Figure 6.14, where there are more severe distortions at 64 kbit/s than at 128 kbit/s. When the sequence is displayed at its normal rate (30 Hz), the positions of the distortions move in different directions over the picture, and the appearance of mosquito noise is quite visible.

Coarse quantisation of the coefficients which results in the loss of high frequency components implies that compression can be modelled as a lowpass filtering process [3,4]. These artefacts are to some extent reduced by using the loop filter (see position of the loop filter in Figure 6.3). The lowpass filter removes the high frequency and

*a*  *b*

*Figure 6.13*  Picture of Claire
  *a*  original
  *b*  H.261 coded at 256 kbit/s

*Figure 6.14* H.261 coded at
    *a*   128 kbit/s
    *b*   64 kbit/s

*Figure 6.15* Coded pictures with loop filter
    *a*   128 kbit/s
    *b*   64 kbit/s

block boundary distortions. The same pictures with the use of a loop filter are shown in Figure 6.15.

Loop filtering is introduced after the motion compensator to improve the prediction. It should be noted that the loop filter has a picture blurring effect. It should be activated only for blocks with motion, otherwise nonmoving parts of the pictures are repeatedly filtered in the following frames, blurring the picture. Since it is motion based, loop filtering is thus carried out on a macroblock basis and it has an impulse response given by:

$$h(x, y) = \frac{1}{16} \begin{bmatrix} 1 & 2 & 1 \\ 2 & 4 & 2 \\ 1 & 2 & 1 \end{bmatrix} \quad (6.2)$$

|   |   |   |   |   |   |   |   |
|---|---|---|---|---|---|---|---|
| 9 | 3 |   |   |   |   |   |   |
| 3 | 1 |   |   |   |   |   |   |
|   |   | 1 | 2 | 1 |   |   |   |
|   |   | 2 | 4 | 2 |   |   |   |
|   |   | 1 | 2 | 1 |   |   |   |
|   |   |   |   |   |   |   |   |
|   |   | 1 | 2 | 1 |   |   |   |
|   |   | 3 | 6 | 3 |   |   |   |

*Figure 6.16    Loop filter impulse response in various parts of the image*

for pixels well inside the picture. For pixels at the image border, or corners, another function may be used. Figure 6.16 shows an example of the filter response in these areas.

The loop filter is only defined for H.261 (no other video codecs use it) and is activated for all six DCT blocks of a macroblock. The filtering should be applied for coding rates of less than $6 \times 64$ kbit/s $= 386$ kbit/s and switched off otherwise. At higher bit rates the filter does not improve the subjective quality of the picture [3]. MPEG-1 does not specify the requirement of a loop filter, because pictures coded with MPEG-1 are at much higher bit rates than 386 kbit/s.

## 6.6  Rate control

The bit rate resulting from the DCT-based coding algorithm fluctuates according to the nature of the video sequence. Variations in the speed of moving objects, their size and texture are the main cause for bit rate variation. The objective of a rate controller is to achieve a constant bit rate for transmission over a circuit switched network. A transmission buffer is usually needed to smooth out the bit rate fluctuations, which are inherent in the interframe coding scheme.

The usual method for bit rate control is to monitor the buffer occupancy and vary the quantiser step size according to the buffer fullness [3,5]. In reference model 8 (RM8) the quantiser step size is calculated as a linear function of the buffer content and expressed by:

$$q = 2 \left\lfloor \frac{buffercontent}{200p} \right\rfloor + 2 \qquad (6.3)$$

where $p$ is the multiplier used in specifying the bit rates as in $p \times 64$ kbit/s, and $\lfloor . \rfloor$ stands for integer division with truncation towards zero.

The buffer control system usually has two additional operating states to prevent buffer underflow or buffer overflow from occurring. If the buffer content reaches the

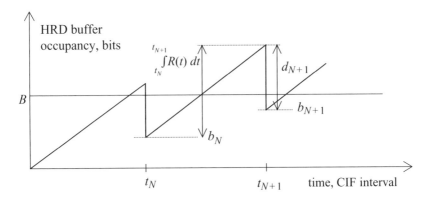

*Figure 6.17    Hypothetical reference buffer occupancy*

trigger point for the overflow state, current and subsequent coded data are not sent to allow the buffer to be emptied. Only trivial side information pertaining to the coded GOB or frame is transmitted.

On the other extreme, bit stuffing is invoked when buffer underflow is threatened. It is essential that buffer underflow is avoided so that the decoder can maintain synchronisation.

In practice, to allow maximum freedom of the H.261 standard codec structure, a hypothetical reference decoder (HRD) buffer is defined. All encoders are required to be compliant with this buffer. The hypothetical reference decoder is best explained with reference to Figure 6.17.

The hypothetical buffer is initially empty. It is examined at CIF intervals ($1/29.97 \cong 33$ ms), and if at least one complete coded picture is in the buffer, then all the data from the earliest picture is instantly removed (e.g. at $t_N$ in Figure 6.17) [6]. Immediately after removing the above data, the buffer occupancy should be less than $B$, with $B = 4R_{max}/29.97$ and $R_{max}$ being the maximum video bit rate to be used in the connection. To meet this requirement, the number of bits for $(N+1)$th coded picture, $d_{N+1}$ must satisfy

$$d_{N+1} > b_N + \int_{t_N}^{t_{N+1}} R(t)\, dt - B \tag{6.4}$$

where $b_N$ is the buffer occupancy just after time $t_N$, $t_N$ is the time at which the $N$th coded picture is removed from the hypothetical reference decoder buffer and $R(t)$ is the video bit rate at time $t$. Note that the time interval $(t_{N+1} - t_N)$ is an integer number of CIF picture periods (1/29.97, 2/29.97, 3/29.97, ...).

This specification constrains all encoders to restrict the picture start lead jitters to four CIF picture period's worth of channel bits. This prevents decoder buffer overflow at the decoder in a correctly designed H.261 codec. Jitters in the opposite direction (phase lag) are not constrained by the H.261 recommendation. Phase lag corresponds to buffer underflow at the decoder, which is simply dealt with by making the decoder wait for sufficient bits to have arrived to continue decoding.

A major deficiency with the RM8/H.261 model rate controls is that bits might be unfairly distributed across the image. For example, in the active parts of the picture sequences, such as in the middle of the head-and-shoulders pictures, the buffer tends to fill up very quickly, and hence the quantiser step size rises rapidly. This causes the rest of the picture to be coded coarsely. One of the key issues in H.261, as well as in any video codec, is the way in which the quantiser step size or the rate control is managed. In the past decade numerous manufacturers have produced H.261 codecs, but they may not perform equally. Because of the need for interoperability, the general structure of H.261 must be based on coding elements as shown in Figure 6.3. Therefore the only part which makes one codec better than the others is the rate control mechanism. This part is kept secret by the manufacturers and is subject to further research.

## 6.7 Problems

1 In a CIF picture, find the number of macroblocks and blocks per

   a GOB
   b picture

2 Calculate the macroblock interval in CIF pictures. How large is this value for a QCIF video at 10 Hz?

3 The absolute value of the motion compensated frame difference per macroblock, $|mfd|$, is normally smaller than that without motion compensation, $|fd|$. In a search for motion compensation, the following values for $|mfd|$ and $|fd|$ have been calculated. Using Figure 6.4, determine in which of the following cases motion compensation should be used:

   a $|fd| = 1200$ and $|mfd| = 1000$
   b $|fd| = 600$ and $|mfd| = 500$
   c $|fd| = 200$ and $|mfd| = 50$

4 Why is nonmotion compensation preferred to motion compensation for very small frame difference images?

5 To decide whether a macroblock should be interframe or intraframe coded, the variances of intra and motion compensated interframe macroblocks are compared, according to Figure 6.5. Find in each of the following whether a macroblock should be intra or interframe coded:

   a $\sigma^2_{intra} = 1500$, $\sigma^2_{inter} = 1450$
   b $\sigma^2_{intra} = 500$, $\sigma^2_{inter} = 600$
   c $\sigma^2_{intra} = 50$, $\sigma^2_{inter} = 60$

6 In macroblocks with small energy (inter or intra), inter macroblock is preferred to intra, why?

7 Calculate the coded block pattern (CBP) indices of the following macroblocks in a H.261 codec if:

   *a* all the blocks in the macroblock are coded
   *b* only the luminance blocks are coded
   *c* only the chrominance blocks are coded

8 Assume the transform coefficients of the block given in problem 3 of Chapter 5 belong to an H.261 codec. These coefficients are linearly quantised with the quantiser of Figure 6.9, with $th = 16$ and $q = 12$. Using Figure 6.12, calculate the number of bits required to code this block.

9 Pixels on the top left corner of a picture have values of:
   200  135  180  210
   75   110  134  230
   62   89   52   14
   and are filtered with the loop filter of Figure 6.16. Find the filtered values of these pixels.

10 The maximum quantiser step size in H.261 is 62 (quantiser parameter is 31, defined with five bits). A 384 kbit/s H.261 encoder with an RM8 type rate control has a smoothing buffer of 5 kbytes. Find the spare capacity of the buffer, when the quantiser step size is at its maximum value.

## 6.8 References

1 H.261: 'Recommendation H.261, video codec for audiovisual services at $p \times 64$ kbit/s'. Geneva, 1990
2 CCITT SG XV WP/1/Q4: 'Specialist group on coding for visual telephony'. Description of reference model 8 (RM8), 1989
3 PLOMPEN, R.H.J.M.: 'Motion video coding for visual telephony' (Proefschrift, 1989)
4 NGAN, K.N.: 'Two-dimensional transform domain decimation technique', IEEE international conference on *Acoust. speech and signal processing, ICASP '86*, 1986, pp.1001–1004
5 CHEN, C.T., and WONG, A.: 'A self-governing rate buffer control strategy for pseudoconsant bit rate video coding', *IEEE Trans. Image processing*, 1993, **2**:1, pp.50–59.
6 CARR, M.D.: 'Video codec hardware to realise a new world standard', *Br. Telecom. J.*, 1990, **8**:3, pp.28–35

*Chapter 7*

# Coding of moving pictures for digital storage media (MPEG-1)

MPEG-1 is the first generation of video codecs proposed by the Motion Picture Experts Group as a standard to provide video coding for digital storage media (DSM), such as CD, DAT, Winchester discs and optical drives [1]. This development was in response to industry needs for an efficient way of storing visual information on storage media other than the conventional analogue video cassette recorders (VCR). At the time the CD-ROMs had the capability of 648 Mbytes, sufficient to accommodate movie programs at a rate of approximately 1.2 Mbit/s, and the MPEG standard aimed to conform roughly with this target. Although in most applications the MPEG-1 video bit rate is in the range of 1–1.5 Mbit/s, the international standard does not limit the bit rate, and higher bit rates might be used for other applications.

It was also envisaged that the stored data would be within both 625 and 525-line television systems and provide flexibility for use with workstations and personal computers. For this reason, the MPEG-1 standard is based on progressively scanned images and does not recognise interlacing. Interlaced sources have to be converted to a noninterlaced format before coding. After decoding, the decoded image may be converted back to provide an interlaced format for display.

Since coding for digital storage can be regarded as a competitor to VCRs, then MPEG-1 video quality at the rate of 1 to 1.5 Mbit/s is expected to be comparable to VCRs. Also, it should provide the viewing conditions associated with VCRs such as forward play, freeze picture, fast forward, fast reverse, slow forward and random access. The ability of the decoder to provide these modes depends to some extent on the nature of digital storage media. However, it should be borne in mind that efficient coding and flexibility in operation are not compatible. Provision of the added functionality of random access necessitates regular intraframe pictures in the coded sequence. Those frames that do not exploit the temporal redundancy in video have poor compression, and as a result the overall bit rate is increased.

Both H.261 [2] and MPEG-1 [1] are standards defined for relatively low bit rate coding of low spatial resolution pictures. Like H.261, MPEG-1 utilises DCT for lossy coding of its intraframe and interframe prediction errors. The MPEG-1 video coding algorithm is largely an extension of H.261, and many of the features are common. Their bit streams are, however, incompatible, although their encoding units are very similar.

The MPEG-1 standard, like H.261, does not specify the design of the decoder, and even less information is given about the encoder. What is expected from MPEG-1, like H.261, is to produce a bit stream which is decodable. Manufacturers are free to choose any algorithms they wish, and to optimise them for better efficiency and functionality. Therefore in this Chapter we again look at the fundamentals of MPEG-1 coding, rather than details of the implementation.

## 7.1 Systems coding outline

The MPEG-1 standard gives the syntax description of how audio, video and data are combined into a single data stream. This sequence is formally termed the ISO 11172 stream [3]. The structure of this ISO 11172 stream is illustrated in Figure 7.1. It consists of a compression layer and a systems layer. In this book we study only the video part of the compression layer, but the systems layer is important for the proper delivery of the coded bit stream to the video decoder, and hence we briefly describe it.

The MPEG-1 systems standard defines a packet structure for multiplexing coded audio and video into one stream and keeping it synchronised. The systems layer is organised into two sublayers known as the pack and packet layers. A pack consists of a pack header that gives the systems clock reference (SCR) and the bit rate of the

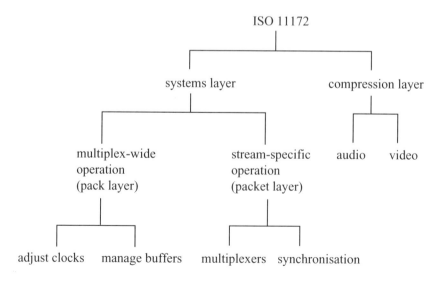

*Figure 7.1   Structure of an ISO 11172 stream*

multiplexed stream followed by one or more packets. Each packet has its own header that conveys essential information about the elementary data that it carries. The aim of the systems layer is to support the combination of video and audio elementary streams. The basic functions are as follows:

- synchronised presentation of decoded streams
- construction of the multiplexed stream
- initialisation of buffering for playback start-up
- continuous buffer management
- time identification.

In the systems layer, elements of direct interest to the video encoding and decoding processes are mainly those of the stream-specific operations, namely multiplexing and synchronisation.

### 7.1.1 *Multiplexing elementary streams*

The multiplexing of elementary audio, video and data is performed at the packet level. Each packet thus contains only one elementary data type. The systems layer syntax allows up to 32 audio, 16 video and two data streams to be multiplexed together. If more than two data streams are needed, substreams may be defined.

### 7.1.2 *Synchronisation*

Multiple elementary streams are synchronised by means of presentation time stamps (PTS) in the ISO 11172 bit stream. End-to-end synchronisation is achieved when the encoders record time stamps during capture of raw data. The receivers will then make use of these PTS in each associated decoded stream to schedule their presentations. Playback synchronisation is pegged onto a master time base, which may be extracted from one of the elementary streams, the digital storage media (DSM), channel or some external source. This prototypical synchronisation arrangement is illustrated in Figure 7.2. The occurrences of PTS and other information such as the systems clock reference (SCR) and systems headers will also be essential for facilitating random access of the MPEG-1 bit stream. This set of access codes should therefore be located near to the part of the elementary stream where decoding can begin. In the case of video, this site will be near the head of an intraframe.

To ensure guaranteed decoder buffer behaviour, the MPEG-1 systems layer employs a systems target decoder (STD) and decoding time stamp (DTS). The DTS differs from PTS only in the case of video pictures that require additional reordering delay during the decoding process.

## 7.2 **Preprocessing**

The source material for video coding may exist in a variety of forms such as computer files or live video in CCIR-601 format [4]. If CCIR-601 is the source, since MPEG-1 is

154  *Standard codecs: image compression to advanced video coding*

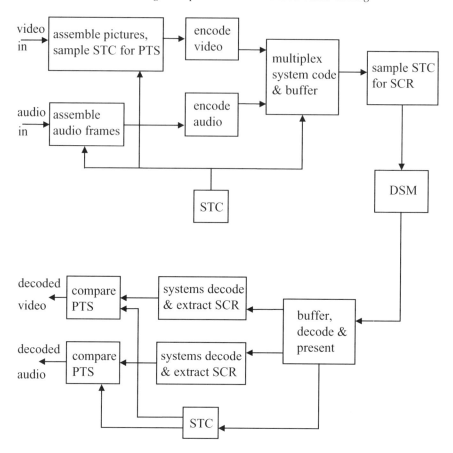

Figure 7.2   *MPEG-1's prototypical encoder and decoder illustrating end-to-end synchronisation (STC: systems time clock; SCR: systems clock reference; PTS: presentation time stamp; DSM: digital storage media)*

for coding of video at VCR resolutions, then SIF format is normally used. These source pictures must be processed prior to coding. In Chapter 2 we explained how CCIR-601 video was converted to SIF format. If the source is film, we also discussed the conversion methodology in that Chapter. However, if computer source files do not have the SIF format, they have to be converted too. In MPEG-1, another preprocessing step is required to reorder the input pictures for coding. This is called picture reordering.

## 7.2.1  Picture reordering

Because of the conflicting requirements of random access and highly efficient coding, the MPEG suggested that not all pictures of a video sequence should be coded in the same way. They identified four types of picture in a video sequence. The first type is called I-pictures, which are coded without reference to the previous picture.

They provide access points to the coded sequence for decoding. These pictures are intraframe coded as for JPEG, with a moderate compression. The second type is the P-pictures, which are predictively coded with reference to the previous I or P-coded pictures. They themselves are used as a reference (anchor) for coding of the future pictures. Coding of these pictures is very similar to H.261. The third type is B-pictures, or bidirectionally coded pictures, which may use past, future or combinations of both pictures in their predictions. This increases the motion compensation efficiency, since occluded parts of moving objects may be better compensated for from the future frame. B-pictures are never used for predictions. This part, which is unique to MPEG, has two important implications:

(i) If B-pictures are not used for predictions of future frames, then they can be coded with the highest possible compression without any side effects. This is because, if one picture is coarsely coded and is used as a prediction, the coding distortions are transferred to the next frame. This frame then needs more bits to clear the previous distortions, and the overall bit rate may increase rather than decrease.

(ii) In applications such as transmission of video over packet networks, B-pictures may be discarded (e.g. due to buffer overflow) without affecting the next decoded pictures [5]. Note that if any part of the H.261 pictures, or I and P-pictures in MPEG, are corrupted during the transmission, the effect will propagate until they are refreshed [6].

Figure 7.3 illustrates the relationship between these three types of picture. Since B-pictures use I and P-pictures as predictions, they have to be coded later. This requires reordering the incoming picture order, which is carried out at the preprocessor.

The fourth picture type is the D-pictures. These are intraframe coded, where only the DC coefficients are retained. Hence the picture quality is poor and normally used for applications like fast forward. D-pictures are not part of the GOP, hence they are not present in a sequence containing any other picture type.

## 7.3 Video structure

### 7.3.1 Group of pictures (GOP)

Since in the H.261 standard successive frames are similarly coded, a picture is the top level of the coding hierarchy. In MPEG-1 due to the existence of several picture types, a group of pictures, called GOP, is the highest level of the hierarchy. A GOP is a series of one or more pictures to assist random access into the picture sequence. The first coded picture in the group is an I-picture. It is followed by an arrangement for P and B-pictures, as shown in Figure 7.3.

The GOP length is normally defined as the distance between I-pictures, which is represented by parameter $N$ in the standard codecs. The distance between the anchor I/P to P-pictures is represented by $M$. In the above Figure $N = 12$ and $M = 3$. The group of pictures may be of any length, but there should be at least one I-picture in each GOP. Applications requiring random access, fast forward play or fast and normal

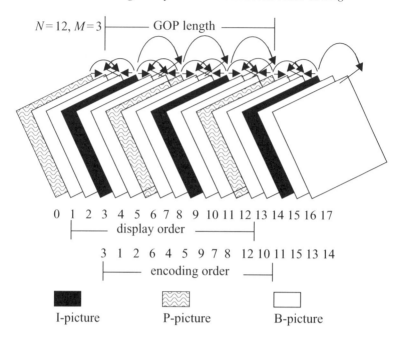

*Figure 7.3    An example of MPEG-1 GOP*

reverse play may use short GOPs. GOP may also start at scene cuts or other cases where motion compensation is not effective. The number of consecutive B-pictures is variable. Neither a P nor a B-picture needs to be present. For most applications, GOP in the SIF-625/50 format has $N = 12$ and $M = 3$. In SIF-525/60, the values are 15 and 3, respectively.

The encoding or transmission order of pictures differs from the display or incoming picture order. In the Figure B-pictures 1 and 2 are encoded after P-picture 0 and I-picture 3 are encoded. Also in this Figure B-pictures 13 and 14 are a part of the next GOP. Although their display order is 0,1,2,...,11, their encoding order is 3,1,2,6,4,5.... This reordering introduces delays amounting to several frames at the encoder (equal to the number of B-pictures between the anchor I and P-pictures). The same amount of delay is introduced at the decoder, in putting the transmission/decoding sequence back to its original. This format inevitably limits the application of MPEG-1 for telecommunications.

### 7.3.2   Picture

All the three main picture types, I, P and B, have the same SIF size with 4 : 2 : 0 format. In SIF-625 the luminance part of each picture has 360 pixels, 288 lines and 25 Hz, and those of each chrominance are 180 pixels, 144 lines and 25 Hz. In SIF-525, these values for luminance are 360 pixels, 240 lines and 30 Hz, and for the chrominance are 180, 120 and 30, respectively. For 4 : 2 : 0 format images, the luminance and chrominance samples are positioned as shown in Figure 2.3.

## 7.3.3 Slice

Each picture is divided into a group of macroblocks, called slices. In H.261 such a group was called GOB. The reason for defining a slice is the same as that for defining a GOB, namely resetting the variable length code to prevent channel error propagation into the picture. Slices can have different sizes within a picture, and the division in one picture need not be the same as the division in any other picture.

The slices can begin and end at any macroblock in a picture, but with some constraints. The first slice must begin at the top left of the picture (the first macroblock) and the end of the last slice must be the bottom right macroblock (the last macroblock) of the picture, as shown in Figure 7.4. Therefore the minimum number of slices per picture is one, and the maximum number is equal to the number of macroblocks (e.g. 396 in SIF-625).

Each slice starts with a slice start code, and is followed by a code that defines its position and a code that sets the quantisation step size. Note that in H.261 the quantisation step sizes were set at each GOB or row of GOBs, but in MPEG-1 they can be set at any macroblock (see below). Therefore, in MPEG-1 the main reason for defining slices is not to reset a new quantiser, but to prevent the effects of channel error propagation. If the coded data is corrupted, and the decoder detects it, then it can search for the new slice, and the decoding starts from that point. Part of the picture slice from the start of the error to the next slice can then be degraded. Therefore in a noisy environment it is desirable to have as many slices as possible. On the other hand each slice has a large overhead, called slice start code (minimum of 32 bits). This creates a large overhead in the total bit rate. For example, if we use the slice structure of Figure 7.4, where there is one slice for each row of MBs, then for SIF-625 video there are 18 slices per picture, and with 25 Hz video, the slice overhead can be $32 \times 18 \times 25 = 14\,400\,\text{bit/s}$.

| 1 begin | end 1 |
|---|---|
| 2 begin | end 2 |
| 3 begin | end 3 |
| 4 begin | end 4 |
| 5 begin | end 5 |
| 6 begin | end 6 |
| 7 begin | end 7 |
| 8 begin | end 8 |
| 9 begin | end 9 |
| 10 begin | end 10 |
| 11 begin | end 11 |
| 12 begin | end 12 |
| 13 begin | end 13 |
| 14 begin | end 14 |
| 15 begin | end 15 |
| 16 begin | end 16 |
| 17 begin | end 17 |
| 18 begin | end 18 |

Figure 7.4   An example of slice structure for SIF-625 pictures

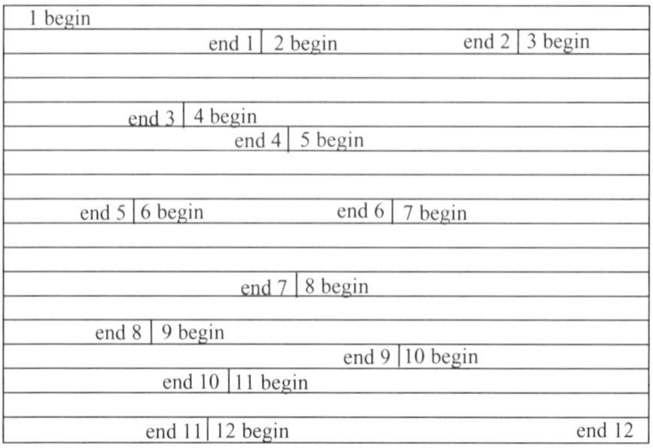

158  *Standard codecs: image compression to advanced video coding*

| 1 begin | | | |
|---|---|---|---|
| | end 1 \| 2 begin | | end 2 \| 3 begin |
| | end 3 \| 4 begin | | |
| | end 4 \| 5 begin | | |
| | end 5 \| 6 begin | end 6 \| 7 begin | |
| | | end 7 \| 8 begin | |
| | end 8 \| 9 begin | | |
| | | end 9 \| 10 begin | |
| | end 10 \| 11 begin | | |
| | end 11 \| 12 begin | | end 12 |

*Figure 7.5   Possible arrangement of slices in SIF-625*

To optimise the slice structure, that is, to give a good immunity from channel errors and at the same time to minimise the slice overhead, one might use short slices for macroblocks with significant energy (such as intra-MB), and long slices for less significant ones (e.g. macroblocks in B-pictures). Figure 7.5 shows a slice structure where in some parts the slice length extends beyond several rows of macroblocks, and in some cases is less than one row.

### 7.3.4  Macroblock

Slices are divided into macroblocks of $16 \times 16$ pixels, similar to the division of GOB into macroblocks in H.261. Macroblocks in turn are divided into blocks, for coding. In Chapter 6, we gave a detailed description of how a macroblock was coded, starting from its type, mode of selection, blocks within the MB, their positional addresses and finally the block pattern. Since MPEG-1 is also a macroblock-based codec, most of these rules are used in MPEG-1. However, due to differences of slice *versus* GOB, picture type *versus* a single picture format in H.261, there are bound to be variations in the coding. We first give a general account of these differences then, in the following section, more details about the macroblocks in the various picture types.

The first difference is that since a slice has a raster scan structure, macroblocks are addressed in a raster scan order. The top left macroblock in a picture has address 0, the next one on the right has address 1 and so on. If there are $M$ macroblocks in a picture (e.g. $M = 396$), then the bottom right macroblock has address $M - 1$. To reduce the address overhead, macroblocks are relatively addressed by transmitting the difference between the current macroblock and the previously coded macroblock. This difference is called the macroblock address increment. In I-pictures, since all the macroblocks are coded, the macroblock address increment is always 1. The exception is that, for the first coded macroblock at the beginning of each slice, the macroblock address is set to that of the right-hand macroblock of the previous row. This address at the beginning of each picture is set to $-1$. If a slice does not start at the left edge of the picture (see the slice structure of Figure 7.5), then the macroblock address increment for the first macroblock in the slice will be larger than one. For example, in the slice structure of Figures 7.4 and 7.5 there are 22 macroblocks per row. For Figure 7.4, at the start of slice two, the macroblock address is set to 21, which is the address of the macroblock at the right-hand edge of the top row of macroblocks. In Figure 7.5, if the first slice contains 30 macroblocks, eight of them would be in the second row, so the address of the first macroblock in the second slice would be 30 and the macroblock increment would be nine. For further reduction of address overhead, macroblock address increments are VLC coded.

There is no code to indicate a macroblock address increment of zero. This is why the macroblock address is set to $-1$ rather than zero at the top of the picture. The first macroblock will have an increment of one, making its address equal to zero.

### 7.3.5  *Block*

Finally, the smallest part of the picture structure is the block of $8 \times 8$ pixels, for both luminance and chrominance components. DCT coding is applied at this block level. Figure 7.6 illustrates the whole structure of partitioning a video sequence, from its GOP level at the top to the smallest unit of block at the bottom.

## 7.4  Encoder

As mentioned, the international standard does not specify the design of the video encoders and decoders. It only specifies the syntax and semantics of the bit stream and signal processing at the encoder/decoder interface. Therefore, options are left open to the video codec manufacturers to trade-off cost, speed, picture quality and coding efficiency. As a guideline, Figure 7.7 shows a block diagram of an MPEG-1 encoder. Again it is similar to the generic codec of Chapter 3 and the H.261 codec of Chapter 6. For simplicity the coding flags shown in the H.261 codec are omitted, although they also exist.

The main differences between this encoder and that defined in H.261 are:

- *Frame reordering*: at the input of the encoder coding of B-pictures is postponed to be carried out after coding the anchor I and P-pictures.
- *Quantisation*: intraframe coded macroblocks are subjectively weighted to emulate perceived coding distortions.
- *Motion estimation*: not only is the search range extended but the search precision is increased to half a pixel. B-pictures use bidirectional motion compensation.
- *No loop filter*.
- *Frame store and predictors*: to hold two anchor pictures for prediction of B-pictures.
- *Rate regulator*: since here there is more than one type of picture, each generating different bit rates.

Before describing how each picture type is coded, and the main differences between this codec and H.261, we can describe the codec on a macroblock basis, as the basic

160  *Standard codecs: image compression to advanced video coding*

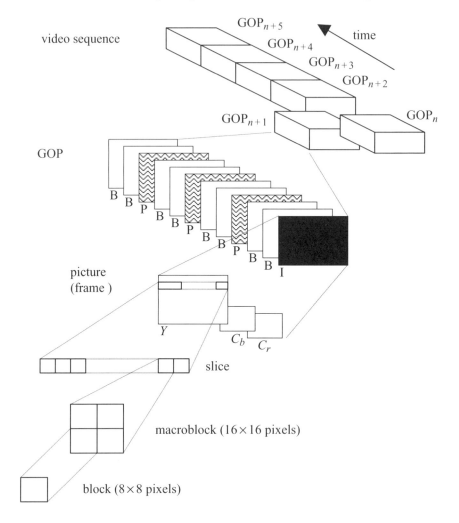

*Figure 7.6  MPEG-1 coded video structure*

unit of coding. Within each picture, macroblocks are coded in a sequence from left to right. Since 4 : 2 : 0 image format is used, then the six blocks of 8 × 8 pixels, four luminance and one of each chrominance component, are coded in turn. Note that the picture area covered by the four luminance blocks is the same as that covered by each of the chrominance blocks.

First, for a given macroblock, the coding mode is chosen. This depends on the picture type, the effectiveness of motion compensated prediction in that local region and the nature of the signal within the block. Secondly, depending on the coding mode, a motion compensated prediction of the contents of the block based on the past and/or future reference pictures is formed. This prediction is subtracted from the actual data in the current macroblock to form an error signal. Thirdly, this error signal

*Coding of moving pictures for digital storage media (MPEG-1)* 161

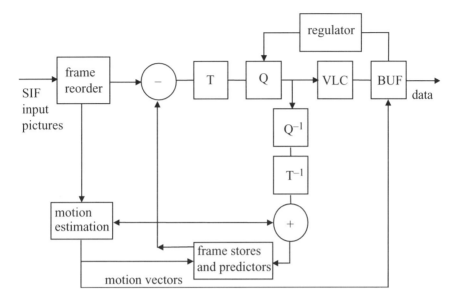

Figure 7.7   A simplified MPEG-1 video encoder

is divided into 8 × 8 blocks and a DCT is performed on each block. The resulting two-dimensional 8 × 8 block of DCT coefficients is quantised and is scanned in zigzag order to convert into a one-dimensional string of quantised DCT coefficients. Fourthly, the side information for the macroblock, including the type, block pattern, motion vector and address alongside the DCT coefficients are coded. For maximum efficiency, all the data are variable length coded. The DCT coefficients are run length coded with the generation of events, as we discussed in H.261.

A consequence of using different picture types and variable length coding is that the overall bit rate is very variable. In applications that involve a fixed rate channel, a FIFO buffer is used to match the encoder output to the channel. The status of this buffer may be monitored to control the number of bits generated by the encoder. Controlling the quantiser parameter is the most direct way of controlling the bit rate. The international standard specifies an abstract model of the buffering system (the video buffering verifier) in order to limit the maximum variability in the number of bits that are used for a given picture. This ensures that a bit stream can be decoded with a buffer of known size (see section 7.8).

## 7.5  Quantisation weighting matrix

The insensitivity of the human visual system to high frequency distortions can be exploited for further bandwidth compression. In this case, the higher orders of DCT coefficients are quantised with coarser quantisation step sizes than the lower frequency ones. Experience has shown that for SIF pictures a suitable distortion weighting matrix

|   |   |   |   |   |   |   |   |   |   |   |   |   |   |   |
|---|---|---|---|---|---|---|---|---|---|---|---|---|---|---|
| 8  | 16 | 19 | 22 | 26 | 27 | 29 | 34 | 16 | 16 | 16 | 16 | 16 | 16 | 16 | 16 |
| 16 | 16 | 22 | 24 | 27 | 29 | 34 | 37 | 16 | 16 | 16 | 16 | 16 | 16 | 16 | 16 |
| 19 | 22 | 26 | 27 | 29 | 34 | 34 | 38 | 16 | 16 | 16 | 16 | 16 | 16 | 16 | 16 |
| 22 | 22 | 26 | 27 | 29 | 34 | 37 | 40 | 16 | 16 | 16 | 16 | 16 | 16 | 16 | 16 |
| 22 | 26 | 27 | 29 | 32 | 35 | 40 | 48 | 16 | 16 | 16 | 16 | 16 | 16 | 16 | 16 |
| 26 | 27 | 29 | 32 | 35 | 40 | 48 | 58 | 16 | 16 | 16 | 16 | 16 | 16 | 16 | 16 |
| 26 | 27 | 29 | 34 | 38 | 46 | 56 | 69 | 16 | 16 | 16 | 16 | 16 | 16 | 16 | 16 |
| 27 | 29 | 35 | 38 | 46 | 56 | 69 | 83 | 16 | 16 | 16 | 16 | 16 | 16 | 16 | 16 |

intra                                  inter

*Figure 7.8    Default intra and inter quantisation weighting matrices*

for the intra DCT coefficients is the one shown in Figure 7.8. This intra matrix is used as the default quantisation matrix for intraframe coded macroblocks.

If the picture resolution departs significantly from the SIF size, then some other matrix may give perceptively better results. The reason is that this matrix is derived from the vision contrast sensitivity curve, for a nominal viewing distance (e.g. viewing distances of 4–6 times the picture height) [7]. For higher or lower picture resolutions, or changing the viewing distance, the spatial frequency will then change, and hence different weighting will be derived.

It should be noted that different weightings may not be used for interframe coded macroblocks. This is because high frequency interframe error does not necessarily mean high spatial frequency. It might be due to poor motion compensation or block boundary artefacts. Hence interframe coded macroblocks use a flat quantisation matrix. This matrix is called the inter or nonintra quantisation weighting matrix.

Note that, since in H.261 all the pictures are interframe coded and a very few macroblocks might be intra coded, then only the nonintra weighting matrix is defined. Little work has been performed to determine the optimum nonintra matrix for MPEG-1, but evidence suggests that the coding performance is more related to the motion and the texture of the scene than the nonintra quantisation matrix. If there is any optimum matrix, it should then be somewhere between the flat default inter matrix and the strongly frequency-dependent values of the default intra matrix.

The DCT coefficients, prior to quantisation, are divided by the weighting matrix. Note that the DCT coefficients prior to weighting have a dynamic range from $-2047$ to $+2047$. Weighted coefficients are then quantised by the quantisation step size and at the decoder, reconstructed quantised coefficients are multiplied to the weighting matrix to reconstruct the coefficients.

## 7.6    Motion estimation

In Chapter 3, block matching motion estimation/compensation and its application in standard codecs was discussed in great detail. We even introduced some fast search

# Coding of moving pictures for digital storage media (MPEG-1)  163

methods for estimation, which can be used in software-based codecs. As we saw, motion estimation in H.261 was optional. This was mainly due to the assumption that, since motion compensation can reduce correlation, then DCT coding may not be efficient. Investigations since the publication of H.261 have proved that this is not the case. What is expected from a DCT is to remove the spatial correlation within a small area of $8 \times 8$ pixels. Measurement of correlations between the adjacent error pixels have shown that there is still strong correlation between the error pixels, which does not impair the potential of DCT for spatial redundancy reduction. Hence motion estimation has become an important integral part of all the later video codecs, such as MPEG-1, MPEG-2, H.263, H.26L and MPEG-4. These will be explained in the relevant chapters.

Considering MPEG-1, the strategy for motion estimation in this codec is different from the H.261 in four main respects:

- motion estimation is an integral part of the codec
- motion search range is much larger
- higher precision of motion compensation is used
- B-pictures can benefit from bidirectional motion compensation.

These features are described in the following sections.

## 7.6.1 Larger search range

In H.261, if motion compensation is used, a search is carried out within every subsequent frame. Also, H.261 is normally used for head-and-shoulders pictures, where the motion speed tends to be very small. In contrast, MPEG-1 is used mainly for coding of films with much larger movements and activities. Moreover, in the search for motion in P-pictures, since they might be several frames apart, the search range becomes many times larger. For example, in a GOP structure with $M = 3$, where there are two B-pictures between the anchor pictures, the motion speed is three times greater than that for consecutive pictures. Thus in MPEG-1 we expect a much larger search range. Considering that in full search block matching the number of search positions for a motion speed of $w$ is $(2w + 1)^2$, then tripling the search range makes motion estimation prohibitively computationally expensive.

In Chapter 3 we introduced some fast search methods such as logarithmic step searches and hierarchical motion estimation. Although the hierarchical method can be used here, of course needing one or more levels of hierarchy, use of a logarithmic search may not be feasible. This is because these methods are very prone to large search ranges, and at these ranges the final minima can be very far away from the local minima, so causing the estimation to fail [8].

One way of alleviating this problem is to use a telescopic search method. This is unique to MPEG with B-pictures. In this method, rather than searching for the motion between the anchor pictures, the search is carried out on all the consecutive pictures, including B-pictures. The final search between the anchor pictures is then the sum of all the intermediate motion vectors, as shown in Figure 7.9. Note that since we

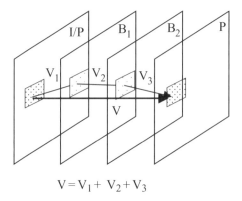

*Figure 7.9  Telescopic motion search*

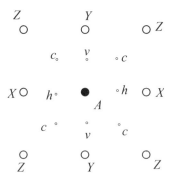

*Figure 7.10  Subpixel search positions, around pixel coordinate A*

are now searching for motion in successive pictures, the search range is smaller, and even fast search methods can be used.

### 7.6.2  Motion estimation with half pixel precision

In the search process with a half pixel resolution, normal block matching with integer pixel positions is carried out first. Then eight new positions, with a distance of half a pixel around the final integer pixel, are tested. Figure 7.10 shows a part of the search area, where the coordinate marked $A$ has been found as the best integer pixel position at the first stage.

In testing the eight subpixel positions, pixels of the macroblock in the previous frame are interpolated, according to the positions to be searched. For subpixel positions, marked with $h$ in the middle of the horizontal pixels, the interpolation is:

$$h = \frac{A + X}{2} \tag{7.1}$$

where the division is truncated. For the subpixels in the vertical midpoints, the interpolated values for the pixels are:

$$v = \frac{A + Y}{2} \tag{7.2}$$

and for subpixels in the corner (centre of four pixels), the interpolation is:

$$c = \frac{A + X + Y + Z}{4} \tag{7.3}$$

Note that in subpixel precision motion estimation, the range of the motion vectors' addresses is increased by 1 bit for each of the horizontal and vertical directions. Thus the motion vector overhead may be increased by two bits per vector (in practice due to variable length coding, this might be less than two bits). Despite this increase in motion vector overhead, the efficiency of motion compensation outweighs the extra bits, and the overall bit rate is reduced. Figure 7.11 shows the motion compensated error, with and without half pixel precision, for two consecutive frames of the Claire sequence. The motion compensated error has been magnified by a factor of four for better representation. It might be seen that half pixel precision has fewer blocking artefacts and, in general, motion compensated errors are smaller.

For further reduction on the motion vector overhead, differential coding is used. The prediction vector at the start of each slice and each intra coded macroblock is set to zero. Note that the predictively coded macroblocks with no motion vectors also set the prediction vector to zero. The motion vector prediction errors are then variable length coded.

*Figure 7.11*   *Motion compensated prediction error*
           *a*   with half pixel precision
           *b*   without half pixel precision

166  *Standard codecs: image compression to advanced video coding*

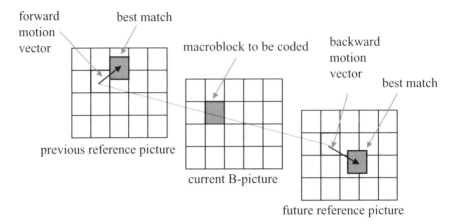

*Figure 7.12  Motion estimation in B-pictures*

### 7.6.3  Bidirectional motion estimation

B-pictures have access to both past and future anchor pictures. They can then use either past frame, called forward motion estimation, or the future frame for backward motion estimation, as shown in Figure 7.12.

Such an option increases the motion compensation efficiency, particularly when there are occluded objects in the scene. In fact, one of the reasons for the introduction of B-pictures was the fact that the forward motion estimation used in H.261 and P-pictures cannot compensate for the uncovered background of moving objects.

From the two forward and backward motion vectors, the coder has a choice of choosing any of the forward, backward or their combined motion compensated predictions. In the latter case, a weighted average of the forward and backward motion compensated pictures is calculated. The weight is inversely proportional to the distance of the B-picture with its anchor pictures. For example, in the GOB structure of I, $B_1$, $B_2$, P, the bidirectionally interpolated motion compensated picture for $B_1$ would be two-thirds of the forward motion compensated pixels from the I-picture and one-third from backward motion compensated pixels of the P-picture. This ratio is reversed for $B_2$. Note that B-pictures do not use motion compensation from each other, since they are not used as predictors. Also note that the motion vector overhead in B-pictures is much more than in P-pictures. The reason is that for B-pictures there are more macroblock types, which increase the macroblock type overhead, and for the bidirectionally motion compensated macroblocks two motion vectors have to be sent.

### 7.6.4  Motion range

When B-pictures are present, due to various distances between a picture and its anchor, it is expected that the search range for motion estimation will be different for different picture types. For example, with $M = 3$, P-pictures are three frames apart from their anchor pictures. $B_1$-pictures are only one frame apart from their past frame and two frames from their future frames, and those of $B_2$-pictures are in reverse order. Hence the motion range for P-pictures is larger than the backward motion range of $B_1$-pictures, which is itself larger than the forward motion vector. For normal scenes, the maximum search range for P-pictures is usually taken as 11 pixels/3 frames, and the forward and backward motion range for $B_1$-pictures are 3 pixels/frame and 7 pixels/2 frames, respectively. These values for $B_2$-pictures become 7 and 3.

It should be noted that, although motion estimation for B-pictures, due to the calculation of forward and backward motion vectors, is more processing demanding than that of the P-pictures nevertheless, due to larger motion range for P-pictures, the latter can be more costly than the former. For example, if the full search method is used, the number of search operations for P-pictures will be $(2 \times 11 + 1)^2 = 529$. This value for the forward and backward motion vectors of $B_1$-pictures will be $(2 \times 3 + 1)^2 = 49$ and $(2 \times 7 + 1)^2 = 225$, respectively. For $B_2$-pictures, the forward and backward motion estimation cost becomes 225 and 49, respectively. Thus, although motion estimation cost for P-pictures in this example is 529, the cost for a B-picture is about $49 + 225 = 274$, which is less. For motion estimation with half pixel accuracy, for P and B-pictures 8 and 16 more operations have to be added to these values, respectively. For more active pictures, where the search ranges for both P and B-pictures are larger, the gap on motion estimation cost becomes wider.

## 7.7 Coding of pictures

Since the encoder was described in terms of the basic unit of a macroblock, then the picture types may be defined in terms of their macroblock types. In the following each of these picture types are defined.

### 7.7.1 I-pictures

In I-pictures all the macroblocks are intra coded. There are two intra macroblock types: one that uses the current quantiser scale, intra-d, and the other that defines a new value for the quantiser scale, intra-q. Intra-d is the default value when the quantiser scale is not changed. Although these two types can be identified with 0 and 1, and no variable length code is required, the standard has foreseen some possible extensions to the macroblock types in the future. For this reason, they are VLC coded and intra-d is assigned with 1, and intra-q with 01. Extensions to the VLC codes with a start code of 0 are then open. The policy of making the coding tables open in this way was adopted by the MPEG group video committee in developing the international standard. The advantage of future extensions was judged to be worth the slight coding inefficiency.

If the macroblock type is intra-q, then the macroblock overhead should contain an extra five bits, to define the new quantiser scale between 1 and 31. For intra-d macroblocks, no quantiser scale is transmitted and the decoder uses the previously set value. Therefore the encoder may prefer to use as many intra-d types as possible. However, when the encoding rate is to be adjusted, which normally causes a new quantiser to be defined, the type is changed to intra-q. Note that, since in H.261 the bit rate is controlled at either the start of GOBs or rows of a GOB, then, if there is any intra-q in a GOB, it must be the first MB in that GOB, or rows of the GOB. In I-pictures of MPEG-1, an intra-q can be any of the macroblocks.

Each block within the MB is DCT coded and the coefficients are divided by the quantiser step size, rounded to the nearest integer. The quantiser step size is derived from the multiplication of the quantisation weighting matrix and the quantiser parameter (1 to 31). Thus the quantiser step size is different for different coefficients and may change from MB to MB. The only exception is the DC coefficients, which are treated differently. This is because the eye is sensitive to large areas of luminance and chrominance errors; then the accuracy of each DC value should be high and fixed. The quantiser step size for the DC coefficient is fixed to eight. Since in the quantisation weighting matrix the DC weighting element is eight, the quantiser parameter for the DC coefficient is always 1, irrespective of the quantisation parameter used for the remaining AC coefficients.

Due to the strong correlation between the DC values of blocks within a picture, the DC indices are coded losslessly by DPCM. Such a correlation does not exist among the AC coefficients, and hence they are coded independently. The prediction for the DC coefficients of luminance blocks follows the coding order of blocks within a macroblock and the raster scan order. For example, in the macroblocks of $4:2:0$ format pictures shown in Figure 7.13, the DC coefficient of block $Y_2$ is used as a prediction for the DC coefficient of block $Y_3$. The DC coefficient of block $Y_3$ is a prediction for the DC coefficient of $Y_0$ of the next macroblock. For the chrominance, we use the DC coefficients of the corresponding value of the block in the previous macroblock.

The differentially coded DC coefficient and the remaining AC coefficients are zigzag scanned, in the same manner as was explained for H.261 coefficients of Chapter 6. A copy of the coded picture is stored in the frame store to be used for the prediction of the next P and the past or future B-pictures.

### 7.7.2 P-pictures

As in I-pictures, each P-picture is divided into slices, which are in turn divided into macroblocks and then blocks for coding. Coding of P-pictures is more complex

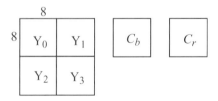

*Figure 7.13*   *Positions of luminance and chrominance blocks within a macroblock in $4:2:0$ format*

than of I-pictures, since motion compensated blocks may be constructed. For inter macroblocks, the difference between the motion compensated macroblock and the current macroblock is partitioned into blocks, and then DCT transformed and coded.

Decisions on the type of a macroblock, or whether motion compensation should be used or not, are similar to those for H.261 (see Chapter 6). Other H.261 coding tools, such as differential encoding of motion vectors, coded block pattern, zigzag scan, nature of variable length coding etc. are similar. In fact, coding of P-pictures is the same as coding each frame in H.261 with two major differences:

1. Motion estimation has a half pixel precision and, due to larger distances between the P-frames, the motion estimation range is much larger.
2. In MPEG-1 all intra-MB use the quantisation weighting matrix, whereas in H.261 all MB use a flat matrix. Also in MPEG-1 the intra-MB of P-pictures are predictively coded like those of I-pictures, with the exception that the prediction value is fixed at $128 \times 8$ if the previous macroblock is not intra coded.

Locally decoded P-pictures are stored in the frame store for further prediction. Note that, if B-pictures are used, two buffer stores are needed to store two prediction pictures.

### 7.7.3 B-pictures

As in I and P-pictures, B-pictures are divided into slices, which in turn are divided into macroblocks for coding. Due to the possibility of bidirectional motion compensation, coding is more complex than for P-pictures. Thus the encoder has more decisions to make than in the case of P-pictures. These are: how to divide the picture into slices, determining the best motion vectors to use, deciding whether to use forward, backward or interpolated motion compensation or to code intra, and how to set the quantiser scale. These make processing of B-pictures computationally very intensive. Note that motion compensation is the most costly operation in the codecs, and for every macroblock both forward and backward motion compensations have to be performed.

The encoder does not need to store decoded B-pictures, since they are not used for prediction. Hence B-pictures can be coded with larger distortions. In this regard, to reduce the slice overhead, larger slices (fewer slices in the picture) may be chosen.

In P-pictures, as for H.261, there are eight different types of macroblock. In B-pictures, due to backward motion compensation and interpolation of forward and backward motion compensation, the number of macroblock types is about 14. Figure 7.14 shows the flow chart for macroblock type decisions in B-pictures.

The decision on the macroblock type starts with the selection of a motion compensation mode based on the minimisation of a cost function. The cost function is the mean-squared/absolute error of the luminance difference between the motion compensated macroblock and the current macroblock. The encoder first calculates the best forward motion compensated macroblock from the previous anchor picture for forward motion compensation. It then calculates the best motion compensated macroblock from the future anchor picture, as the backward motion compensation. Finally, the average of the two motion compensated errors is calculated to produce the

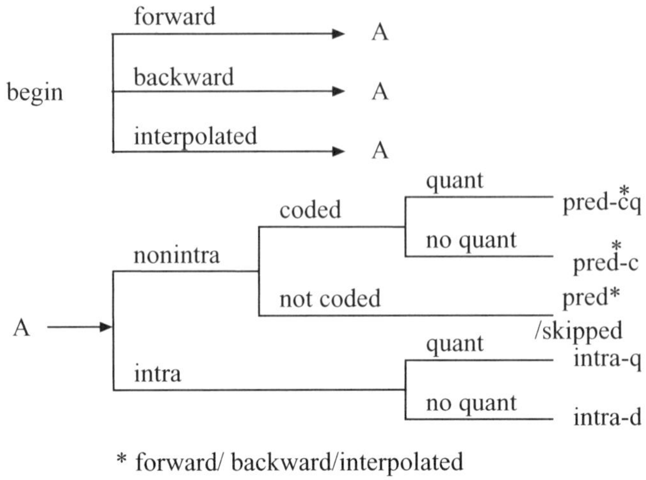

170  *Standard codecs: image compression to advanced video coding*

```
                forward
                ─────────────▶ A
        begin   backward
                ─────────────▶ A
                interpolated
                ─────────────▶ A

                                      quant
                             coded ───────── pred-c̃q
                   nonintra       no quant
                                      ─────── pred-c̃
                             not coded        pred*
        A ──▶                                 /skipped
                                      quant
                   intra             ─────── intra-q
                                   no quant
                                      ─────── intra-d

        * forward/ backward/interpolated
```

*Figure 7.14  Selection of macroblock types in B-pictures*

interpolated macroblock. It then selects the one that had the smallest error difference with the current macroblock. In the event of a tie, an interpolated mode is chosen.

Another difference between macroblock types in B and P-pictures is in the definition of noncoded and skipped macroblocks. In P-pictures, the skipped MB is the one in which none of its blocks has any significant DCT coefficient (cbp (coded block pattern) = 0), and the motion vector is also zero. The first and the last MB in a slice cannot be declared skipped. They are treated as noncoded.

A noncoded MB in P-pictures is the one in which none of its blocks has any significant DCT coefficient (cbp = 0), but the motion vector is nonzero. Thus the first and the last MB in a slice, which could be skipped, is noncoded with motion vector set to zero! In H.261 the noncoded MB was called motion vector only (MC).

In B-pictures, the skipped MB has again all zero DCT coefficients, but the motion vector and the type of prediction mode (forward, backward or interpolated) is exactly the same as that of its previous MB. Similar to P-pictures, the first and the last MB in a slice cannot be declared skipped, and is in fact called noncoded.

The noncoded MB in B-pictures has all of its DCT coefficients zero (cbp = 0), but either its motion vector or its prediction (or both) is different from its previous MB.

### 7.7.4  D-pictures

D-pictures contain only low frequency information, and are coded as the DC coefficients of the blocks. They are intended to be used for fast visible search modes. A bit is transmitted for the macroblock type, although there is only one type. In addition there is a bit denoting the end of the macroblock. D-pictures are not part of the constrained bit stream.

## 7.8 Video buffer verifier

A coded bit stream contains different types of pictures, and each type ideally requires a different number of bits to encode. In addition, the video sequence may vary in complexity with time, and it may be desirable to devote more coding bits to one part of a sequence than to another. For constant bit rate coding, varying the number of bits allocated to each picture requires that the decoder has a buffer to store the bits not needed to decode the immediate picture. The extent to which an encoder can vary the number of bits allocated to each picture depends on the size of this buffer. If the buffer is large an encoder can use greater variations, increasing the picture quality, but at the cost of increasing the decoding delay. The delay is the time taken to fill the input buffer from empty to its current level. An encoder needs to know the size of the decoder's input buffer in order to determine to what extent it can vary the distribution of coding bits among the pictures in the sequence.

In constant bit rate applications (for example decoding a bit stream from a CD-ROM), problems of synchronisation may occur. In these applications, the encoder should generate a bit stream that is perfectly matched to the device. The decoder will display the decoded pictures at their specific rate. If the display clock is not locked to the channel data rate, and this is typically the case, then any mismatch between the encoder and channel clock, and the display clock will eventually cause a buffer overflow or underflow. For example, assume that the display clock runs one part per million too slow with respect to the channel clock. If the data rate is 1 Mbit/s, then the input buffer will fill at an average rate of one bit per second, eventually causing an overflow. If the decoder uses the entire buffer to allocate bits between pictures, the overflow could occur more quickly. For example, suppose the encoder fills the buffer completely except for one byte at the start of each picture. Then overflow will occur after only 8 s!

The model decoder is defined to resolve three problems: to constrain the variability in the number of bits that may be allocated to different pictures; it allows a decoder to initialise its buffer when the system is started; it allows the decoder to maintain synchronisation while the stream is played. At the beginning of this Chapter we mentioned multiplexing and synchronisation of audio and video streams. The tools defined in the international standard for the maintenance of synchronisation should be used by decoders when multiplexed streams are being played.

The definition of the parameterised model decoder is known as the video buffer verifier (VBV). The parameters used by a particular encoder are defined in the bit stream. This really defines a model decoder that is needed if encoders are to be assured that the coded bit stream they produce will be decodable. The model decoder looks like Figure 7.15.

A fixed rate channel is assumed to put bits at a constant rate into the buffer, at regular intervals, set by the picture rate. The picture decoder instantaneously removes all the bits pertaining to the next picture from the input buffer. If there are too few bits in the input buffer, that is all the bits for the next picture have been received, then the input buffer underflows and there is an underflow error. If during the time between the picture starts the capacity of the input buffer is exceeded, then there is an overflow error.

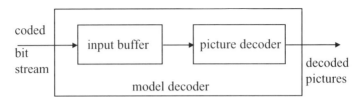

*Figure 7.15  Model decoder*

Practical decoders may differ from this model in several important ways. They may not remove all the bits required to decode a picture from the input buffer instantaneously. They may not be able to control the start of decoding very precisely as required by the buffer fullness parameters in the picture header, and they take a finite time to decode. They may also be able to delay decoding for a short time to reduce the chance of underflow occurring. But these differences depend in degree and kind on the exact method of implementation. To satisfy the requirements of different implementations, the MPEG video committee chose a very simple model for the decoder. Practical implementations of decoders must ensure that they can decode the bit stream constrained in this model. In many cases this will be achieved by using an input buffer that is larger than the minimum required, and by using a decoding delay that is larger than the value derived from the buffer fullness parameter. The designer must compensate for any differences between the actual design and the model in order to guarantee that the decoder can handle any bit stream that satisfies the model.

Encoders monitor the status of the model to control the encoder so that overflow does not occur. The calculated buffer fullness is transmitted at the start of each picture so that the decoder can maintain synchronisation.

### 7.8.1  Buffer size and delay

For constant bit rate operation each picture header contains a variable delay parameter (*vbv_delay*) to enable decoders to synchronise their decoding correctly. This parameter defines the time needed to fill the input buffer of Figure 7.15 from an empty state to the current level immediately before the picture decoder removes all the bits from the picture. This time thus represents a delay and is measured in units of 1/90 000 s. This number was chosen because it is almost an exact factor of the picture duration in various original video formats: 1/24, 1/25, 1/29.97 and 1/30 s, and because it is comparable in duration to an audio sample. The delay is given by:

$$D = \frac{vbv\_delay}{90\,000} \text{ s} \qquad (7.4)$$

For example, if *vbv_delay* was 9000, then the delay would be 0.1 s. This means that at the start of a picture the input buffer of the model decoder should contain exactly 0.1 s worth of data from the input bit stream.

The bit rate, $R$, is defined in the sequence header. The number of bits in the input buffer at the beginning of the picture is thus given by

$$B = D \times R = \frac{vbv\_delay}{90\,000} \times R \text{ bits} \tag{7.5}$$

For example, if *vbv_delay* and $R$ were 9000 and 1.2 Mbit/s, respectively, then the number of bits in the input buffer would be 120 kbits. The constrained parameter bit stream requires that the input buffer have a capacity of 327 680 bits, and $B$ should never exceed this value [3].

## 7.8.2 Rate control and adaptive quantisation

The encoder must make sure that the input buffer of the model decoder is neither overflowed nor underflowed by the bit stream. Since the model decoder removes all the bits associated with a picture from its input buffer instantaneously, it is necessary to control the total number of bits per picture. In H.261 we saw that the encoder could control the bit rate by simply checking its output buffer content. As the buffer fills up, so the quantiser step size is raised to reduce the generated bit rate, and *vice versa*. The situation in MPEG-1, due to the existence of three different picture types, where each generates a different bit rate, is slightly more complex. First, the encoder should allocate the total number of bits among the various types of picture within a GOP, so that the perceived image quality is suitably balanced. The distribution will vary with the scene content and the particular distribution of I, P and B-pictures within a GOP.

Investigations have shown that for most natural scenes, each P-picture might generate as many as 2–5 times the number of bits of a B-picture, and an I-picture three times those of the P-picture. If there is little motion and high texture, then a greater proportion of the bits should be assigned to I-pictures. Similarly, if there is strong motion, then a proportion of bits assigned to P-pictures should be increased. In both cases lower quality from the B-pictures is expected, to permit the anchor I and P-pictures to be coded at their best possible quality.

Our investigations with variable bit rate (VBR) video, where the quantiser step size is kept constant (no rate control), show that the ratios of generated bits are 6 : 3 : 2, for I, P and B-pictures, respectively [9]. Of course at these ratios, due to the fixed quantiser step size, the image quality is almost constant, not only for each picture (in fact slightly better for B-pictures, due to better motion compensation), but throughout the image. Again, if we lower the expected quality for B-pictures, we can change that ratio in favour of I and P-pictures.

Although these ratios appear to be very important for a suitable balance in picture quality, one should not worry very much about their exact values. The reason is that it is possible to make the encoder intelligent enough to learn the best ratio. For example, after coding each GOP, one can multiply the average value of the quantiser scale in each picture by the bit rate generated at that picture. Such a quantity can be used as the complexity index, since larger complexity indices should be due to both larger quantiser step sizes and larger bit rates. Therefore, based on the complexity index one can derive a new coding ratio, and the target bit rate for each picture in the next GOP is based on this new ratio.

As an example, let us assume that SIF-625 video is to be coded at 1.2 Mbit/s. Let us also assume that the GOP structure of $N = 12$ and $M = 3$ is used. Therefore there will be one I-picture, three P-pictures and eight B-pictures in each GOP. First of all, the target bit rate for each GOP is $1200 \times \frac{12}{25} = 576$ kbit/GOP. If we assume a coding ratio of $6:3:2$, then the target bit rate for each of the I, P and B-pictures will be:

I-picture $\qquad \frac{6}{6+3\times3+2\times8} \times 576 = \frac{6}{31} \times 576 = 112$ kbits

P-picture $\qquad \frac{3}{31} \times 576 = 56$ kbits

B-picture $\qquad \frac{2}{31} \times 576 = 37$ kbits

Therefore, each picture is aiming for its own target bit rate. Similar to H.261, one can control the quantiser step size for that picture, such that the required bit rate is achieved. At the end of the GOP, the complexity index for each picture type is calculated. Note that for P and B-pictures, the complexity index is the average of three and eight complexity indices, respectively. These ratios are used to define new coding ratios between the picture type for coding of the next GOP. Also, bits generated in that GOP are added together and the extra bit rate, or the deficit, from the GOP target bit rate is transferred to the next GOP.

In practice, the target bit rates for B-pictures compared with other picture types are deliberately reduced by a factor of 1.4. This is done for two reasons. First, due to efficient bidirectional motion estimation in B-pictures, their transform coefficients are normally small. Increasing the quantiser step size hardly affects these naturally small value coefficients' distortions, but the overall bit rate can be reduced significantly. Second, since B-pictures are not used in the prediction loop of the encoder, even if they are coarsely coded, the encoding error is not transferred to the subsequent frames. This is not the case with the anchor I and P-pictures, since through the prediction loop, any saving in one frame due to coarser coding has to be paid back in the following frames.

Experimental results indicate that by reducing the target bit rates of the B-pictures by a factor of 1.4, the average quantiser step size for these pictures rises almost by the same factor, but its quality (PSNR) only slightly deteriorates, which makes it worth doing.

Note also that, although in the above example (which is typical) the bits per B-picture are fewer than those of I and P-pictures, nevertheless the eight B-pictures in a GOP generate almost $8 \times 37 = 296$ kbits, which is more than 50 per cent of the bits in a GOP. The first implication is that use of the factor 1.4 can have a significant reduction in the overall bit rate. The second implication of this is in the transmission of video over packet networks, where during periods of congestion, if only B-pictures are discarded, so reducing the network load by 50 per cent, congestion can be eased without significantly affecting the picture quality. Note that B-pictures are not used for predictions, so their loss will result in only a very brief (480 ms) reduction in quality.

## 7.9 Decoder

The decoder block diagram is based on the same principle as the local decoder associated with the encoder as shown in Figure 7.16.

The incoming bit stream is stored in the buffer, and is demultiplexed into the coding parameters such as DCT coefficients, motion vectors, macroblock types, addresses etc. These are then variable length decoded using the locally provided tables. The DCT coefficients after inverse quantisation are inverse DCT transformed and added to the motion compensated prediction (as required) to reconstruct the pictures. The frame stores are updated by the decoded I and P-pictures. Finally, the decoded pictures are reordered to their original scanned form.

At the beginning of the sequence, the decoder will decode the sequence header, including the sequence parameters. If the bit stream is not constrained, and a parameter exceeds the capability of the decoder, then the decoder should be able to detect this. If the decoder determines that it can decode the bit stream, then it will set up its parameters to match those defined in the sequence header. This will include horizontal and vertical resolutions and aspect ratio, the bit rate and the quantisation weighting matrices.

Next, the decoder will decode the group of picture header field, to determine the GOP structure. It will then decode the first picture header in the group of pictures and, for constant bit rate operation, determine the buffer fullness. It will delay de-coding the rest of the sequence until the input buffer is filled to the correct level. By doing this, the decoder can be sure that no buffer overflow or underflow will occur during decoding. Normally, the input buffer size will be larger than the minimum required by the bit stream, giving a range of fullness at which the decoder may start to decode.

If it is required to play a recorded sequence from a random point in the bit stream, the decoder should discard all the bits until it finds a sequence start code, a group of pictures start code, or a picture start code which introduces an I-picture. The slices and macroblocks in the picture are decoded and written into a display buffer, and perhaps into another buffer. The decoded pictures may be postprocessed and displayed in the order defined by the temporal reference at the picture rate defined in the sequence header. Subsequent pictures are processed at the appropriate times to avoid buffer overflow and underflow.

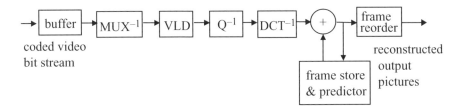

*Figure 7.16   A block diagram of an MPEG-1 decoder*

### 7.9.1 Decoding for fast play

Fast forward can be supported by D-pictures. It can also be supported by an appropriate spacing of I-pictures in a sequence. For example, if I-pictures were spaced regularly every 12 pictures, then the decoder might be able to play the sequence at 12 times the normal speed by decoding and displaying only the I-pictures. Even this simple concept places a considerable burden on the storage media and the decoder. The media must be capable of speeding up and delivering 12 times the data rate. The decoder must be capable of accepting this higher data rate and decoding the I-pictures. Since I-pictures typically require significantly more bits to code than P and B-pictures, the decoder will have to decode significantly more than the $\frac{1}{12}$ of the data rate. In addition, it has to search for picture start codes and discard the data for P and B-pictures. For example, consider a sequence with $N = 12$ and $M = 3$, such as:

I  B  B  P  B  B  P  B  B  P  B  B  I  B  B

Assume that the average bit rate is $C$, each B-picture requires $0.6C$, each P-picture requires $1.4C$ and the remaining $3C$ are assigned to the I-picture in the GOP. Then the I-pictures should code $\frac{3}{12} \times 100 = 25$ per cent of the total bit rate in just $\frac{1}{12}$ of the display time.

Another way to achieve fast forward in a constant bit rate application is for the medium itself to sort out the I-pictures and transmit them. This would allow the data rate to remain constant. Since this selection process can be made to produce a valid MPEG-1 video bit stream, the decoder should be able to decode it. If every I-picture of the preceding example were selected, then one I-picture would be transmitted every three picture periods, and the speed up rate would be $\frac{12}{3} = 4$ times.

If alternate I-pictures of the preceding example were selected, then one I-picture would again be transmitted every three picture periods, but the speed up rate would be $\frac{24}{3} = 8$ times. If one in $N$ I-pictures of the preceding example were selected, then the speed up rate would be $4N$.

### 7.9.2 Decoding for pause and step mode

Decoding for pause requires the decoder to be able to control the incoming bit stream, and display a decoded picture without decoding any additional pictures. If the decoder has full control over the bit stream, then it can be stopped for pause and resumed when play begins. If the decoder has less control, as in the case of a CD-ROM, there may be a delay before play can be resumed.

### 7.9.3 Decoding for reverse play

To decode a bit stream and play in reverse, the decoder must decode each group of pictures in the forward direction, store the entire decoded pictures, then display them in reverse order. This imposes severe storage requirements on the decoder in addition to any problems in gaining access to the decoded bit stream in the correct order.

To reduce decoder memory requirements, groups of pictures should be small. Unfortunately, there is no mechanism in the syntax for the encoder to state what the

*Coding of moving pictures for digital storage media (MPEG-1)* 177

```
B  B  I  B  B  P  B  B  P  B  B  P     pictures in display order
0  1  2  3  4  5  6  7  8  9  10 11    temporal reference

I  B  B  P  B  B  P  B  B  P  B  B     pictures in decoding order
2  0  1  5  3  4  8  6  7  11 9  10    temporal reference

I  P  P  P  B  B  B  B  B  B  B  B     pictures in new order
2  5  8  11 10 9  7  6  4  3  1  0     temporal reference
```

*Figure 7.17  Example of group of pictures, in the display, decoding and new orders*

decoder requirements are in order to play in reverse. The amount of display buffer storage may be reduced by reordering the pictures, either by having the storage unit read and transmit them in another order, or by reordering the coded pictures in a decoder buffer. To illustrate this, consider the typical group of pictures shown in Figure 7.17.

The decoder would decode pictures in the new order, and display them in the reverse of the normal display. Since the B-pictures are not decoded until they are ready to be displayed, the display buffer storage is minimised. The first two B-pictures, 0 and 1, would remain stored in the input buffer until the last P-picture in the previous group of pictures was decoded.

## 7.10 Postprocessing

### 7.10.1 Editing

Editing of a video sequence is best performed before compression, but situations may arise where only the coded bit stream is available. One possible method would be to decode the bit stream, perform the required editing on the pixels and recode the bit stream. This usually leads to a loss in video quality, and it is better, if possible, to edit the coded bit stream itself.

Although editing may take several forms, the following discussion pertains only to editing at the picture level, that is deletion of the coded video material from a bit stream and insertion of coded video material into a bit stream, or rearrangement of coded video material within a bit stream.

If a requirement for editing is expected (e.g. clip video is provided analogous to clip art for still pictures), then the video can be encoded with well defined cutting points. These cutting points are places at which the bit stream may be broken apart or joined. Each cutting point should be followed by a closed group of pictures (e.g. a GOP that starts with an I-picture). This allows smooth play after editing.

To allow the decoder to play the edited video without having to adopt any unusual strategy to avoid overflow and underflow, the encoder should make the buffer fullness take the same value at the first I-picture following every cutting point. This value

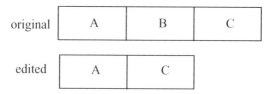

*Figure 7.18   Edited sequences*

should be the same as that of the first picture in the sequence. If this suggestion is not followed, then the editor may make an adjustment either by padding (stuffing bits or macroblocks) or by recording a few images to make them smaller.

If the buffer fullness is mismatched and the editor makes no correction, then the decoder will have to make some adjustment when playing over an edited cut. For example, consider a coded sequence consisting of three clips, A, B and C, in order. Assume that clip B is completely removed by editing, so that the edited sequence consists only of clip A followed immediately by clip C, as illustrated in Figure 7.18.

Assume that in the original sequence the buffer is three-quarters full at the beginning of clip B, and one-quarter full at the beginning of clip C. A decoder playing the edited sequence will encounter the beginning of clip C with its buffer three-quarters full, but the first picture in clip C will contain a buffer fullness value corresponding to a quarter full buffer. In order to avoid buffer overflow, the decoder may try to pause the input bit stream, or discard pictures without displaying them (preferably B-pictures), or change the decoder timing.

For another example, assume that in the original sequence the buffer is one-quarter full at the beginning of clip B, and three-quarters full at the beginning of clip C. A decoder playing the edited sequence will encounter the beginning of clip C with its buffer one-quarter full, but the first picture in clip C will contain a buffer fullness value corresponding to a three-quarters full buffer. In order to avoid buffer underflow, the decoder may display one or more pictures for longer than the normal time.

If provision for editing was not specifically provided in the coded bit stream, or if it must be available at any picture, then the editing task is more complex, and places a greater burden on the decoder to manage buffer overflow and underflow problems. The easiest task is to cut at the beginning of a group of pictures. If the group of pictures following the cut is open (e.g. GOP starts with two B-pictures), which can be detected by examining the closed GOP flag in the group of pictures header, then editing must set the broken link bit to 1 to indicate to the decoder that the previous group of pictures cannot be used for decoding any B-pictures.

### 7.10.2   Resampling and up conversion

The decoded sequence may not match the picture rate or the spatial resolution of the display device. In such situations (which occur frequently), the decoded video must be resampled or scaled. In Chapter 2 we saw that CCIR-601 video was subsampled into SIF format for coding, hence for display it is appropriate to upsample it back into its original format. Similarly, it has to be temporally converted for proper display, as was discussed in Chapter 2. This is particularly important for cases where video was converted from film.

## 7.11 Problems

1 In MPEG-1, with the group of picture structure of $N = 12$ and $M = 3$, the maximum motion speed is assumed to be 15.5 pixels/frame (half pixel precision), calculate the number of search operations required to estimate the motion in P-pictures with:

  a telescopic method
  b direct on the P-pictures

2 In an MPEG-1 encoder, for head-and-shoulders-type pictures, the maximum motion speed for P-pictures is set to 13 pixels and those of the forward and backward for the first B-picture in the subgroup are set to five pixels and nine pixels, respectively.

  a explain why the search range for the B-picture is smaller than that of the P-picture
  b what would be the forward and backward search ranges for the second B-picture in the subgroup?
  c calculate the number of search operations with half pixel precision for the P and B-pictures

3 An I-picture is coded at 50 kbits. If the quantiser step size is linearly distributed between 10 and 16, find the complexity index for this picture.

4 In coding of SIF-625 video at 1.2 Mbit/s, with a group of pictures (GOP) structure $N = 12, M = 3$, the ratios of complexity indices of I, P and B are 20:10:7, respectively. Calculate the target bit rate for coding of each frame in the next GOP.

5 If in problem 4 the allocated bits to B-pictures were reduced by a factor of 1.4, find the new complexity indices and the target bits to each picture type.

6 In problem 4 if due to scene change the average quantiser step size in the last P-picture of the GOP was doubled, but those of other pictures did not change significantly:

  a how do the complexity index ratios change?
  b what is the new target bit rate for each picture type?

7 If in problem 4 the complexity indices ratios were wrongly set to 1 : 1 : 1, but after coding the average quantiser step sizes for I, P and B were 60, 20 and 15 respectively, find:

   a the target bit rate for each picture type before coding
   b the target bit rate for each picture type of the next GOP.

## 7.12 References

1 MPEG-1: 'Coding of moving pictures and associated audio for digital storage media at up to about 1.5 Mbit/s'. ISO/IEC 1117–2: video, November 1991
2 H.261: 'ITU-T Recommendation H.261, video codec for audiovisual services at p × 64 kbit/s'. Geneva, 1990
3 'Coding of moving pictures and associated audio for digital storage media at up to about 1.5 Mbit/s'. ISO/IEC 1117–2: Systems, November 1991
4 CCIR Recommendation 601: 'Digital methods of transmitting television information'. Recommendation 601, encoding parameters of digital television for studios
5 WILSON, D., and GHANBARI, M.: 'Frame sequence partitioning of video for efficient multiplexing', *Electron. Lett.*, 1998, **34:15**, pp.1480–1481
6 GHANBARI, M.: 'An adapted H.261 two-layer video codec for ATM networks', *IEEE Trans. Commun.*, 1992, **40:9**, pp.1481–1490
7 PEARSON, D.E.: 'Transmission and display of pictorial information' (Pentech Press, 1975)
8 SEFERIDIS, V., and GHANBARI, M.: 'Adaptive motion estimation based on texture analysis', *IEEE Trans. Commun.*, 1994, **42:2/3/4**, pp.1277–1287
9 ALDRIDGE, R.P., GHANBARI, M., and PEARSON, D.E.: 'Exploiting the structure of MPEG-2 for statistically multiplexing video'. Proceedings of 1996 international *Picture coding* symposium, *PCS '96*, Melbourne, Australia, 13–15 March 1996, pp.111–113

*Chapter 8*
# Coding of high quality moving pictures (MPEG-2)

Following the universal success of the H.261 and MPEG-1 video codecs, there was a growing need for a video codec to address a wide variety of applications. Considering the similarity between H.261 and MPEG-1, ITU-T, and ISO/IEC made a joint effort to devise a generic video codec. Joining the study was a special group in ITU-T, Study Group 15, SG15, who were interested in coding of video for transmission over the future broadband integrated services digital networks (B-ISDN) using asynchronous transfer mode (ATM) transport. The devised generic codec was finalised in 1995, and takes the name of MPEG-2/H.262, although it is more commonly known as MPEG-2 [1].

At the time of the development, the following applications for the generic codec were foreseen:

- BSS       broadcasting satellite service (to the home)
- CATV      cable TV distribution on optical networks, copper etc.
- CDAD      cable digital audio distribution
- DAB       digital audio broadcasting (terrestrial and satellite)
- DTTB      digital terrestrial television broadcast
- EC        electronic cinema
- ENG       electronic news gathering (including SNG, satellite news gathering)
- FSS       fixed satellite service (e.g. to head ends)
- HTT       home television theatre
- IPC       interpersonal communications (videoconferencing, videophone, ...)
- ISM       interactive storage media (optical discs etc.)
- MMM       multimedia mailing
- NCA       news and current affairs
- NDS       networked database services (*via* ATM etc.)
- RVS       remote video surveillance
- SSM       serial storage media (digital VTR etc.).

182  *Standard codecs: image compression to advanced video coding*

Of particular importance is the application to satellite systems where the limitations of radio spectrum and satellite parking orbit result in pressure to provide acceptable quality TV signals at relatively low bit rates. As we will see at the end of this Chapter, today we can accommodate about 6–8 MPEG-2 coded TV programs in the same satellite channel that used to carry only one analogue TV program. Numerous new applications have been added to the list. In particular high definition television (HDTV) and digital versatile disc (DVD) for home storage systems appear to be the main beneficiaries of further MPEG-2 development.

## 8.1  MPEG-2 systems

The MPEG-1 standard targeted coding of audio and video for storage, where the media error rate is negligible [2]. Hence the MPEG-1 system is not designed to be robust to bit error rates. Also, MPEG-1 was aimed at software oriented image processing, where large and variable length packets could reduce the software overhead [3].

The MPEG-2 standard on the other hand is more generic for a variety of audiovisual coding applications. It has to include error resilience for broadcasting, and ATM networks. Moreover, it has to deliver multiple programs simultaneously without requiring them to have a common time base. These require that the MPEG-2 transport packet length should be short and fixed.

MPEG-2 defines two types of stream: the program stream and the transport stream. The program stream is similar to the MPEG-1 systems stream, but uses a modified syntax and new functions to support advanced functionalities (e.g. scalability). It also provides compatibility with the MPEG-1 systems stream, that is MPEG-2 should be capable of decoding an MPEG-1 bit stream. Like the MPEG-1 decoder, program stream decoders typically employ long and variable length packets. Such packets are well suited for software-based processing and error-free transmission environments, such as coding for storage of video on a disc. Here the packet sizes are usually 1–2 kbytes long, chosen to match the disc sector sizes (typically 2 kbytes). However, packet sizes as long as 64 kbytes are also supported.

The program stream also includes features not supported by MPEG-1 systems. These include scrambling of data, assignment of different priorities to packets, information to assist alignment of elementary stream packets, indication of copyright, indication of fast forward, fast reverse and other trick modes for storage devices. An optional field in the packets is provided for testing the network performance, and optional numbering of a sequence of packets is used to detect lost packets.

In the transport stream, MPEG-2 significantly differs from MPEG-1 [4]. The transport stream offers robustness for noisy channels as well as the ability to assemble multiple programs into a single stream. The transport stream uses fixed length packets of size 188 bytes with a new header syntax. This can be segmented into four 47-bytes to be accommodated in the payload of four ATM cells, with of the AAL1 adaptation scheme [5]. It is therefore more suitable for hardware processing and for error correction schemes, such as those required in television broadcasting,

## Coding of high quality moving pictures (MPEG-2)

satellite/cable TV and ATM networks. Furthermore, multiple programs with independent time bases can be multiplexed in one transport stream. The transport stream also allows synchronous multiplexing of programs, fast access to the desired program for channel hopping, multiplexing of programs with clocks unrelated to transport clock and correct synchronisation of elementary streams for playback. It also allows control of the decoder buffers during start up and playback for both constant and variable bit rate programs.

A basic data structure that is common to the organisation of both the program stream and transport stream is called the packetised elementary stream (PES) packet. Packetising the continuous streams of compressed video and audio bit streams (elementary streams) generates PES packets. Simply stringing together PES packets from the various encoders with other packets containing necessary data to generate a single bit stream generates a program stream. A transport stream consists of packets of fixed length containing four bytes of header followed by 184 bytes of data, where the data is obtained by segmenting the PES packets.

Figure 8.1 illustrates both types of program and transport stream multiplexes of MPEG-2 systems. Like MPEG-1, the MPEG-2 systems layer is also capable of combining multiple sources of user data along with encoded audio and video. The audio and video streams are packetised to form PES packets that are sent to either a program multiplexer or a transport multiplexer, resulting in a program stream or transport stream, respectively. As mentioned earlier, program streams

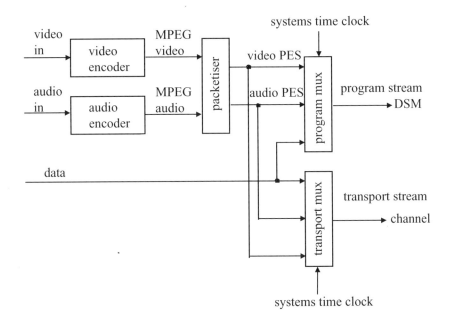

*Figure 8.1  MPEG-2 systems multiplex of program and transport streams*

184　*Standard codecs: image compression to advanced video coding*

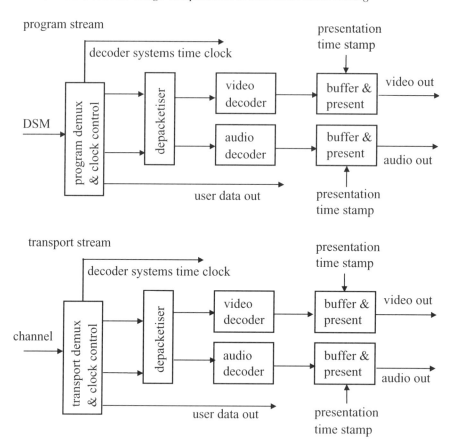

*Figure 8.2　MPEG-2 systems demultiplexing of program and transport streams*

are intended for an error-free environment such as digital storage media (DSM). Transport streams are intended for noisier environments such as terrestrial broadcast channels.

At the receiver the transport streams are decoded by a transport demultiplexer (which includes a clock extraction mechanism), unpacketised by a depacketiser, and sent to audio and video decoders for decoding, as shown in Figure 8.2.

The decoded signals are sent to the receiver buffer and presentation units, which output them to a display device and a speaker at the appropriate time. Similarly, if the program streams are used, they are decoded by the program stream demultiplexer, depacketiser and sent to the audio and video decoders. The decoded signals are sent to the respective buffer to await presentation. Also similar to MPEG-1 systems, the information about systems timing is carried by the clock reference field in the bit stream that is used to synchronise the decoder systems clock (STC). Presentation time stamps (PTS) that are also carried by the bit stream control the presentation of the decoded output.

## 8.2 Level and profile

MPEG-2 is intended to be generic in the sense that it serves a wide range of applications, bit rates, resolutions, qualities and services. Applications should cover, among other things, digital storage media, television broadcasting and communications. In the course of the development, various requirements from typical applications were considered, and they were integrated into a single syntax. Hence MPEG-2 is expected to facilitate the interchange of bit streams among different applications. Considering the practicality of implementing the full syntax of the bit stream, however, a limited number of subsets of the syntax are also stipulated by means of profile and level [6].

A profile is a subset of the entire bit stream syntax that is defined by the MPEG-2 specification. Within the bounds imposed by the syntax of a given profile it is still possible to encompass very large variations in the performance of encoders and decoders depending upon the values taken by parameters in the bit stream. For instance, it is possible to specify frame sizes as large as (approximately) $2^{14}$ samples wide by $2^{14}$ lines high. It is currently neither practical nor economical to implement a decoder capable of dealing with all possible frame sizes. In order to deal with this problem levels are defined within each profile. A level is a defined set of constraints imposed on parameters in the bit stream. These constraints may be simple limits on numbers. Alternatively, they may take the form of constraints on arithmetic combinations of the parameters (e.g. frame width multiplied by frame height multiplied by frame rate). Both profiles and levels have a hierarchical relationship, and the syntax supported by a higher profile or level must also support all the syntactical elements of the lower profiles or levels.

Bit streams complying with the MPEG-2 specification use a common syntax. In order to achieve a subset of the complete syntax, flags and parameters are included in the bit stream which signal the presence or otherwise of syntactic elements that occur later in the bit stream. Then to specify constraints on the syntax (and hence define a profile) it is only necessary to constrain the values of these flags and parameters that specify the presence of later syntactic elements.

In order to parse the bit stream into specific applications, they are ordered into layers. If there is only one layer, the coded video data is called a nonscalable video bit stream. For two or more layers, the bit stream is called a scalable hierarchy. In the scalable mode, the first layer, called the base layer, is always decoded independently. Other layers are called enhancement layers, and can only be decoded together with the lower layers.

Before describing how various scalabilities are introduced in MPEG-2, let us see how levels and profiles are defined. MPEG-2 initially defined five hierarchical structure profiles and later on added two profiles that do not fit the hierarchical structure. The profile mainly deals with the supporting tool for coding, such as the group of picture structure, the picture format and scalability. The seven known profiles and their current applications are summarised in Table 8.1.

The level deals with the picture resolutions such as the number of pixels per line, lines per frame, frame per seconds (fps) and bits per second or the bit rate (e.g. Mbps). Table 8.2 summarises the levels that are most suitable for each profile.

Table 8.1  Various profiles defined for MPEG-2

| Type | Supporting tools | Application |
| --- | --- | --- |
| Simple | I & P picture, 4 : 2 : 0 format; nonscalable | currently not used |
| Main | simple profile + B-pictures | broadcast TV |
| SNR scalable | main profile + SNR scalability | currently not used |
| Spatial | SNR profile + Spatial scalability | currently not used |
| High | spatial profile + 4 : 2 : 2 format | currently not used |
| 4 : 2 : 2 | IBIBIB... pictures, extension of main profile to high bit rates | studio postproduction; high quality video for storage (VTR) & Video distribution |
| Multiview | main profile + temporal scalability | several video streams; stereo presentation |

Table 8.2  The levels defined for each profile

| Level | resolutions | Simple I,P 4:2:0 | Main I,P,B 4:2:0 | SNR I,P,B 4:2:0 | Spatial I,P,B 4:2:0 | High I,P,B 4:2:0 4:2:2 | 4:2:2 I,P,B 4:2:0 4:2:2 | Multiview I,P,B 4:2:0 |
| --- | --- | --- | --- | --- | --- | --- | --- | --- |
| Low | pel/line |  |  | 352 | 352 |  |  | 352 |
|  | line/fr |  |  | 288 | 288 |  |  | 288 |
|  | fps |  |  | 30/25 | 30/15 |  |  | 30/25 |
|  | Mbps |  |  | 4 | 4 |  |  | 8 |
| Main | pel/line | 720 | 720 | 720 |  | 720 | 720 | 720 |
|  | line/fr | 576 | 576 | 576 |  | 576 | 512/608 | 576 |
|  | fps | 30/25 | 30/25 | 30/25 |  | 30/25 | 30/25 | 30/25 |
|  | Mbps | 15 | 15 | 15 |  | 20 | 50 | 25 |
| High 1440 | pel/line | 1440 |  | 1440 | 1440 | 1440 |  |  |
|  | line/fr | 1152 |  | 1152 | 1152 | 1152 |  |  |
|  | fps | 60 |  | 60 | 60 | 60 |  |  |
|  | Mbps | 60 |  | 60 | 80 | 100 |  |  |
| High | pel/line | 1920 |  |  | 1920 | 1920 |  | 1920 |
|  | line/fr | 1152 |  |  | 1152 | 1152 |  | 1152 |
|  | fps | 60 |  |  | 60 | 60 |  | 60 |
|  | Mbps | 80 |  |  | 100 | 130 |  | 300 |

*Coding of high quality moving pictures (MPEG-2)* 187

The main profile main line (MP@ML) is the most widely used pair for broadcast TV and the 4 : 2 : 2 profile and main line (4 : 2 : 2@ML) is for studio video production and recording.

## 8.3 How does the MPEG-2 video encoder differ from MPEG-1?

*8.3.1 Major differences*

From the profile and level we see that the picture resolutions in MPEG-2 can vary from SIF (352 × 288 × 25 or 30) to HDTV with 1920 × 1250 × 60. Moreover, most of these pictures are interlaced, whereas in MPEG-1 pictures are noninterlaced (progressive). Coding of interlaced pictures is the first difference between the two coding schemes.

In the MPEG-2 standard, combinations of various picture formats and the interlaced/progressive option create a new range of macroblock types. Although each macroblock in a progressive mode has six blocks in the 4 : 2 : 0 format, the number of blocks in the 4 : 4 : 4 image format is 12. Also, the dimensions of the unit of blocks used for motion estimation/compensation can change. In the interlaced pictures, since the number of lines per field is half the number of lines per frame then, with equal horizontal and vertical resolutions, for motion estimation it might be appropriate to choose blocks of 16 × 8, i.e. 16 pixels over eight lines. These types of submacroblock have half the number of blocks of the progressive mode.

The second significant difference between the MPEG-1 and the MPEG-2 video encoders is the new function of scalability. The scalable modes of MPEG-2 are intended to offer interoperability among different services or to accommodate the varying capabilities of different receivers and networks upon which a single service may operate. They allow a receiver to decode a subset of the full bit stream in order to display an image sequence at a reduced quality, spatial and temporal resolution.

*8.3.2 Minor differences*

Apart from the two major distinctions there are some other minor differences, which have been introduced to increase the coding efficiency of MPEG-2. They are again due to the picture interlacing used in MPEG-2. The first one is the scanning order of DCT coefficients. In MPEG-1, like H.261, zigzag scanning is used. MPEG-2 has the choice of using alternate scan, as shown in Figure 8.3*b*. For interlaced pictures, since the vertical correlation in the field pictures is greatly reduced, should field prediction be used, an alternate scan may perform better than a zigzag scan.

The second minor difference is on the nature of quantisation of the DCT coefficients. MPEG-2 supports both linear and nonlinear quantisation of the DCT coefficients. The nonlinear quantisation increases the precision of quantisation at high bit rates by employing lower quantiser scale values. This improves picture quality at low contrast areas. At lower bit rates, where larger step sizes are needed, again the nonlinear behaviour of the quantiser provides a larger dynamic range for quantisation of the coefficients.

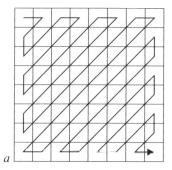 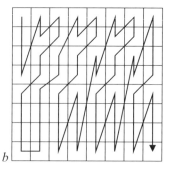

Figure 8.3    Two types of scanning method
   *a*   zigzag scan
   *b*   alternate scan

### 8.3.3  MPEG-1 and MPEG-2 syntax differences

The IDCT mismatch control in MPEG-1 is slightly different from that in MPEG-2. After the inverse quantisation process of the DCT coefficients, the MPEG-1 standard requires that all the nonzero coefficients are added with 1 or $-1$. In the MPEG-2 standard, only the last coefficient need be added with 1 or $-1$ provided that the sum of all coefficients is even after inverse quantisation. Another significant variance is the run level values. In MPEG-1, those that cannot be coded with a variable length code (VLC) are coded with the escape code, followed by either a 14-bit or 22-bit fixed length coding (FLC), whereas for MPEG-2 they are followed by an 18-bit FLC.

The constraint parameter flag mechanism in MPEG-1 has been replaced by the profile and level structures in MPEG-2. The additional chroma formats (4:2:2 and 4:4:4) and the interlaced related operations (field prediction and scalable coding modes) make MPEG-2 bit stream syntax different from that of MPEG-1.

The concept of the group of pictures (GOP) layer is slightly different. GOP in MPEG-2 may indicate that certain B-pictures at the beginning of an edited sequence comprise a broken link which occurs if the forward reference picture needed to predict the current B-pictures is removed from the bit stream by an editing process. It is an optional structure for MPEG-2 but mandatory for MPEG-1. The final point is that slices in MPEG-2 must always start and end on the same horizontal row of macroblocks. This is to assist the implementations in which the decoding process is split into some parallel operations along horizontal strips within the same pictures.

Although these differences may make direct decoding of the MPEG-1 bit stream by an MPEG-2 decoder infeasible, the fundamentals of video coding in the two codecs remain the same. In fact, as we mentioned, there is a need for backward compatibility, such that the MPEG-2 decoder should be able to decode the MPEG-1 encoded bit stream. Thus MPEG-1 is a subset of MPEG-2. They employ the same concept of a group of pictures, and the interlaced field pictures now become I, P and B-fields, and all the macroblock types have to be identified as field or frame based. Therefore in describing the MPEG-2 video codec, we will avoid repeating what has already been said about MPEG-1 in Chapter 7. Instead we concentrate on those parts which have risen due to interlacing and scalability of MPEG-2. However, for information on the difference between MPEG-1 and MPEG-2 refer to [7].

## 8.4 MPEG-2 nonscalable coding modes

This simple nonscalable mode of the MPEG-2 standard is the direct extension of the MPEG-1 coding scheme with the additional feature of accommodating interlaced video coding. The impact of the interlaced video on the coding methodology is that interpicture prediction may be carried out between the fields, as they are closer to each other. Furthermore, for slow moving objects, vertical pixels in the same frame are closer, making frame prediction more efficient.

As usual we define the prediction modes on a macroblock basis. Also, to be in line with the MPEG-2 definitions, we define the odd and the even fields as the top and bottom fields, respectively. A field macroblock, similar to the frame macroblock, consists of $16 \times 16$ pixels. In the following, five modes of prediction are described [3]. They can be equally applied to P and B-pictures, unless specified otherwise.

Similar to the reference model in H.261, software-based reference codecs for laboratory testing have also been considered for MPEG-1 and 2. For these codecs, the reference codec is called the test model (TM), and the latest version of the test model is TM5 [8].

### 8.4.1 Frame prediction for frame pictures

Frame prediction for frame pictures is exactly identical to the predictions used in MPEG-1. Each P-frame can make a prediction from the previous anchor frame, and there is one motion vector for each motion compensated macroblock. B-frames may use previous, future or interpolated past and future anchor frames. There will be up to two motion vectors (forward and backward) for each B-frame motion compensated macroblock. Frame prediction works well for slow to moderate motion as well as panning over a detailed background.

### 8.4.2 Field prediction for field pictures

Field prediction is similar to frame prediction, except that pixels of the target macroblock (MB to be coded) belong to the same field. Prediction macroblocks also should belong to one field, either from the top or the bottom field. Thus for P-pictures the prediction macroblock comes from the two most recent fields, as shown in Figure 8.4. For example, the prediction for the target macroblocks in the top field of a P-frame, $T_P$, may come either from the top field, $T_R$, or the bottom field, $B_R$, of the reference frame.

The prediction for the target macroblocks in the bottom field, $B_P$, can be made from its two recent fields, the top field of the same frame, $T_P$, or the bottom field of the reference frame, $B_R$.

190  Standard codecs: image compression to advanced video coding

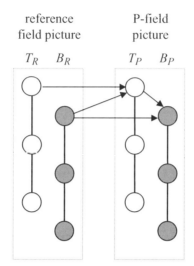

Figure 8.4  Field prediction of field pictures for P-picture MBs

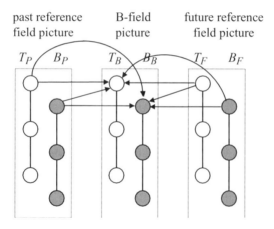

Figure 8.5  Field prediction of field pictures for B-picture MBs

For B-pictures the prediction MBs are taken from the two most recent anchor pictures (I/P or P/P). Each target macroblock can make a forward or a backward prediction from either of the fields.

For example in Figure 8.5 the forward prediction for the bottom field of a B-picture, $B_B$, is either $T_P$ or $B_P$, and the backward prediction is taken from $T_F$ or $B_F$. There will be one motion vector for each P-field target macroblock, and two motion vectors for those of B-fields.

# Coding of high quality moving pictures (MPEG-2)

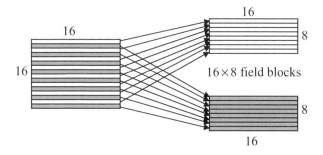

*Figure 8.6  A target macroblock is split into two 16 × 8 field blocks*

## 8.4.3 Field prediction for frame pictures

In this case the target macroblock in a frame picture is split into two top field and bottom field pixels, as shown in Figure 8.6. Field prediction is then carried out independently for each of the 16 × 8 pixel target macroblocks.

For P-pictures, two motion vectors are assigned for each 16 × 16 pixel target macroblock. The 16 × 8 predictions may be taken from either of the two most recently decoded anchor pictures. Note that the 16 × 8 field prediction cannot come from the same frame, as was the case in field prediction for field pictures.

For B-pictures, due to the forward and the backward motion, there can be two or four motion vectors for each target macroblock. The 16 × 8 predictions may be taken from either field of the two most recently decoded anchor pictures.

## 8.4.4 Dual prime for P-pictures

Dual prime is only used in P-pictures where there are no B-pictures in the GOP. Here only one motion vector is encoded (in its full format) in the bit stream together with a small differential motion vector correction. In the case of the field pictures two motion vectors are then derived from this information. These are used to form predictions from the two reference fields (one top, one bottom) which are averaged to form the final prediction. In the case of frame pictures this process is repeated for the two fields so that a total of four field predictions are made.

Figure 8.7 shows an example of dual prime motion compensated prediction for the case of frame pictures. The transmitted motion vector has a vertical displacement of three pixels. From the transmitted motion vector two preliminary predictions are computed, which are then averaged to form the final prediction.

The first preliminary prediction is identical to the field prediction, except that the reference pixels should come from the previously coded fields of the same parity (top or bottom fields) as the target pixels. The reference pixels, which are obtained from the transmitted motion vector, are taken from two fields (taken from one field for field pictures). In the Figure the predictions for target pixels in the top field, $T_P$, are

192  *Standard codecs: image compression to advanced video coding*

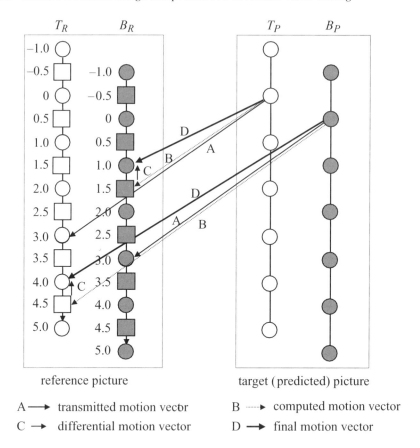

*Figure 8.7*  *Dual prime motion compensated prediction for P-pictures*

taken from the top reference field, $T_R$. Target pixels in the bottom field, $B_P$, take their predictions from the bottom reference field, $B_R$.

The second preliminary prediction is derived using a computed motion vector plus a small differential motion vector correction. For this prediction, reference pixels are taken from the parity field opposite to the first parity preliminary prediction. For the target pixels in the top field $T_P$, pixels are taken from the bottom reference field $B_R$. Similarly, for the target pixels in the bottom field $B_P$, prediction pixels are taken from the top reference field $T_R$.

The computed motion vectors are obtained by a temporal scaling of the transmitted motion vector to match the field in which the reference pixels lie, as shown in Figure 8.7. For example, for the transmitted motion vector of value 3, the computed motion vector for $T_P$ would be $3 \times 1/2 = 1.5$, since the reference field $B_R$ is mid way between the top reference field and the top target field. The computed motion vector for the bottom field is $3 \times 3/2 = 4.5$, as the distance between the reference top field and the bottom target field is three fields (3/2 frames). The differential motion vector correction, which can have up to one half pixel precision, is then added to the computed motion vector to give the final corrected motion vector.

In the Figure the differential motion vector correction has a vertical displacement of $-0.5$ pixel. Therefore the corrected motion vector for the top target field, $T_P$, would be $1.5 - 0.5 = 1$, and for the bottom target field it is $4.5 - 0.5 = 4$, as shown with thicker lines in the Figure.

For interlaced video the performance of dual prime prediction can, under some circumstances, be comparable to that of B-picture prediction and has the advantage of low encoding delay. However, for dual prime, unlike B-pictures, the decoded pixels should be stored to be used as reference pixels.

### 8.4.5  16 × 8 motion compensation for field pictures

In motion compensation mode, a field of $16 \times 16$ pixel macroblocks is split into upper half and lower half $16 \times 8$ pixel blocks, and a separate field prediction is carried out for each. Two motion vectors are transmitted for each P-picture macroblock and two or four motion vectors for the B-picture macroblock. This mode of motion compensation may be useful in field pictures that contain irregular motion. Note the difference between this mode and the field prediction for frame pictures in section 8.4.3. Here a field macroblock is split into two halves, and in the field prediction for frame pictures a frame macroblock is split into two top and bottom field blocks.

Thus the five modes of motion compensation in MPEG-2 in relation to field and frame predictions can be summarised in Table 8.3.

### 8.4.6  Restrictions on field pictures

It should be noted that field pictures have some restrictions on I, P and B-picture coding type and motion compensation. Normally, the second field picture of a frame must be of the same coding type as the first field. However, if the first field picture of a frame is an I-picture, then the second field can be either I or P. If it is a P-picture, the prediction macroblocks must all come from the previous I-picture, and dual prime cannot be used.

Table 8.3  *Five motion compensation modes in MPEG-2*

| Motion compensation mode | Use in field pictures | Use in frame pictures |
|---|---|---|
| Frame prediction for frame pictures | no | yes |
| Field prediction for field pictures | yes | no |
| Field prediction for frame pictures | no | yes |
| Dual prime for P-pictures | yes | yes |
| 16 × 8 motion compensation for field pictures | yes | no |

### 8.4.7 Motion vectors for chrominance components

As explained, the motion vectors are estimated based on the luminance pixels, hence they are used for the compensation of the luminance component. For each of the two chrominance components the luminance motion vectors are scaled according to the image format:

- 4 : 2 : 0 both the horizontal and vertical components of the motion vector are scaled by dividing by two
- 4 : 2 : 2 the horizontal component of the motion vector is scaled by dividing by two; the vertical component is not altered
- 4 : 4 : 4 the motion vector is unmodified.

### 8.4.8 Concealment motion vectors

Concealment motion vectors are motion vectors that may be carried by the intra macroblocks for the purpose of concealing errors should transmission error result in loss of information. A concealment motion vector is present for all intra macroblocks if, and only if, the *concealment_motion_vectors* flag in the picture header is set. In the normal course of events no prediction is formed for such macroblocks, since they are of intra type. The specification does not specify how error recovery shall be performed. However, it is a recommendation that concealment motion vectors should be suitable for use by a decoder that is capable of performing the function. If concealment is used in an I-picture then the decoder should perform prediction in a similar way to a P-picture.

Concealment motion vectors are intended for use in the case where a data error results in information being lost. There is therefore little point in encoding the concealment motion vector in the macroblock for which it is intended to be used. This is because, if the data error results in the need for error recovery, it is very likely that the concealment motion vector itself would be lost or corrupted. As a result the following semantic rules are appropriate:

- For all macroblocks except those in the bottom row of macroblocks concealment motion vectors should be appropriate for use in the macroblock that lies vertically below the macroblock in which the motion vector occurs.
- When the motion vector is used with respect to the macroblock identified in the previous rule a decoder must assume that the motion vector may refer to samples outside of the slices encoded in the reference frame or reference field.
- For all macroblocks in the bottom row of macroblocks the reconstructed concealment motion vectors will not be used. Therefore the motion vector (0,0) may be used to reduce unnecessary overhead.

## 8.5 Scalability

Scalable video coding is often regarded as being synonymous with layered video coding, which was originally proposed by the author to increase robustness of video codecs against packet (cell) loss in ATM networks [9]. At the time (late 1980s), H.261 was under development and it was clear that purely interframe coded video by this codec was very vulnerable to loss of information. The idea behind layered coding was that the codec should generate two bit streams, one carrying the most vital video information and called the base layer, and the other to carry the residual information to enhance the base layer image quality, named the enhancement layer. In the event of network congestion, only the less important enhancement data should be discarded, and the space made available for the base layer data. Such a methodology had an influence on the formation of ATM cell structure, to provide two levels of priority for protecting base layer data [5]. This form of two-layer coding is now known as SNR scalability in the MPEG-2 standard, and currently a variety of new two-layer coding techniques have been devised. They now form the basic scalability functions of the MPEG-2 standard.

Before describing various forms of scalability in some detail, it will be useful to know the similarity and more importantly any dissimilarity between these two coding methods. In the next section this will be dealt with in some depth, but since scalability is the commonly adopted name for all the video coding standards, throughout the book we use scalability to address both methods.

The scalability tools defined in the MPEG-2 specifications are designed to support applications beyond that supported by the single-layer video. Among the noteworthy applications areas addressed are video telecommunications, video on asynchronous transfer mode networks (ATM), interworking of video standards, video service hierarchies with multiple spatial, temporal and quality resolutions, HDTV with embedded TV, systems allowing migration to higher temporal resolution HDTV etc. Although a simple solution to scalable video is the simulcast technique, which is based on transmission/storage of multiple independently coded reproductions of video, a more efficient alternative is scalable video coding, in which the bandwidth allocated to a given reproduction of video can be partially reutilised in coding of the next reproduction of video. In scalable video coding, it is assumed that given an encoded bit stream, decoders of various complexities can decode and display appropriate reproductions of the coded video. A scalable video encoder is likely to have increased complexity when compared with a single-layer encoder. However, the standard provides several different forms of scalability which address nonoverlapping applications with corresponding complexities. The basic scalability tools offered are: data partitioning, SNR scalability, spatial scalability and temporal scalability. Moreover, combinations of these basic scalability tools are also supported and are referred to as hybrid scalability. In the case of basic scalability, two layers of video, referred to as the base layer and the enhancement layer, are allowed, whereas in hybrid scalability up to three layers are supported.

### 8.5.1 Layering versus scalability

Considering the MPEG-1 and MPEG-2 systems functions, defined in section 8.1, we see that MPEG-2 puts special emphasis on the transport of the bit stream. This is because MPEG-1 video is mainly for storage and software-based decoding applications in an almost error free environment, whereas MPEG-2, or H.262 in the ITU-T standard, is for transmission and distribution of video in various networks. Depending on the application, the emphasis can be put on either transmission or distribution. In the introduction to MPEG-2/H262 potential applications for this codec were listed. The major part of the application is the transmission of video over networks, such as: satellite and terrestrial broadcasting, news gathering, personal communications, video over ATM networks etc., where for better quality of service the bit stream should be protected against channel misbehaviour, which is very common in these environments. One way of protecting data against channel errors is to add some redundancy, like forward error correcting bits into the bit stream. The overhead is a percentage of the bit rate (depending on the amount of protection) and will be minimal if the needed protection part had a small channel rate requirement. Hence it is logical to design the codec in such a way as to generate more than one bit stream, and protect the most vital bit stream against the error to produce a basic quality picture. The remaining bit streams should be such that their presence enhances the video quality, but their absence or corruption should not degrade the video quality significantly. Similarly, in ATM networks, partitioning the bit stream into various parts of importance, and then providing a guaranteed channel capacity for a small part of the bit stream, is much easier than dealing with the entire bit stream. This is the fundamental concept behind the layered video coding. The most notable point is that the receiver is always expecting to receive the entire bit stream, but if some parts are not received, the picture will not break up, or the quality will not degrade significantly. Increasing the number of layers, and unequally protecting the layers against errors (cell loss in ATM networks) according to their importance in their contribution to video quality, can give a graceful degradation of video quality.

On the other hand, some applications set forth for MPEG-2 are mainly for distribution of digital video to the receivers of various capabilities. For example, in cable TV distribution over optical networks (CATV), the prime importance is to be able to decode various quality pictures from a single bit stream. In this network, there might be receivers with various decoding capability (processing powers) or customers with different requirements for video quality. Then, according to the need, a portion of the bit stream is decoded for that specific service. Here, it is assumed that the error rate in the optical channels is negligible and sufficient channel capacity for the bit stream exists, both assumptions are plausible.

Thus comparing layered coding with scalability, the fundamental difference is that in layered coding the receiver expects to receive the entire bit stream, but occasionally some parts might be in error or missing, while in scalable coding, a receiver expects to decode a portion of the bit stream, but when that is available, it remains so for the entire communication session. Of course, in scalable coding, for efficient compression, generation of the bit stream is made in a hierarchical structure such that the basic portion of the bit stream gives a video of minimum acceptable quality, similar to the base layer video. The subsequent segments of the bit stream enhance the video quality accordingly, similar to the enhancement layers. If the channel requires any protection, then the base layer bit stream should be guarded, or in ATM networks the required channel capacity be provided. Now in this respect we can say scalable and layered video coding are the same. This means that scalable coding can be used as a layering technique, but, however, layered coded data may not be scalable. Thus, scalability is a more generic name for layering, and throughout the book we use scalability to address both. Those parts where the video codec acts as a layered encoder but not as a scalable encoder, will be particularly identified.

### 8.5.2 Data partitioning

Data partitioning is a tool intended for use when two channels are available for the transmission and/or storage of a video bit stream, as may be the case in ATM networks, terrestrial broadcasting, magnetic media etc. Data partitioning in fact is not true scalable coding, but as we will see it is a layered coding technique. It is a means of dividing the bit stream of a single-layer MPEG-2 into two parts or two layers. The first layer comprises the critical parts of the bit stream (such as headers, motion vectors, lower order DCT coefficients) which are transmitted in the channel with the better error performance. The second layer is made of less critical data (such as higher DCT coefficients) and is transmitted in the channel with poorer error performance. Thus, degradations to channel errors are minimised since the critical parts of a bit stream are better protected. Data from neither channel may be decoded on a decoder that is not intended for decoding data partitioned bit streams. Even with the proper decoder, data extracted from the second layer decoder cannot be used unless the decoded base layer data is available.

A block diagram of a data partitioning encoder is shown in Figure 8.8. The single-layer encoder is in fact a nonscalable MPEG-2 video encoder that may or may not include B-pictures. At the encoder, during the quantisation and zigzag scanning of each $8 \times 8$ DCT coefficient, the scanning is broken at the priority break point (PBP), as shown in Figure 8.9.

The first part of the scanned quantised coefficients after variable length coding, with the other overhead information such as motion vectors, macroblock types and addresses etc., including the priority break point (PBP), is taken as the base layer bit stream. The remaining scanned and quantised coefficients plus the end of block

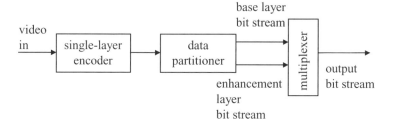

*Figure 8.8   Block diagram of a data partitioning encoder*

198  *Standard codecs: image compression to advanced video coding*

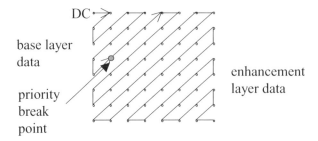

*Figure 8.9  Position of the priority break point in a block of DCT coefficients*

(EOB) code constitute the enhancement layer bit stream. Figure 8.9 also shows the position of the priority break point in the DCT coefficients.

The base and the enhancement layer bit streams are then multiplexed for transmission into the channel. For prioritised transmission such as ATM networks, each bit stream is first packetised into high and low priority cells and the cells are multiplexed. At the decoder, knowing the position of PBP, a block of DCT coefficients is reconstructed from the two bit streams. Note that PBP indicates the last DCT coefficient of the base. Its position at the encoder is determined based on the portion of channel rate from the total bit rate allocated to the base layer.

Figure 8.10 shows single shots of an 8 Mbit/s data partitioning MPEG-2 coded video and its associated base layer picture. The priority break point is adjusted for a base layer bit rate of 2 Mbit/s. At this bit rate, the quality of the base layer is almost acceptable. However, some areas in the base layer show blocking artefacts and in others the picture is blurred. Blockiness is due to the reconstruction of some macroblocks from only the DC and/or from a few AC coefficients. Blurriness is due to loss of high frequency DCT coefficients.

It should be noted that, since the encoder is a single-layer interframe coder, then at the encoder both the base and the enhancement layer coefficients are used at the encoding prediction loop. Thus reconstruction of the picture from the base layer only can result in a mismatch between the encoder and decoder prediction loops. This causes picture drift on the reconstructed picture, that is a loss of enhancement data at the decoder is accumulated and appears as mosquito-like noise. Picture drift only occurs on P-pictures, but since B-pictures may use P-pictures for prediction, they suffer from picture drift too. Also, I-pictures reset the feedback prediction, hence they clean up the drift. The more frequent the I-pictures, the less is the appearance of picture drift, but at the expense of higher bit rates.

In summary to have a drift-free video, the receiver should receive the entire bit stream. That is, a receiver that decodes only the base layer portion of the bit stream cannot produce a stable video. Therefore data partitioned bit stream is not scalable, but it is layered coded. It is in fact the simplest form of layering technique, which has no extra complexity over the single-layer encoder.

*Coding of high quality moving pictures (MPEG-2)* 199

*a*

*b*

*Figure 8.10  Data partitioning*
    *a* enhanced
    *b* base picture

Although the base picture suffers from picture drift and may not be usable alone, that of the enhanced (base layer plus the enhancement layer) picture with occasional losses is quite acceptable. This is due to normally low loss rates in most networks (e.g. less than $10^{-4}$ in ATM networks), such that before the accumulation of loss becomes significant, the loss area is cleaned up by I-pictures.

### 8.5.3 SNR scalability

SNR scalability is a tool intended for use in video applications involving telecommunications and multiple quality video services with standard TV and enhanced TV, i.e. video systems with the common feature that a minimum of two layers of video quality is necessary. SNR scalability involves generating two video layers of the same spatio–temporal resolution but different video qualities from a single video source such that the base layer is coded by itself to provide the basic video quality and the enhancement layer is coded to enhance the base layer. The enhancement layer when added back to the base layer regenerates a higher quality reproduction of the input video. Since the enhancement layer is said to enhance the signal-to-noise ratio (SNR) of the base layer, this type of scalability is called SNR. Alternatively, as we will see later, SNR scalability could have been called coefficient amplitude scalability or quantisation noise scalability. These types, although a bit wordy, may better describe the nature of this encoder.

Figure 8.11 shows a block diagram of a two-layer SNR scalable encoder. First, the input video is coded at a low bit rate (lower image quality), to generate the base layer bit stream. The difference between the input video and the decoded output of the base layer is coded by a second encoder, with a higher precision, to generate the enhancement layer bit stream. These bit streams are multiplexed for transmission over the channel. At the decoder, decoding of the base layer bit stream results in the base picture. When the decoded enhancement layer bit stream is added to the base layer, the result is an enhanced image. The base and the enhancement layers may either use the MPEG-2 standard encoder or the MPEG-1 standard for the base layer

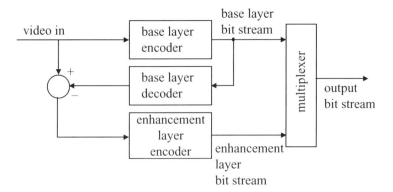

*Figure 8.11*   *Block diagram of a two-layer SNR scalable coder*

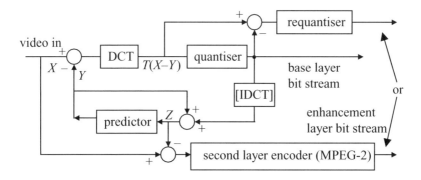

*Figure 8.12   A DCT based base layer encoder*

and MPEG-2 for the enhancement layer. That is, in the latter a 4 : 2 : 0 format picture is generated at the base layer, but a 4 : 2 : 0 or 4 : 2 : 2 format picture at the second layer.

It may appear that the SNR scalable encoder is much more complex than is the data partitioning encoder. The former requires at least two nonscalable encoders whereas data partitioning is a simple single-layer encoder, and partitioning is just carried out on the bit stream. The fact is that if both layer encoders in the SNR coder are of the same type, e.g. both nonscalable MPEG-2 encoders, then the two-layer encoder can be simplified. Consider Figure 8.12, which represents a simplified nonscalable MPEG-2 of the base layer [10].

According to the Figure, the difference between the input pixels block $X$ and their motion compensated predictions $Y$ are transformed into coefficients $T(X-Y)$. These coefficients after quantisation can be represented with $T(X-Y) - Q$, where $Q$ is the introduced quantisation distortion. The quantised coefficients after the inverse DCT (IDCT) reconstruct the prediction error. They are then added to the motion compensated prediction to reconstruct a locally decoded pixel block $Z$.

Thus the interframe error signal $X - Y$ after transform coding becomes:

$$T(X - Y) \tag{8.1}$$

and after quantisation, a quantisation distortion $Q$ is introduced to the transform coefficients. Then eqn. 8.1 becomes:

$$T(X - Y) - Q \tag{8.2}$$

After the inverse DCT the reconstruction error can be formulated as:

$$T^{-1}[T(X - Y) - Q] \tag{8.3}$$

where $T^{-1}$ is the inverse transformation operation. Because transformation is a linear operator, then the reconstruction error can be written as:

$$T^{-1}T(X - Y) - T^{-1}(Q) \tag{8.4}$$

Also, due to the orthonormality of the transform, where $T^{-1}T = 1$, eqn. 8.4 is simplified to:

$$X - Y - T^{-1}(Q) \qquad (8.5)$$

When this error is added to the motion compensated prediction $Y$, the locally decoded block becomes:

$$Z = Y + X - Y - T^{-1}(Q) = X - T^{-1}(Q) \qquad (8.6)$$

Thus, according to Figure 8.12, what is coded by the second layer encoder is:

$$X - Z = X - X + T^{-1}(Q) = T^{-1}(Q) \qquad (8.7)$$

that is, the inverse transform of the base layer quantisation distortion. Since the second layer encoder is also an MPEG encoder (e.g. a DCT-based encoder), then DCT transformation of $X - Z$ in eqn. 8.7 would result in:

$$T(X - Z) = TT^{-1}(Q) = Q \qquad (8.8)$$

where again the orthonormality of the transform is employed. Thus the second layer transform coefficients are in fact the quantisation distortions of the base layer transform coefficients, $Q$. For this reason the codec can also be called a coefficient amplitude scalability or quantisation noise scalability unit.

Therefore the second layer of an SNR scalable encoder can be a simple requantiser, as shown in Figure 8.12, without much more complexity than a data partitioning encoder. The only problem with this method of coding is that, since normally the base layer is poor, or at least worse than the enhanced image (base plus the second layer), then the used prediction is not good. A better prediction would be a picture of the sum of both layers, as shown in Figure 8.13. Note that the second layer is still encoding the quantisation distortion of the base layer.

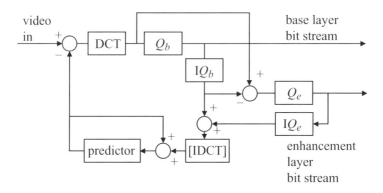

Figure 8.13  A two-layer SNR scalable encoder with drift at the base layer

In this encoder, for simplicity, the motion compensation, variable length coding of both layers and the channel buffer have been omitted. In the Figure $Q_b$ and $Q_e$ are the base and the enhancement layer quantisation step sizes, respectively. The quantisation distortion of the base layer is requantised with a finer precision ($Q_e < Q_b$), and then it is fed back to the prediction loop, to represent the coding loop of the enhancement layer. Now compared with data partitioning, this encoder only requires a second quantiser, and so the complexity is not so great.

Note the tight coupling between the two-layer bit streams. For freedom from drift in the enhanced picture, both bit streams should be made available to the decoder. For this reason this type of encoder is called an SNR scalable encoder with drift at the base layer or no drift in the enhancement layer. If the base layer bit stream is decoded by itself then, due to loss of differential refinement coefficients, the decoded picture in this layer will suffer from picture drift. Thus, this encoder is not a true scalable encoder, but is in fact a layered encoder. Again, although the drift should only appear in P-pictures, since B-pictures use P-pictures for predictions, this drift is transferred into B-pictures too. I-pictures reset the distortion and drift is cleaned up.

For applications with the occasional loss of information in the enhancement layer, parts of the picture have the base layer quality, and other parts that of the enhancement layer. Therefore picture drift can be noticed in these areas.

If a true SNR scalable encoder with drift-free pictures at both layers is the requirement, then the coupling between the two layers must be loosened. Applications such as simulcasting of video with two different qualities from the same source need such a feature. One way to prevent picture drift is not to feed back the enhancement data into the base layer prediction loop. In this case the enhancement layer will be intra coded and bit rate will be very high.

In order to reduce the second layer bit rate, the difference between the input to and the output of the base layer (see Figure 8.11) can be coded by another MPEG encoder [11]. However, here we need two encoders and the complexity is much higher than that for data partitioning. To reduce the complexity, we need to code only the quantisation distortion. However, since the transformation operator and most importantly in SNR scalability the temporal and spatial resolutions of the base and enhancement layers pictures are identical, then the motion estimation and compensation can be shared between them. Following the previous eqns 8.1–8.8, we can also simplify the two independent encoders into one encoder generating two bit streams, such that each bit stream is drift-free decodable. Figure 8.14 shows a block diagram of a three-layer truly SNR scalable encoder, where the generated picture of each layer is drift free and can be used for simulcasting [12].

In the Figure T, B, IT and EC represent transformation, prediction buffer, inverse transformation and entropy coding and $Q_i$, is the $i$th layer quantiser. A common motion vector is used at all layers. Note that although this encoder looks to be made of several single-layer encoders, since motion estimation and many coding decisions are common to all the layers, and the motion estimation comprises about 55–70 per cent of encoding complexity of an encoder, the increase in complexity is moderate. Figure 8.15 shows the block diagram of the corresponding three-layer SNR decoder.

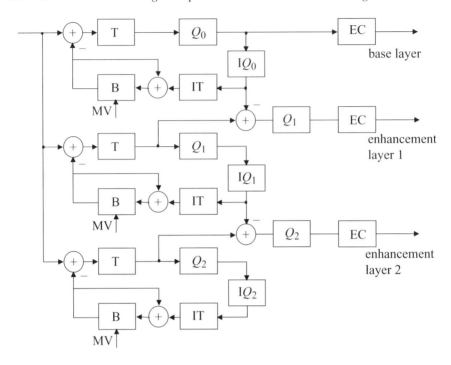

*Figure 8.14    A three-layer drift-free SNR scalable encoder*

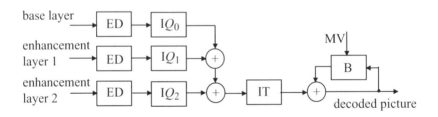

*Figure 8.15    A block diagram of a three-layer SNR decoder*

After entropy decoding (ED) each layer is inverse quantised and then all are added together to represent the final DCT coefficients. These coefficients are inverse transformed and are added to the motion compensated previous picture to reconstruct the final picture. Note that there is only one motion vector, which is transmitted at the base layer. Note also, the decoder of Figure 8.13, with drift in the base layer, is also similar to this Figure, but with only two layers of decoding.

Figure 8.16 shows the picture quality of the base layer at 2 Mbit/s. That of the base plus the enhancement layer would be similar to those of the data partitioning,

*Coding of high quality moving pictures (MPEG-2)* 205

*Figure 8.16    Picture quality of the base layer of SNR encoder at 2 Mbit/s*

albeit with slightly higher bit rate. At this bit rate the extra bits would be of the order of 15–20 per cent, due to the overhead of the second layer data [11] (also see Figure 8.26). Due to coarser quantisation, some parts of the picture are blocky, as was the case in data partitioning. However, since any significant coefficient can be included at the base layer, the base layer picture of this encoder, unlike that of data partitioning, does not suffer from loss of high frequency information.

Experimental results show that the picture quality of the base layer of an SNR scalable coder is much superior to that of data partitioning, especially at lower bit rates [13]. This is because, at lower base layer bit rates, data partitioning can only retain DC and possibly one or two AC coefficients. Reconstructed pictures with these few coefficients are very blocky.

### 8.5.4    Spatial scalability

Spatial scalability involves generating two spatial resolution video streams from a single video source such that the base layer is coded by itself to provide the basic spatial resolution and the enhancement layer employs the spatially interpolated base layer which carries the full spatial resolution of the input video source [14]. The base and

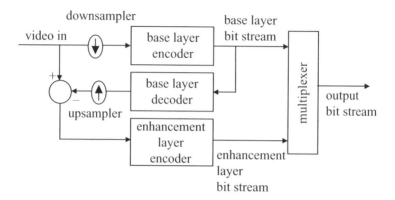

*Figure 8.17  Block diagram of a two-layer spatial scalable encoder*

the enhancement layers may either use both the coding tools in the MPEG-2 standard, or the MPEG-1 standard for the base layer and MPEG-2 for the enhancement layer, or even an H.261 encoder at the base layer and an MPEG-2 encoder at the second layer. Use of MPEG-2 for both layers achieves a further advantage by facilitating interworking between video coding standards. Moreover, spatial scalability offers flexibility in the choice of video formats to be employed in each layer. The base layer can use SIF or even lower resolution pictures at 4:2:0, 4:2:2 or 4:1:1 formats, and the second layer can be kept at CCIR-601 with 4:2:0 or 4:2:2 format. Like the other two scalable coders, spatial scalability is able to provide resilience to transmission errors as the more important data of the lower layer can be sent over a channel with better error performance, and the less critical enhancement layer data can be sent over a channel with poorer error performance. Figure 8.17 shows a block diagram of a two-layer spatial scalable encoder.

An incoming video is first spatially reduced in both the horizontal and vertical directions to produce a reduced picture resolution. For 2:1 reduction, normally a CCIR-601 video is converted into an SIF image format. The filters for the luminance and the chrominance colour components are the 7 and 4 tap filters, respectively, described in section 2.3. The SIF image sequence is coded at the base layer by an MPEG-1 or MPEG-2 standard encoder, generating the base layer bit stream. The bit stream is decoded and upsampled to produce an enlarged version of the base layer decoded video at CCIR-601 resolution. The upsampling is carried out by inserting zero level samples between the luminance and chrominance pixels, and interpolating with the 7 and 4 tap filters, similar to those described in section 2.3. An MPEG-2 encoder at the enhancement layer codes the difference between the input video and the interpolated video from the base layer. Finally, the base and enhancement layer bit streams are multiplexed for transmission into the channel.

If the base and the enhancement layer encoders are of the same type (e.g. both MPEG-2), then the two encoders can interact. This is not only to simplify the two-layer encoder, as was the case for the SNR scalable encoder, but also to make the coding more efficient. Consider a macroblock at the base layer. Due to 2:1 picture resolution between the enhancement and the base layers, the base layer macroblock corresponds to four macroblocks at the enhancement layer. Similarly, a macroblock at the enhancement layer corresponds to a block of 8 × 8 pixels at the base layer. The interaction would be in the form of upsampling the base layer block of 8 × 8 pixels into a macroblock of 16 × 16 pixels, and using it as a part of the prediction in the enhancement layer coding loop.

Figure 8.18 shows a block of 8 × 8 pixels from the base layer that is upsampled and is combined with the prediction of the enhancement layer to form the final prediction for a macroblock at the enhancement layer. In the Figure the base layer upsampled macroblock is weighted by $w$ and that of the enhancement layer by $1 - w$.

More details of the spatial scalable encoder are shown in Figure 8.19. The base layer is a nonscalable MPEG-2 encoder, where each block of this encoder is upsampled, interpolated and fed to a weighting table (WT). The coding elements of the enhancement layer are shown without the motion compensation, variable length code

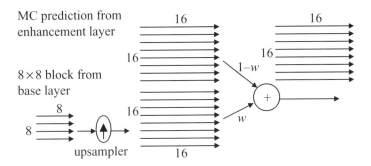

*Figure 8.18*   *Principle of spatio–temporal prediction in the spatial scalable encoder*

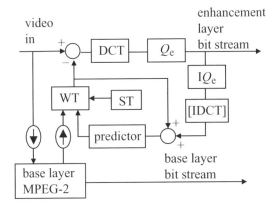

*Figure 8.19*   *Details of spatial scalability encoder*

and the other coding tools of the MPEG-2 standard. A statistical table (ST) sets the weighting table elements. Note that the weighted base layer macroblocks are used in the prediction loop, which will be subtracted from the input macroblocks. This part is similar to taking the difference between the input and the decoded base layer video and coding their differences by a second layer encoder, as was illustrated in the general block diagram of this encoder in Figure 8.17.

Figure 8.20 shows a single shot of the base layer picture at 2 Mbit/s. The picture produced by the base plus the enhancement layer at 8 Mbit/s would be similar to that resulting for data partitioning, shown in Figure 8.10. Note that since picture size is one quarter of the original, the 2 Mbit/s allocated to the base layer would be sufficient to code the base layer pictures at almost identical quality to the base plus the enhancement layer at 8 Mbit/s. An upsampled version of the base layer picture to fill the display at the CCIR-601 size is also shown in the Figure. Comparing this picture with those of data partitioning and the simple version of the SNR scalable coders, it can be seen that the picture is almost free from blockiness. However, still some very high frequency information is missing and aliasing distortions due to the upsampling will be introduced into the picture. Note that the base layer picture can be used alone without picture drift. This was not the case for data partitioning and the simple SNR scalable encoders. However, the price paid is that this encoder is made up of two MPEG encoders, and is more complex than data partitioning and SNR scalable encoders. Note that unlike the true SNR scalable encoder of Figure 8.14, here due to differences in the picture resolutions of the base and enhancement layers, the same motion vector cannot be used for both layers.

### 8.5.5 Temporal scalability

Temporal scalability is a tool intended for use in a range of diverse video applications from telecommunications to HDTV. In such systems migration to higher temporal resolution systems from that of lower temporal resolution systems may be necessary. In many cases, the lower temporal resolution video systems may be either the existing systems or the less expensive early generation systems. The more sophisticated systems may then be introduced gradually.

Temporal scalability involves partitioning of video frames into layers, in which the base layer is coded by itself to provide the basic temporal rate and the enhancement layer is coded with temporal prediction with respect to the base layer. The layers may have either the same or different temporal resolutions, which, when combined, provide full temporal resolution at the decoder. The spatial resolution of frames in each layer is assumed to be identical to that of the input video. The video encoders of the two layers may not be identical. The lower temporal resolution systems may only decode the base layer to provide basic temporal resolution, whereas more sophisticated systems of the future may decode both layers and provide high temporal resolution video while maintaining interworking capability with earlier generation systems.

Since in temporal scalability the input video frames are simply partitioned between the base and the enhancement layer encoders, the encoder need not be more complex than a single-layer encoder. For example, a single-layer encoder may be switched

*Coding of high quality moving pictures (MPEG-2)* 209

*Figure 8.20*  *a*  Base layer picture of a spatial scalable encoder at 2 Mbit/s
             *b*  its enlarged version

between the two base and enhancement modes to generate the base and the enhancement bit streams alternately. Similarly, a decoder can be reconfigured to decode the two bit streams alternately. In fact, the B-pictures in MPEG-1 and MPEG-2 provide a very simple temporal scalability that is encoded and decoded alongside the anchor I and P-pictures within a single codec. I and P-pictures are regarded as the base layer,

210  *Standard codecs: image compression to advanced video coding*

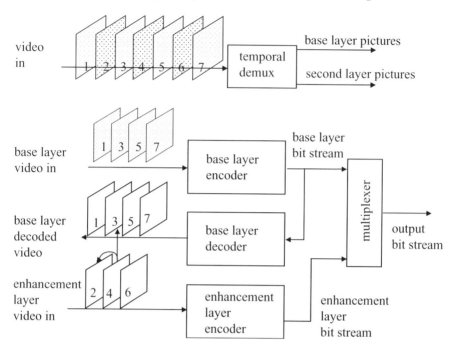

*Figure 8.21  A block diagram of a two-layer temporal scalable encoder*

and the B-pictures become the enhancement layer. Decoding of I and P-pictures alone will result in the base pictures with low temporal resolution, and when added to the decoded B-pictures the temporal resolution is enhanced to its full size. Note that, since the enhancement data does not affect the base layer prediction loop, both the base and the enhanced pictures are free from picture drift.

Figure 8.21 shows the block diagram of a two-layer temporal scalable encoder. In the Figure a temporal demultiplexer partitions the input video into the base and enhancement layer, input pictures. For the 2:1 temporal scalability shown in the Figure, the odd numbered pictures are fed to the base layer encoder and the even numbered pictures become inputs to the second layer encoder. The encoder at the base layer is a normal MPEG-1, MPEG-2, or any other encoder. Again for greater interaction between the two layers, either to make encoding simple or more efficient, both layers may employ the same type of coding scheme.

At the base layer the lower temporal resolution input pictures are encoded in the normal way. Since these pictures can be decoded independently of the enhancement layer, they do not suffer from picture drift. The second layer may use prediction from the base layer pictures, or from its own picture, as shown for frame 4 in the Figure. Note that at the base layer some pictures might be coded as B-pictures, using their own previous, future or their interpolation as prediction, but it is essential that some pictures should be coded as anchor pictures. On the other hand, in the enhancement layer, pictures can be coded in any mode. Of course, for greater compression, at the enhancement layer, most if not all the pictures are coded as B-pictures. These B-pictures have the choice of using past, future and their interpolated values, either from the base or the enhancement layer.

### 8.5.6 Hybrid scalability

MPEG-2 allows combination of individual scalabilities such as spatial, SNR or temporal scalability to form hybrid scalability for certain applications. If two scalabilities are combined, then three layers are generated and they are called the base layer, enhancement layer 1 and enhancement layer 2. Here enhancement layer 1 is a lower layer relative to enhancement layer 2, and hence decoding of enhancement layer 2 requires the availability of enhancement layer 1. In the following some examples of hybrid scalability are shown.

#### 8.5.6.1 Spatial and temporal hybrid scalability

Spatial and temporal scalability is perhaps the most common use of hybrid scalability. In this mode the three-layer bit streams are formed by using spatial scalability between the base and enhancement layer 1, while temporal scalability is used between enhancement layer 2 and the combined base and enhancement layer 1, as shown in Figure 8.22.

In this Figure, the input video is temporally partitioned into two lower temporal resolution image sequences In-1 and In-2. The image sequence In-1 is fed to the spatial scalable encoder, where its reduced version, In-0, is the input to the base layer encoder. The spatial encoder then generates two bit streams, for the base and enhancement layer 1. The In-2 image sequence is fed to the temporal enhancement encoder to generate the third bit stream, enhancement layer 2. The temporal enhancement encoder can use the locally decoded pictures of a spatial scalable encoder as predictions, as was explained in section 8.5.4.

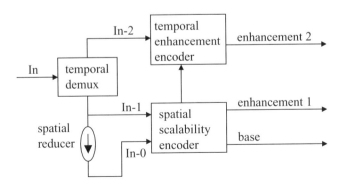

*Figure 8.22*   *Spatial and temporal hybrid scalability encoder*

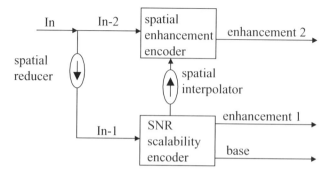

*Figure 8.23   SNR and spatial hybrid scalability encoder*

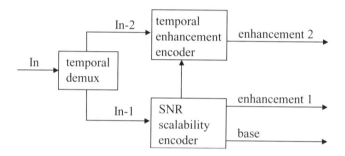

*Figure 8.24   SNR and temporal hybrid scalability encoder*

### 8.5.6.2   SNR and spatial hybrid scalability

Figure 8.23 shows a three-layer hybrid encoder employing SNR scalability and spatial scalability. In this encoder the SNR scalability is used between the base and enhancement layer 1 and the spatial scalability is used between layer 2 and the combined base and enhancement layer 1. The input video is spatially downsampled (reduced) to lower resolution as In-1 is to be fed to the SNR scalable encoder. The output of this encoder forms the base and enhancement layer 1 bit streams. The locally decoded pictures from the SNR scalable coder are upsampled to full resolution to form prediction for the spatial enhancement encoder.

### 8.5.6.3   SNR and temporal hybrid scalability

Figure 8.24 shows an example of an SNR and temporal hybrid scalability encoder. The SNR scalability is performed between the base layer and the first enhancement layer. The temporal scalability is used between the second enhancement layer and the locally decoded picture of the SNR scalable coder. The input image sequence through a temporal demultiplexer is partitioned into two sets of image sequences, and these are fed to each individual encoder.

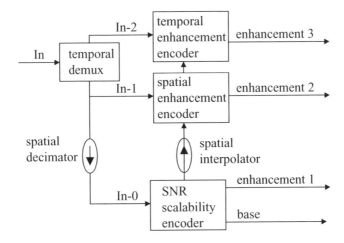

*Figure 8.25   SNR, spatial and temporal hybrid scalability encoder*

### 8.5.6.4  SNR, spatial and temporal hybrid scalability

The three scalable encoders might be combined to form a hybrid coder with a larger number of levels. Figure 8.25 shows an example of four levels of scalability, using all the three scalability tools mentioned.

The temporal demultiplexer partitions the input video into image sequences In-1 and In-2. Image sequence In-2 is coded at the highest enhancement layer (enhancement 3), with the prediction from the lower levels. The image sequence In-1 is first downsampled to produce a lower resolution image sequence, In-0. This sequence is then SNR scalable coded, to provide the base and the first enhancement layer bit streams. An upsampled and interpolated version of the SNR scalable decoded video forms the prediction for the spatial enhancement encoder. The output of this encoder results in the second enhancement layer bit stream (enhancement 2).

Figure 8.25 was just an example of how various scalability tools can be combined to produce bit streams of various degrees of importance. Of course, depending on the application, formation of the base and the level of the hierarchy of the higher enhancement layers might be defined in a different way to suit the application. For example, when the above scalability methods are applied to each of the I, P and B-pictures, since these pictures have different levels of importance, then their layered versions can increase the number of layers even further.

### 8.5.7  *Overhead due to scalability*

Although scalability or layering techniques provide a means of delivering a better video quality to the receivers than do single-layer encoders, this is done at the expense of higher encoder complexity and higher bit rate. We have seen that data partitioning is the simplest form of layering, and spatial scalability the most complex one. The amount of extra bits generated by these scalability techniques is also different.

Data partitioning is a single-layer encoder, but inclusion of the priority break point (PBP) in the zigzag scanning path and the fact that the zero run of the zigzag scan is now broken into two parts, incur some additional bits. These extra bits, along with redundant declaration of the macroblock addresses at both layers, generate some overhead over the single-layer coder. Our investigations show that the overhead bit is of the order of 3–4 per cent of the single-layer counterpart, almost irrespective of the percentage of the bits from the total bit rate assigned to the base layer.

In SNR scalability the second layer codes the quantisation distortions of the base layer, plus the other addressing information. The additional bits over the single layer depend on the relationship between the quantiser step sizes of the base and enhancement layers, and consequently on the percentage of the total bits allocated to the base layer. At the lower percentages, the quantiser step size of the base layer is large and hence the second layer efficiently codes any residual base layer quantisation distortions. This is very similar to successive approximation (two sets of bit planes), hence it is not expected that the SNR scalable coding efficiency will be much worse than the single layer.

At the higher percentages, the quantiser step sizes of the base and enhancement layers become close to each other. Considering that for a base layer quantiser step size of $Q_b$, the maximum quantisation distortion of the quantised coefficients is $Q_b/2$ and the nonquantised ones that fall in the dead zone is $Q_b$, then as long as the enhancement quantiser step size $Q_b > Q_e > Q_b/2$, none of the significant base layer coefficients is coded by the enhancement layer, except of course the ones in the dead zone of the base layer. Thus, again both layers code the data efficiently, that is the coefficient is either coded at the base layer or the enhancement layer, and the overall coding efficiency is not worse than the single layer. Reducing the base layer bit rate from its maximum value, means increasing $Q_b$. As long as $Q_b/2 < Q_e$, none of the base layer quantisation distortions (except the ones on the dead zone) can be coded by the enhancement layer. Hence, the enhancement layer does not improve the picture quality noticeably, and since the base layer is coded at a lower bit rate, the overall quality will be worse than that for the single layer. The worst quality occurs when $Q_e = Q_b/2$.

If the aim was to produce the same picture quality, then the bit rate of the SNR scalable coder had to be increased, as shown in Figure 8.26. In this Figure the overall bit rate of the SNR scalable coder is increased over the single layer such that the picture quality under both encoders is identical. The percentage of the bits assigned to the base layer from the total bits is varied from its minimum value to its maximum value. As we see the poorest performance of the SNR scalable (the highest overhead) is when the base layer is allocated about 40–50 per cent of the total bit rate. In fact, at this bit rate, the average quantiser step size of the enhancement layer is half of that of the base layer. This maximum overhead is 30 per cent and that of the data partitioning is also shown, which reads about three per cent irrespective of the bits assigned to the base layer.

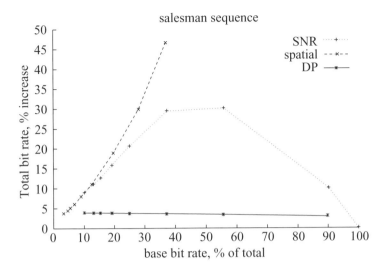

*Figure 8.26   Increase in bit rate due to scalability*

In spatial scalability the smaller size picture of the base layer is upsampled and its difference with the input picture is coded by the enhancement layer. Hence the enhancement layer, in addition to the usual redundant addressing, has to code two new items of information. One is the aliasing distortion of the base layer due to upsampling and the other is the quantisation distortion of the base layer, if there is any.

At the very low percentage of the bits assigned to the base layer, both of these distortions are coded efficiently by the enhancement layer. As the percentage of bits assigned to the base layer increases, similar to SNR scalability the overhead increases too. At the time where the quantiser step sizes of both layers become equal, $Q_b = Q_e$, any increase in the base layer bit rate (making $Q_b < Q_e$), means that the enhancement layer cannot improve the distortions of the base layer further. Beyond this point, aliasing distortion will be the dominant distortion and any increase in the base layer bit rate will be wasted. Thus, as the bit rate budget of the base layer increases, the overhead increases too, as shown in Figure 8.26. This differs from the behaviour of the SNR scalability.

In fact, in spatial scalability with a fixed total bit rate, increasing the base layer bit rate beyond the critical point of $Q_b = Q_e$ will reduce the enhancement layer bit rate budget. In this case, increasing the base layer bit rate will increase the aliasing distortion and hence, as the base layer bit rate increases, the overall quality decreases!!

In temporal scalability, in contrast to the other scalability methods, in fact the bit rate can be less than that for a single-layer encoder!! This is because there is no redundant addressing to create overhead. Moreover, since the enhancement layer pictures have more choice for their optimum prediction, from either base or enhancement layers, they are coded more efficiently than the single layer. A good example is the B-pictures in MPEG-2, that can be coded at much lower bit rate than the P-pictures. Thus temporal scalability in fact can be slightly more efficient than for single-layer coding.

### 8.5.8 Applications of scalability

Considering the nature of the basic scalability of data partitioning, SNR, spatial and temporal scalability and their behaviour with regard to picture drift and the overhead, suitable applications for each method may be summarised as:

a  *Data partitioning*: this mode is the simplest of all but since it has a poor base layer quality and is sensitive to picture drift, it can be used in the environment where there is rarely any loss of enhancement data (e.g. loss rate $<10^{-6}$). Hence the best application would be video over ATM networks, where through admission control, the loss ratio can be maintained at low levels [15].

b  *SNR scalability*: in this method two pictures of the same spatio–temporal resolutions are generated, but one has a lower picture quality than the other. SNR scalability has generally a higher bit rate over nonscalable encoders, but can have a good base picture quality and can be drift free. Hence suitable applications can be:

- transmission of video at different quality of interest, such as multiquality video, video on demand, broadcasting of TV and enhanced TV
- video over networks with a high error or packet loss rates, such as Internet, or heavily congested ATM networks.

c  *Spatial scalability*: this is the most complex form of scalability, where each layer requires a complete encoder/decoder. Such a loose dependency between the layers has the advantage that each layer is free to use any codec, with different spatio–temporal and quality resolutions. Hence there can be numerous applications for this mode, such as:

- interworking between two different standard video codecs (e.g. H.263 and MPEG-2)
- simulcasting of drift free good quality video at two spatial resolutions, such as standard TV and HDTV
- distribution of video over computer networks
- video browsing
- reception of good quality low spatial resolution pictures over mobile networks
- similar to other scalable coders, transmission of error resilience video over packet networks.

d  *Temporal scalability*: this is a moderately complex encoder, where either a single-layer coder encodes both layers, such as coding of B and the anchor I and P-pictures in MPEG-1 and 2, or two separate encoders operating at two different temporal rates. The major applications can then be:

- migration to progressive (HDTV) from the current interlaced broadcast TV
- internetworking between lower bit rate mobile and higher bit rate fixed networks
- video over LANs, Internet and ATM for computer workstations
- video over packet (Internet/ATM) networks for loss resilience.

Tables 8.4, 8.5 and 8.6 summarise a few applications of various scalability techniques that can be applied to broadcast TV. In each application, parameters of the base and enhancement layers are also shown.

*Table 8.4   Applications of SNR scalability*

| Base layer | Enhancement layer | Application |
|---|---|---|
| ITU-R-601 | same resolution and format as lower layer | two quality service for standard TV |
| high definition | same resolution and format as lower layer | two quality service for HDTV |
| 4:2:0 high definition | 4:2:2 chroma simulcast | video production/distribution |

*Table 8.5   Applications of spatial scalability*

| Base | Enhancement | Application |
|---|---|---|
| progressive (30 Hz) | progressive (30 Hz) | CIF/QCIF compatibility or scalability |
| interlace (30 Hz) | interlace (30 Hz) | HDTV/SDTV scalability |
| progressive (30 Hz) | interlace (30 Hz) | ISO/IECE11172-2/compatibility with this specification |
| interlace (30 Hz) | progressive (60 Hz) | migration to HR progressive HDTV |

*Table 8.6   Applications of temporal scalability*

| Base | Enhancement | Higher | Application |
|---|---|---|---|
| progressive (30 Hz) | progressive (30 Hz) | progressive (60 Hz) | migration to HR progressive HDTV |
| interlace (30 Hz) | interlace (30 Hz) | progressive (60 Hz) | migration to HR progressive HDTV |

## 8.6 Video broadcasting

Currently more than ninety-five percent of MPEG-2 coded video is for broadcasting applications, carried *via* terrestrial, satellite and cable TV networks to homes. In Europe, the standard ITU-R 601 video is encoded in the range of 3–8 Mbit/s, depending on the scene content of the video. The lower end of the bit rate is for head-and-shoulders type video such as the video clip of a news reader, and the higher bit rates are required for the critical scenes such as sports programs, similar to the snap shot shown in Figure 8.10. Normally, scenes with grass and tree leaves, which have detailed texture and random motion due to wind, if they appear alongside a plain scene, like a lake or stream, are the most difficult to code. Random motion of the detailed area makes motion estimation useless, and for a limited bit rate budget, increase in quantiser step size will cause blocking artefacts in the plain areas. For HDTV video, the required bit rate is of the order of 20 Mbit/s.

In both terrestrial and satellite TV, for better channel utilisation, several TV programs may be multiplexed and then digitally modulated on to a carrier. At the destination, the receiver, known as the set top box, separates the channels, decodes each program and feeds the individual analogue signals to the television set for display. Although the same multiplexing technique can be used, because digital modulation techniques for terrestrial and satellite are different, unfortunately the same set top box cannot be used for both.

For multiplexing of TV programs, the individual bit streams are first decoded and are then reencoded to a new target bit rate. For optimum multiplexing, the target bit rate for each TV channel is made dependent on the content (statistics) of each program, and hence it is called statistical multiplexing. Here, more complex video might be assigned higher bit rates and since video complexity may vary over time, then for optimum statistical multiplexing we need to monitor the video complexity continuously.

One way of calculating the complexity of a scene in a video program is to define the scene complexity as the sum of the complexity indices of its I, P and B-pictures in a group of pictures (GOP) [16]. For each picture type, the complexity index is the product of its average quantiser step size to its bit rate in that frame. For example, the scene complexity index (SCI) of a video with a GOP structure of $N = 12$ and $M = 3$, which has one I, three P and eight B-pictures is:

$$SCI = \frac{1}{12}\left[ IQ_I + \sum_{j=1}^{3} P_j Q_{Pj} + \sum_{j=1}^{8} B_j Q_{Bj} \right] \tag{8.9}$$

where $I$, $P$ and $B$ are the target bit rates for the I, P and B pictures, and $Q_I$, $Q_P$ and $Q_B$ are their respective average quantiser step sizes. After calculating SCI for each TV program, the total bit rate is divided between the TV channels in proportion to their SCI. Values of the SCI can be continuously calculated on a frame by frame basis (within a window of a GOP), to provide optimum statistical multiplexing.

One of the main attractions of digital satellite TV is the benefit of broadcasting many TV programmes from a single transponder. In the analogue era, one satellite transponder with a bandwidth of 36 MHz could accommodate only one frequency modulated (FM) TV programme, whereas currently about 6–8 digital TV programmes can be multiplexed into 27 Msymbol/s and are accommodated in the same transponder. There are even stations that squeeze about 10–15 digital TV programmes into a transponder, albeit at a slightly lower video quality. In addition to this increase of the number of TV channels, the required transmitted power for digital can be of the order of 10–20 per cent of analogue, or for the same power, the satellite dishes can be made much smaller (45–60 cm diameter dishes compared to 80 cm used in analogue). Digital terrestrial TV also benefits from the low power transmitters.

In digital terrestrial TV, normally one programme is digitally modulated into an 8 MHz (European) UHF channel [17]. The bit stream prior to channel modulation is orthogonal frequency division multiplexed (OFDM) into 1705 carriers (2000 carriers is also an option) and the channel modulation is a 64-QAM. At a higher modulation rate (e.g. 256-QAM) it is even possible to accommodate an 18–24 Mbit/s bit stream into the same 8 MHz UHF channel, thus being able to multiplex 2–3 digital programmes (or even higher for poorer quality) into one existing analogue UHF terrestrial channel.

Since at the baseband a 4–8 Mbit/s MPEG-2 video is OFDM modulated into almost 2000 carriers, then each bit of the video signals is transmitted at a rate of 2–4 kbit/s. Such a low data rate (large interval) is very robust against interference, similar to a high frequency burst of noise, and can be cleaned up easily. Thus OFDM is particularly attractive for ghost-free TV broadcasting in big cities, where multiple reflections from tall buildings can create interference (a common problem with analogue TV). Moreover, it is possible to cover the whole broadcast TV network (nationwide programmes not regional programmes) with a single frequency, since interference is not a problem. This will release a lot of wireless bandwidth for other communication services. Finally, similar to satellite, the transmitter power can be reduced by a factor of ten.

The price paid for all these benefits of digital TV is the sensitivity of the digital TV to channel errors. During heavy rain or snow, pictures become blocky or in the more severe cases, a complete loss of picture (picture freeze). This is the main disadvantage of digital TV since, in analogue TV, weaker reception may cause snowy pictures, which is better than picture break up or freeze in digital TV. To alleviate this problem, layered video coding with unequal error protection to various layers may be used. Or one may use the more intelligent technique of distributing the transmitter power among the layers, such that the picture quality gradually degrades, closer to quality degradation in analogue TV.

## 8.7 Digital versatile disc (DVD)

Digital versatile/video disc (DVD) is a new storage medium for MPEG-2 coded high quality video. DVD discs with 9 Gbytes storage capacity (in two tracks of 4.5 Gbytes) are introduced to replace the 648 (or 700) Mbytes CD-ROMs. The main reason for the introduction of this new product is that viewers' expectations of video quality have grown over time. CD-ROMs could only store MPEG-1 compressed video of SIF format at about a target rate of 1.2 Mbyte/s. When SIF pictures are enlarged to the standard size (e.g. 720 pixels by 576 lines), to be displayed on TV sets, for certain scenes the enlarged pictures look blocky. This is usually not sufficient for home movies or HDTV programmes.

In DVD, video of CCIR-601 standard size is MPEG-2 compressed. Considering the double-track DVD discs of total capacity of 9 Gbytes, the nominal movie of 90 minutes long can be coded at an average bit rate of 6–12 Mbit/s, depending whether one or both tracks are used.

To increase the video quality and at the same time to optimise the storage capacity, the MPEG-2 encoder is set to encode the video at a variable bit rate (VBR). This is done by fixing the quantiser step size at a constant value, producing video of almost constant quality over the entire programme, irrespective of scene complexity. Due to constant quantiser step size, during high picture activity, the instantaneous bit rate of the encoder can be very high (e.g. 30 Mbit/s). However, these events only occur for a short period, and there are occasions when the scenes might be very quiet, producing lower bit rates (e.g. 2 Mbit/s). Depending on the proportions of the scene activities in the video, its peak-to-mean bit rate ratio, even smoothed over a GOP can be of the order of 3–5 (peak/mean ratio smoothed over one frame can easily rise above 10). Thus had the video been coded at a constant bit rate (CBR), then for the same picture quality as VBR, the target bit rate would have to be set to the peak bit rate. Hence for quiet scenes, the storage capacity of the disc can be wasted. In fact the advantage of VBR over CBR is the saving in storage capacity by the ratio of the peak bit rate to the mean bit rate, which can be considerable.

The main problem with VBR is that the chunk of compressed data read from the disc decodes a variable number of pictures per given time unit (e.g. seconds). For a uniform and smooth display (e.g. 25 pictures per second), the read data from the disc has to be smoothed. This is done by writing it into a random access memory (RAM) and reading it at the desired rate of the decoder. Considering that today electronic notebooks are equipped with 256–512 Mbytes RAM, they are not too expensive to be included in the DVD decoders. These are sufficient to store about 5–10 minutes of the programme, well over what is needed to produce pictures without interruptions.

## 8.8 Video over ATM networks

MPEG-2 and in particular layered video coding and the asynchronous transfer mode (ATM) networks have a very strong link. They were introduced at about the same time (early 1990) and influenced each other's development. The cell loss priority in ATM is the direct product of the success of layered two-layer video coding in delivering a minimum acceptable picture quality [9]. Selection of 188-byte packet size for the MPEG-2 transport stream was influenced by the ATM cell size. An ATM cell (packet) is 53 bytes long with a five-byte header and a 48-byte payload. One of the ATM adaptation layers (AAL) (where the data to be transported is interfaced to the channel) called AAL1 accepts 47-byte raw data, and adds one-byte synchronisation and other information to the payload [15]. Hence each MPEG-2 packet can be transported with four ATM cells.

ATM is a slotted channel, where each 53-byte cell is seized by the server to insert its data for transmission. If the server has nothing to send, the cell is left empty. Hence, the source can send its data at a variable rate. Variable bit rate (VBR) video transmission is particularly attractive since compressed video is variable in rate by nature. With a constant quantiser step size, the video is coded at almost constant quality. At low picture activity (low motion and low texture) less bits are generated and at high picture activity more bits are generated. Such variable bit rate transmission makes statistical multiplexing even more effective than the one used with fixed bit rate broadcast TV. It is easy to show that more variable bit rate services can be accommodated in a given channel than the fixed bit rate services.

The main problem with VBR transmission is that, if bursts of data from various services occur at the same time, there will be more traffic than the network can handle and it will be congested. In this case, cells carrying visual information might be excessively delayed. There is a maximum tolerable delay beyond which late arrival cells will be of no use. Either the switching nodes or the receiver can discard these cells. In the former case, the cell discard is due to the limited capacity of the switching multiplex buffer, and in the latter, the received information is too late to be of any use by the decoder. In both cases, loss of cells leads to degradation in picture quality.

The cell loss priority bit in ATM cells coupled with two-layer video coding can enhance the video quality significantly. Here the base layer video is assigned high priority and the enhancement layer lower priority. In the event of network congestion low priority cells (enhancement data) can be discarded and room made available to the high priority cells (base layer). For example, even in the normal MPEG-2 with a GOP structure of $N = 12$ and $M = 3$, which can be regarded as temporal scalability, during the network congestion all the B-pictures can be temporarily discarded to make room for the I and P-pictures.

In ATM networks, in addition to layering, the packetisation strategy also plays an important role in the video quality. One form of packetisation may confine the effect of a lost packet to a small area of the picture, while other methods may spread degradation to a larger area. With AAL1 packetisation [15], where every 47 bytes of the bit stream are packed into the ATM cell payload without any further processing, if a cell is lost, the following cells may not be recoverable until the next slice or GOB. Thus a large part of a picture slice may be degraded, depending on the location of the lost macroblock. This problem can be overcome by making the first macroblock of each cell absolutely addressed, hence the loss can be confined to a smaller area of the picture [18]. Let us call this method of packing AAL$x$, as shown in Figure 8.27.

In AAL$x$, where the first macroblock in each ATM cell is absolutely addressed, the lost area could be confined to the area covered by the lost cell. All following cells could then be decodable. For the decoder to be able to recognise the absolute address, an additional 11-bit header (absolute address header) must be inserted before the address. Also, the average length of the relative addressing is normally two bits, whereas the absolute address can be nine bits long, resulting in an additional seven bits [18]. Thus AAL$x$ has an almost five per cent extra overhead compared to AAL1.

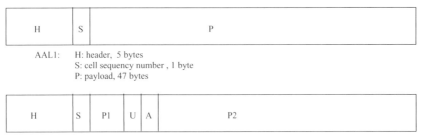

*Figure 8.27    Structure of ALL1 and AALx cells*

Referring to the multiplex cell discard graphs, this can result in five to ten times more cell loss, depending on the network load and the number of channels in the multiplex [19].

In an experiment, 90 frames of the Salesman image sequence were MPEG-2 coded with the first frame being intra (I-frame) coded and the remaining frames predictively (P-frame) coded ($N = \infty$, $M = 1$). Two types of packetisation method, AALx and AAL1 were used. The AALx-type cells were discarded with the ITU-T cell loss model with a cell loss rate of $10^{-2}$ and a mean burst length of 1 (see Appendix E) [20]. Those of AAL1 were discarded at cell loss rates of $10^{-3}$ (10 times lower) and $10^{-4}$ (100 times lower) with the same mean burst length. From Figure 8.28, it can be seen that AALx outperforms AAL1 at ten times lower cell loss rate, but is inferior to AAL1 with a cell loss rate of 100 times lower. Considering that in the experiment AALx is likely to experience five to ten times more loss than AAL1, AALx is a better packetisation scheme for this type of image format (e.g. H.261or H.263).

In another experiment the same 90 frames of the Salesman image sequence were MPEG-2 coded with a GOP structure of $N = 12$, $M = 1$. The packetisation techniques were similar to those of the previous experiment. In this case, shown in Figure 8.29, AALx does not show the same improvement over AAL1, as was the case for Figure 8.28. In fact its performance, due to higher overhead, is worse than AAL1 with ten times lower cell loss rate.

The implications of these two experiments are that with MPEG-1 and 2 structures, where there are regular I-pictures every $N$ frames, AAL1 outperforms AALx. But for very large $N$ (e.g. in H.261) AALx is better than AAL1.

It should be noted the video quality can be improved by concealing the effect of packet loss or channel errors. In MPEG-2 there is an option that additional motion vectors for I-pictures are derived and they are transmitted in the following slice. In case some of the macroblocks are damaged, these motion vectors are used to copy pixels from the previous frame, displaced by the amount of motion vectors to replace the damaged macroblocks. In the next Chapter more general forms of concealing side effects of packet losses and channel errors will be discussed in greater depth.

Coding of high quality moving pictures (MPEG-2)  223

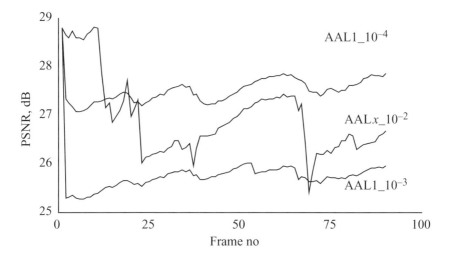

Figure 8.28  PSNR of MPEG-2 coded video sequence GOP (IPPPPPP...)
   a   AALx with error rate of $10^{-2}$
   b   AAL1 with error rate of $10^{-3}$
   c   AAL1 with error rate of $10^{-4}$

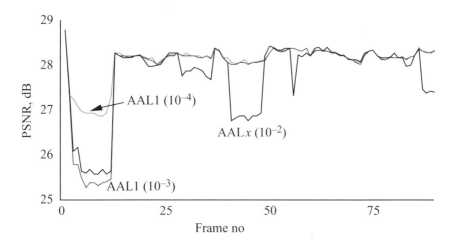

Figure 8.29  PSNR of MPEG-2 coded video sequence with 12 frames per GOP (IPP...IPPP...IP...)
   a   AALx with error rate of $10^{-2}$
   b   AAL1 with error rate of $10^{-3}$
   c   AAL1 with error rate of $10^{-4}$

## 8.9 Problems

1 Why are the systems in MPEG-2 different from those in MPEG-1?

2 Which of the following represents level and profile?

   a  1.5 Mbit/s
   b  SIF
   c  SNR scalability
   d  720 × 576 pixels.

3 The DCT coefficients of a motion compensated picture block are given as:

   | 33  | −10 | −41 | 3  | 17  | 2  | 7  | −13 |
   |-----|-----|-----|----|-----|----|----|-----|
   | 61  | −5  | 23  | 12 | −11 | 5  | 6  | −9  |
   | −3  | 11  | 3   | 9  | −15 | 6  | 3  | −1  |
   | 2   | −34 | 6   | 4  | 0   | 1  | 3  | 1   |
   | −21 | −3  | 0   | 5  | 12  | 3  | 0  | 1   |
   | −7  | −5  | 9   | 3  | 2   | 7  | −1 | −2  |
   | 6   | 3   | 2   | 5  | 7   | −2 | −3 | 1   |
   | −5  | 4   | −2  | 6  | 3   | 1  | 2  | 1   |

   They are linearly quantised with $th = q$, zigzag scanned and the assigned bits are calculated from Figure 6.12. For $q = 8$, identify the two-dimensional events of (run, index) and the number of bits required to code the block.

4 The block of problem 3 is partitioned into two, and the priority break point (PBP) is set at coefficient $(2, 2)$. Assuming that PBP can be identified with six bits, and the quantiser step size is $q = 8$, calculate the number of bits generated in each layer and the total number of bits. (Note the first DCT coefficient is defined at $(0, 0)$.)

5 The block in problem 3 is SNR scalable coded with the base and enhancement quantiser step sizes of 14 and 8, respectively. What are the number of generated bits in each layer, and the total number of bits (assume in each layer $th = q$)?

6 An MPEG-2 coded video with its associated audio and forward error correcting codes comprises 8 Mbit/s. With a 64-QAM modulation, determine how many such videos can be accommodated in a UHF channel of 8 MHz bandwidth, with 2 MHz guard band. Assume each modulated symbol occupies 1.25 Hz of the channel.

7 Draw a two-state channel error model and determine the transition probabilities for each of the following conditions:

   a  bit error rate of $P = 10^{-5}$ and burst length of $B = 5$
   b  bit error rate of $P = 10^{-5}$ and burst length of $B = 1$.

8 Table 8.7 shows the duration of various parts of a 90 minute VBR-MPEG-2 coded video stored on a DVD. The given bit rate is smoothed over a GOP, but is

presented in Mbit/s:

a calculate the required storage capacity
b calculate the peak-to-mean bit rate ratio
c calculate the storage required if the video was coded in CBR at a quality not poorer than the VBR.

Table 8.7 Duration of various picture activity in a DVD program

| Duration [min] | 0.5 | 5 | 10 | 20 | 30 | 24.5 |
|---|---|---|---|---|---|---|
| Bit rate [Mb/s] | 20 | 15 | 10 | 7.5 | 5 | 4 |

9 The ATM cells with AAL1 adaptation layer have a five-byte header and 48-byte payload, of which 47 bytes are used for packing the video data. If the channel bit error rate is $10^{-7}$, calculate the probability that:

a video is decoded erroneously
b the cell is lost.

10 The stored DVD video in problem 8 is to be streamed *via* an ATM network, with a maximum channel capacity of 50 Mbit/s. Due to the other users on the link, on the average only 30 per cent of the link capacity can be used by the DVD server. With an AAL1 packetisation, calculate the time required to download the entire DVD video stream over the link.

11 The cell loss rate of an ATM link can be modelled with $P = 10^{-10(1-\rho^2)}$, where $0 \le \rho \le 1$ is the load of the link. Twenty-five video sources, each coded at an average bit rate of 4 Mbit/s, are streamed *via* a 155 Mbit/s ATM link, with AAL1 adaptation layer. Calculate:

a the network load, $\rho$
b the loss rate that each ATM cell may experience.

12 The video sources in problem 11 were two-layer coded with SNR scalability, but the overall video quality was assumed to remain the same. If in each source 50 per cent of its data is assigned to the base layer, and the base layer cells are always served in preference to the enhancement layers cells, calculate the cell loss probability at the:

a base layer
b enhancement layer

(hint, use Figure 8.25 for the additional overhead due to scalability).

13 Repeat problem 12 for data partitioning.

14 Repeat problem 12 for spatial scalability.

## 8.10 References

1 MPEG-2: 'Generic coding of moving pictures and associated audio information'. ISO/IEC 13818–2 Video, draft international standard, November 1994
2 MPEG-1: 'Coding of moving pictures and associated audio for digital storage media at up to about 1.5 Mbit/s'. ISO/IEC 1117-2: video, November 1991
3 HASKEL, B.G., PURI, A., and NETRAVALI, A.N.: 'Digital video: an introduction to MPEG-2' (Chapman and Hall, 1997)
4 'Generic coding of moving pictures and associated audio information', ISO/IEC 13818-1 Systems, draft international standard, November 1994
5 ITU-T recommendation I.363: 'B-ISDN ATM adaptation layer (AAL) specification'. June 1992
6 OKUBA, S., McCANN, K., and LIPPMAN, A.: 'MPEG-2 requirements, profile and performance verification', *Signal Process., Image Commun.*, 1995, **7**:3 pp.201–209
7 SAVATIER, T.: 'Difference between MPEG-1 and MPEG-2 video'. ISO/IEC JTC1/SC29/WG11 MPEG94/37, March 1994
8 Test model editing committee: 'MPEG-2 video test model 5'. ISO/IEC JTC1/SC29/WG11 doc. N0400, April 1993
9 GHANBARI, M.: 'Two-layer coding of video signals for VBR networks', *IEEE J. Sel. Areas Commun.*, 1989, **7**:5, pp.771–781
10 GHANBARI, M.: 'An adapted H.261 two-layer video codec for ATM networks', *IEEE Trans. Commun.*, 1992, **40**:9, pp.1481–1490
11 GHANBARI, M., and SEFERIDIS, V.: 'Efficient H.261 based two-layer video codecs for ATM networks', *IEEE Trans. Circuits Syst. Video Technol.*, 1995, **5**:2, pp.171–175
12 ITU-T study group XVI: 'Efficient coding of synchronised H.26L streams'. Document VCG-N35, September 2001
13 HERPEL, C.: 'SNR scalability vs data partitioning for high error rate channels'. ISO/IEC JTC1/SC29/WG11 doc. MPEG 93/658, July 1993
14 MORRISON, G., and PARKE, I.: 'A spatially layered hierarchical approach to video coding', *Signal Process., Image Commun.*, 1995, **5:5–6**, pp.445–462
15 ITU-T draft recommendation I.371: 'Traffic control and congestion control in B-ISDN'. Geneva, 1992
16 ROSDIANA, E., and GHANBARI, M.: 'Picture complexity based rate allocation algorithm for transcoded video over ABR networks', *Electron. Lett.*, 2000, **36**:6, pp.521–522
17 dvb blue book, ftp://dvbftp@ftp.dvb.org/Blue_Books/
18 GHANBARI, M., and HUGHES, C.J.: 'Packing coded video signals into ATM cells', *IEEE ACM Trans. Networking*, 1993, **1**:5, pp.505–509
19 HUGHES, C.J., GHANBARI, M., PEARSON, D.E., SEFERIDIS, V., and XIONG, J.: 'Modelling and subjective assessment of cell discard in ATM Video', *IEEE Trans. Image Process.*, 1993, **2**:2, pp.212–222
20 ITU SGXV working party XV/I, Experts Group for ATM video coding, working document AVC-205, January 1992

## Chapter 9
# Video coding for low bit rate communications (H.263)

The H.263 Recommendation specifies a coded representation that can be used for compressing the moving picture components of audio-visual services at low bit rates. Detailed specifications of the first generation of this codec under the test model (TM) to verify the performance and compliance of this codec were finalised in 1995 [1]. The basic configuration of the video source algorithm in this codec is based on ITU-T Recommendation H.261, which is a hybrid of interpicture prediction to utilise temporal redundancy and transform coding of the residual signal to reduce spatial redundancy. However, during the course of the development of H.261 and the subsequent advances on video coding in MPEG-1 and MPEG-2 video codecs, substantial experience was gained, which has been exploited to make H.263 an efficient encoder [2–4]. In this Chapter those parts of the H.263 standard that make this codec more efficient than its predecessors will be explained.

It should be noted that the primary goal in the H.263 standard codec was coding of video at low or very low bit rates for applications such as mobile networks, the public switched telephone network (PSTN) and narrowband ISDN. This goal could only be achieved with small image sizes such as subQCIF and QCIF, at low frame rates. Today, this codec has been found so attractive that higher resolution pictures can also be coded at relatively low bit rates. The current standard recommends operation on five standard pictures of the CIF family, known as subQCIF, QCIF, CIF, 4CIF and 16CIF.

Soon after the finalisation of the H.263 in 1995, work began to improve the coding performance of this codec further. The H.263+ was the first set of extensions to this family, which was intended for near-term standardisation of enhancements of H.263 video coding algorithms for real-time telecommunications [5]. Work on improving the encoding performance is still an ongoing process under H.263++ and every now and then a new extension called an annex is added to the family [6]. The codec for long-term standardisation is called H.26L [7]. The H.26L project has the mandate from ITU-T to develop a very low bit rate (less than 64 kbit/s with emphasis on less than

24 kbit/s) video coding recommendations achieving better video quality, lower delay, lower complexity and better error resilience than are currently available. The project also has an objective to work closely with the MPEG-4 committee in investigating new video coding techniques and technologies as candidates for recommendation [8].

## 9.1 How does H.263 differ from H.261 and MPEG-1?

The source encoder of H.263 follows the general structure of the generic DCT-based interframe coding technique used in the H.261 and MPEG-1 codecs (see Figure 3.18). The core H.263 employs a hybrid interpicture prediction to utilise temporal redundancy and transform coding of the residual signal to reduce spatial redundancy. The decoder has motion compensation capability, allowing optional incorporation of this technique at the encoder. Half pixel precision is used for the motion compensation, as opposed to the optional full pixel precision and loop filter used in Recommendation H.261. In the new versions of H.263, use of quarter and even one eighth of pixel precision are recommended [6].

Perhaps the most significant differences between the core H.263 and H.261/MPEG-1 are in the coding of the transform coefficients and motion vectors. In the following sections these and some other notable differences such as the additional optional modes are explained.

### 9.1.1 Coding of H.263 coefficients

In H.261 and MPEG-1, we saw that the transform coefficients are converted *via* a zigzag scanning process into two-dimensional, run and index events (see section 6.4). In H.263 these coefficients are represented as a three-dimensional event of (last, run, level). Similar to the two-dimensional event, the run indicates the number of zero-valued coefficients preceding a nonzero coefficient in the zigzag scan, and level is the normalised magnitude of the nonzero coefficient which is sometimes called index. last is a new variable to replace the end of block (EOB) code of H.261 and MPEG-1. last takes only two values, 0 and 1. last 0 means that there are more nonzero coefficients in the block, and 1 means that this is the last nonzero coefficient in the block.

The most likely events of (last, run, level) are then variable length coded. The remaining combinations of (last, run, level) are coded with a fixed 22-bit word consisting of seven bits escape, one bit last, six bits run and eight bits level.

### 9.1.2 Coding of motion vectors

The motion compensation in the core H.263 is based on one motion vector per macroblock of 16 × 16 pixels, with half pixel precision. The macroblock motion vector is then differentially coded with predictions taken from three surrounding macroblocks, as indicated in Figure 9.1. The predictors are calculated separately for the horizontal

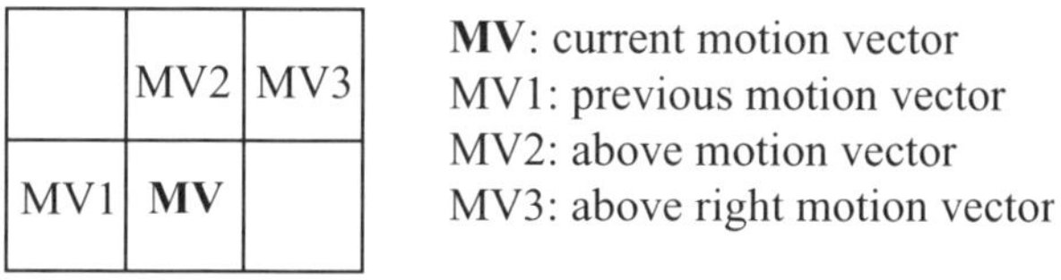

*Video coding for low bit rate communications (H.263)* 229

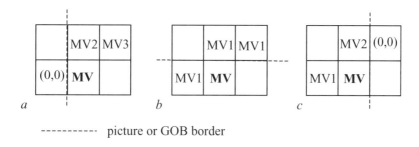

*Figure 9.1  Motion vector prediction*

*Figure 9.2  Motion vector prediction for the border macroblocks*

and vertical components of the motion vectors MV1, MV2 and MV3. For each component, the predictor is the median[1] value of the three candidate predictors for this component:

$$\text{pred}_x = median(MV1_x, MV2_x, MV3_x)$$
$$\text{pred}_y = median(MV1_y, MV2_y, MV3_y) \tag{9.1}$$

The difference between the components of the current motion vector and their predictions are variable length coded. The vector differences are defined by:

$$MVD_x = MV_x - \text{pred}_x$$
$$MVD_y = MV_y - \text{pred}_y \tag{9.2}$$

In the special cases at the borders of the current group of blocks (GOB) or picture, the following decision rules are applied in order as follows:

1. The candidate predictor MV1 is set to zero if the corresponding macroblock is outside the picture at the left side (Figure 9.2a).
2. The candidate predictors MV2 and MV3 are set to MV1 if the corresponding macroblocks are outside the picture at the top, or if the GOB header of the current GOB is nonempty (Figure 9.2b).

---

[1] To find the median value, the components are rank ordered and the middle value is chosen.

3   The candidate predictor MV3 is set to zero if the corresponding macroblock is outside the picture at the right side (Figure 9.2c).
4   When the corresponding macroblock is coded intra or is not coded, the candidate predictor is set to zero.

The values of the difference components are limited to the range [−16 to 15.5]. Since, in H.263, the source images are of the CIF family with the 4 : 2 : 0 format, each macroblock comprises four luminance and two chrominance components, $C_b$ and $C_r$. Hence the motion vector of the macroblock is used for all four luminance blocks in the macroblock. Motion vectors for both chrominance blocks are derived by dividing the component values of the macroblock vector by two, due to the lower chrominance resolution. The resulting values of the quarter pixel resolution vectors are modified towards the nearest half pixel position (note: the macroblock motion vector has half pixel resolution).

### 9.1.3  Source pictures

The source encoder operates on noninterlaced pictures at approximately 29.97 frames per second. These pictures can be one of the five standard picture formats of the CIF family: subQCIF, QCIF, CIF, 4CIF and 16CIF. Since in CIF the luminance and chrominance sampling format is 4 : 2 : 0, then for either of these pictures, the horizontal and vertical resolutions of the chrominance components are half the luminance. Table 9.1 summarises pixel resolutions of the CIF family used in H.263.

Each picture is divided into a group of blocks (GOBs). A GOB comprises $k \times 16$ lines, depending on the picture format ($k = 1$ for subQCIF, QCIF and CIF; $k = 2$ for 4CIF; $k = 4$ for 16CIF). The number of GOBs per picture is six for subQCIF, nine for QCIF, and 18 for CIF, 4CIF and 16CIF. Each GOB is divided into 16 × 16 pixel macroblocks, of which there are four luminance and one each of chrominance blocks of 8 × 8 pixels.

Table 9.1   Number of pixels per line and number of lines per picture for each of the H.263 picture formats

| Picture format | Number of pixels for luminance per line | Number of lines for luminance per picture | Number of pixels for chrominance per line | Number of lines for chrominance per picture |
|---|---|---|---|---|
| SubQCIF | 128 | 96 | 64 | 48 |
| QCIF | 176 | 144 | 88 | 72 |
| CIF | 352 | 288 | 176 | 144 |
| 4CIF | 704 | 576 | 352 | 288 |
| 16CIF | 1408 | 1152 | 704 | 576 |

### 9.1.4 Picture layer

The picture layer contains the picture header, the GOB header together with various coding decisions on macroblocks in a GOB and finally the coded transform coefficients, which are also used in H.261 and MPEG-1 and 2. The most notable difference in the header information for H.263 is in the type information, called PTYPE. For the first generation of H.263, this is a 13-bit code that gives information about the complete picture, in the form of [1]:

bit 1     always 1, in order to avoid start code emulation
bit 2     always 0, for distinction with H.261
bit 3     split screen indicator, 0 off, 1 on
bit 4     document camera indicator, 0 off, 1 on
bit 5     freeze picture release, 0 off, 1 on
bit 6–8     source format, 000 forbidden, 001 subQCIF, 010 QCIF, 011 CIF, 100 4CIF, 101 16CIF, 110 reserved, 111 extended PTYPE
bit 9     picture coding type, 0 intra, 1 inter
bit 10     optional unrestricted motion vector mode, 0 off, 1 on
bit 11     optional syntax-based arithmetic coding mode, 0 off, 1 on
bit 12     optional advanced prediction mode, 0 off, 1 on
bit 13     optional PB frame mode, 0 normal picture, 1 PB frame

The split screen indicator is a signal that indicates that the upper and lower half of the decoded picture could be displayed side by side. This has no direct effect on the encoding and decoding of the picture.

The freeze picture release is a signal from an encoder which responds to a request for packet retransmission (if not acknowledged) or fast update request, and allows a decoder to exit from its freeze picture mode and display decoded picture in the normal manner.

Bits 10–13 refer to the early four optional modes of H.263. Since 1995 more options as annexes have then been added to the extensions of this codec. These optional modes are activated when bits 6–8 of the PTYPE header are in the extended mode of 111, and necessarily some additional bits define the new options. Hence extensions of H.263 have a longer PTYPE header and also a different picture layer than the above 13 bits. All these optional modes are only used after negotiation between the encoder and the decoder *via* the control protocol Recommendation H.245 [9].

Also for further reduction in the overhead, the code for macroblock type and coded block pattern are combined. For example, the combined code of macroblock type and coded block pattern is called MCBPC. MCBPC is always present for each macroblock, irrespective of its type and the options used. Note that in H.261, the MPEG-1 and 2 code block pattern is defined separately from the macroblock type.

## 9.2 Switched multipoint

One of the initial aims in the design of a new low bit rate video codec was to replace H.261 with a more efficient one, that took the name of H.263. Hence the functionalities of H.261, but with some improvements, need to be included in this codec; they appear in Annex C of the H.263 specification that can be activated or disabled as desired [25-C]. Since this annex was introduced prior to all the other annexes, which later on were called optionalities, we introduce this annex prior to all the other options.

In H.263 the decoder can be instructed to alter its normal decoding mode and provide some extra display functions. Instructions for the alterations may be issued by an external device such as Recommendation H.245, which is a control protocol for multimedia communications [9]. Some of the commands and the actions are as follows.

### 9.2.1 Freeze picture request

This signal causes the decoder to freeze its displayed picture until a freeze release signal is received or a time-out period of at least six seconds has expired. A frozen picture is much better perceived by the viewer than, say, a broken picture due to channel errors, or if the encoder cannot deliver the compressed bit stream on time for a continuous display.

### 9.2.2 Fast update request

This command causes the encoder to encode its next picture in intra mode, with coding parameters to avoid buffer overflow. This mode in conjunction with the back channel reduces the probability of error propagation into the subsequent pictures. This mode improves the resilience of the codec to channel errors.

### 9.2.3 Freeze picture release

Freeze picture release is a signal from the encoder, which has responded to a fast update request, and allows a decoder to exit from its freeze mode and display decoded pictures in the normal manner. This signal is transmitted by the PTYPE in the picture header of the first picture coded in response to the fast update request.

### 9.2.4 Continuous presence multipoint

In a multipoint connection, a multipoint control unit (MCU) can assemble two to four video bit streams into one video bit stream, so that at the receiver up to four different video signals can be displayed simultaneously. In H.261, this can be done on a quad screen by only editing the GOB header, but in H.263 it is more complex due to a different GOB structure, overlap motion estimation, multiple motion vectors etc. Therefore, in H.263 a special continuous presence multipoint mode is provided in which four independent video bit streams are transmitted in the four logical channels of a single H.263 video bit stream.

## 9.3 Extensions of H.263

In late 1990s the Video Coding Experts Group (VCEG) of the ITU-Telecommunications standardisation sector set up two activities. The aim was to develop very low bit rate video coding at bit rates less than 64 kbit/s and more specifically at less than 24 kbit/s. One activity is looking at the video coding for very low bit rates, under the name of H.263+ [5]. Recently work of this activity has continued under the name of H.263++, indicating further improvements on H.263+ [6]. The other activity, which has more in common with MPEG-4, is work on advanced low bit rate video coding, under the name of H.26L [7].

The H.263+/H.263++ development effort is intended for short term standardisation of enhancements of the H.263 video coding algorithm for real-time telecommunication and related nonconversational services. The H.26L development effort is aimed at identifying new video coding technology beyond the capabilities of enhancements to H.263 by the H.263+/H.263++ coding algorithms.

These two subgroups also have a close cooperation in the development of their codecs, since the core codec is still H.263. They also work closely with the other bodies of ITU. For example, the collaboration between H.263+ and the mobile group has led to consideration for greater video error resilience capability. The back channel error resilience in H.263+ is especially designed to address the needs of mobile video and other such unreliable bit stream transport environments. The H.26L group work very closely with the MPEG-4 group, as this group has the mandate of developing advanced video coding for storage and broadcasting applications [7].

One of the key features of the H.263+, H.263++ and H.26L is the real-time audio-visual conversational services. In a real-time application, information is simultaneously acquired, processed and transmitted and is usually used immediately at the receiver. This feature implies critical delay and complexity constraints on the codec algorithm.

An important component in any application is the transmission media over which it will need to operate. The transmission media for H.263+/H.26L applications include PSTN, ISDN (1B), dial-up switched-56/64 kbit/s service, LANs, mobile networks (including GSM, DECT, UMTS, FLMPTS, NADC, PCS etc.), microwave and satellite networks, digital storage media (i.e. for immediate recording) and concatenation of the above media. Due to the large number of likely transmission media and the wide variations in the media error and channel characteristics, error resiliency and recovery are critical requirements for this application class.

### 9.3.1 Scope and goals of H.263+

The expected enhancements of H.263+ over H.263 fall into two basic categories:

- enhancing quality within existing applications
- broadening the current range of applications.

A few examples of the enhancements are:

- improving perceptual compression efficiency
- reducing video coding delay
- providing greater resilience to bit errors and data losses.

Note that H.263+ has all the features of H.263, and further tools are added to this codec to increase its coding efficiency and its robustness to errors. This is an ongoing process and more tools are added every year. Recently this codec has been designated H.263++, to emphasise the ongoing improvement in the coding efficiency [6].

### 9.3.2 Scopes and goals of H.26L

The long-term objective of the ITU-U video experts group, under the Advanced Video Coding project, is to provide a video coding recommendation which at very low bit rates can perform substantially better than that achievable with the existing standards (e.g. H.263+). The adopted technology should provide for:

- enhanced visual quality at very low bit rates and particularly at PSTN rates (e.g. at rates below 24 kbit/s)
- enhanced error robustness in order to accommodate the higher error rates experienced when operating for example over mobile links
- low complexity appropriate for small, relatively inexpensive, audio-visual terminals
- low end-to-end delay as required in bidirectional personal communications.

In addition, the group is closely working with the MPEG-4 experts group, to include new coding methods and promote interoperability. Advances in this direction will be discussed in detail at the end of this Chapter.

### 9.3.3 Optional modes of H.263

In the course of development of H.263 numerous optional modes have been added as annexes to the main specifications to improve the visual communication efficiency of this codec. Some of the annexes were introduced along with the introduction of the core H.263, and many more were added gradually under H.263+ and H.263++. It is not the intention to introduce here all these annexes, nor to specify which annex belongs to which generation of the codec. Instead, we try to classify them into groups, without specifying in which generation they were introduced, rather to give a better appreciation of these optional modes in improving coding efficiency. As we will see, since the H.263 video codec is primarily aimed for mobile video communications, and UHF channels are particularly prone to channel errors, the majority of the annexes deal with the protection of visual data against channel errors.

## 9.4 Advanced motion estimation/compensation

Motion estimation/compensation is probably the most evolutionary coding tool in the history of video coding. For every previous generation of video codecs, motion estimation has been considered as a means of improving coding efficiency. In the first video codec (H.120) under COST211, which was a DPCM-based codec, working on pixel-by-pixel, motion estimation for each pixel would have been costly and hence it was never used. Motion estimation was made optional for the H.261 block-based codec, on the grounds that the DCT of this codec is to decorrelate interframe pixels, and since motion compensation reduces this correlation, then nothing is left for DCT! Motion compensation in MPEG-1 was considered seriously, since for B-pictures, which refer to both past and future, the motion of objects even those hidden in the background can be compensated. It was so efficient that it was also recommended for P-pictures, and even with a half pixel precision. In MPEG-2, due to interlacing, a larger variety of motion estimation/compensation between fields and frames, or their combinations, was introduced. The improvement in coding efficiency was at the cost of additional overhead for delivering the motion vectors to the receiver. However, for MPEG-1 and 2, coding at a rate of 1–5 Mbit/s, this overhead is negligible, but the question is how this overhead can be justified for H.263 at a rate of 24 kbit/s or less? In fact, as we will see below, some extensions of H.263 recommend using smaller block sizes, which imply more motion vectors per picture and they even suggest motion estimation precision should be at a quarter or one eighth of a pixel. All these increase the motion vector overhead, which is very significant at a very low bit rate of 24 kbit/s.

The fact is that, if motion compensation is efficient, the motion compensated pictures may not need to be coded by the DCT. That is, these blocks are only represented by their motion vectors. In H.261 and MPEG1, we have seen a form of macroblock that was coded only by the motion vector, without coding the motion compensated error. However, if the expected video quality is low and the motion estimation is efficient, then we can see more of these macroblocks in a codec. This is in fact what is happening with motion compensation in H.263. In the following sections some improvements to motion estimation/compensation, in addition to those introduced for H.261, MPEG-1 and 2, will be discussed. At the end of this section a form of motion estimation that warps the picture for better compensation of complex motion will be introduced. Although this method is not a part of any form of H.263 nor is recommended for other video coding standards, there is no reason why we cannot have this form of motion estimation and compensation in the future video codecs.

### 9.4.1   Unrestricted motion vector

In the default prediction mode of H.263, motion vectors are restricted so that all pixels referenced by them are within the coded picture area. In the optional unrestricted motion vector mode this restriction is removed and therefore motion vectors are allowed to point outside the picture [25-D]. When a pixel referenced by a motion vector is outside of the coded picture area, an edge pixel is used instead. This edge pixel is found by limiting the motion vector to the last full pixel position inside the coded picture area. Limitation of the motion vector is performed on a pixel-by-pixel basis and separately for each component of the motion vector.

### 9.4.2   Advanced prediction

The optional advanced prediction mode of H.263 employs overlapped block matching motion compensation and may have four motion vectors per macroblock [25-F]. The use of this mode is indicated in the macroblock-type header. This mode is only used in combination with the unrestricted motion vector mode [25-D], described above.

### 9.4.2.1 Four motion vectors per macroblock

In H.263, one motion vector per macroblock is used except in the advanced prediction mode, where either one or four motion vectors per macroblock are employed. In this mode, the motion vectors are defined for each $8 \times 8$ pixel block. If only one motion vector for a certain macroblock is transmitted, this is represented as four vectors with the same value. When there are four motion vectors, the information for the first motion vector is transmitted as the codeword MVD (motion vector data), and the information for the three additional vectors in the macroblock is transmitted as the codeword $MVD_{2-4}$.

The vectors are obtained by adding predictors to the vector differences indicated by MVD and $MVD_{2-4}$, as was the case when only one motion vector per macroblock was present (see section 9.1.2). Again, the predictors are calculated separately for the horizontal and vertical components. However, the candidate predictors MV1, MV2 and MV3 are redefined as indicated in Figure 9.3.

As Figure 9.3 shows, the neighbouring $8 \times 8$ blocks that form the candidates for the prediction of the motion vector MV take different forms, depending on the position of the block in the macroblock. Note, if only one motion vector in the neighbouring macroblocks is used, then MV1, MV2 and MV3 are defined as $8 \times 8$ block motion vectors, which possess the same motion vector as the macroblock.

### 9.4.2.2 Overlapped motion compensation

Overlapped motion compensation is only used for $8 \times 8$ luminance blocks. Each pixel in an $8 \times 8$ luminance prediction block is the weighted sum of three prediction values, divided by eight (with rounding). To obtain the prediction values, three motion vectors are used. They are the motion vector of the current luminance block and two

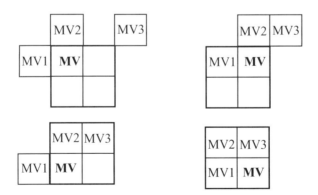

*Figure 9.3  Redefinition of the candidate predictors MV1, MV2 and MV3 for each luminance block in a macroblock*

| 2 | 2 | 2 | 2 | 2 | 2 | 2 | 2 | bottom of the |
|---|---|---|---|---|---|---|---|---|
| 1 | 2 | 2 | 2 | 2 | 2 | 2 | 1 | current block |
| 1 | 1 | 1 | 1 | 1 | 1 | 1 | 1 | |
| 1 | 1 | 1 | 1 | 1 | 1 | 1 | 1 | |
| 1 | 1 | 1 | 1 | 1 | 1 | 1 | 1 | |
| 1 | 1 | 1 | 1 | 1 | 1 | 1 | 1 | top of the |
| 1 | 2 | 2 | 2 | 2 | 2 | 2 | 1 | current block |
| 2 | 2 | 2 | 2 | 2 | 2 | 2 | 2 | |

*Figure 9.4* Weighting values for prediction with motion vectors of the luminance blocks on top or bottom of the current luminance block, $H_1(i, j)$

out of four remote vectors, as follows:

- the motion vector of the block at the left or right side of the current luminance block
- the motion vector of the block above or below the current luminance block.

The remote motion vectors from other groups of blocks (GOBs) are treated the same way as the remote motion vectors inside the GOB.

For each pixel, the remote motion vectors of the block at the two nearest block borders are used. This means that for the upper half of the block the motion vector corresponding to the block above the current block is used, and for the lower half of the block the motion vector corresponding to the block below the current block is used, as shown in Figure 9.4. In this Figure, the neighbouring pixels closer to the pixels in the current block take greater weights.

Similarly, for the left half of the block, the motion vector corresponding to the block at the left side of the current block is used, and for the right half of the block the motion vector corresponding to the block at the right side of the current block is used, as shown in Figure 9.5.

The creation of each interpolated (overlapped) pixel, $p(i, j)$, in an $8 \times 8$ reference luminance block is governed by:

$$p(i, j) = [q(i, j) \times H_0(i, j) + r(i, j) \times H_1(i, j) \\ + s(i, j) \times H_2(i, j) + 4]//8 \quad (9.3)$$

where $q(i, j), r(i, j)$ and $s(i, j)$ are the motion compensated pixels from the reference picture with the three motion vectors defined by:

$$q(i, j) = p(i + MV_x^0, j + MV_y^0)$$
$$r(i, j) = p(i + MV_x^1, j + MV_y^1)$$
$$s(i, j) = p(i + MV_x^2, j + MV_y^2)$$

|   |   |   |   |   |   |   |   |
|---|---|---|---|---|---|---|---|
| 2 | 1 | 1 | 1 | 1 | 1 | 1 | 2 |
| 2 | 2 | 1 | 1 | 1 | 1 | 2 | 2 |
| 2 | 2 | 1 | 1 | 1 | 1 | 2 | 2 |
| 2 | 2 | 1 | 1 | 1 | 1 | 2 | 2 |
| 2 | 2 | 1 | 1 | 1 | 1 | 2 | 2 |
| 2 | 2 | 1 | 1 | 1 | 1 | 2 | 2 |
| 2 | 2 | 1 | 1 | 1 | 1 | 1 | 2 |
| 2 | 1 | 1 | 1 | 1 | 1 | 1 | 2 |

right of the current block ← → left of the current block

*Figure 9.5* Weighting values for prediction with motion vectors of luminance blocks to the left or right of current luminance block, $H_2(i, j)$

|   |   |   |   |   |   |   |   |
|---|---|---|---|---|---|---|---|
| 4 | 5 | 5 | 5 | 5 | 5 | 5 | 4 |
| 5 | 5 | 5 | 5 | 5 | 5 | 5 | 5 |
| 5 | 5 | 6 | 6 | 6 | 6 | 5 | 5 |
| 5 | 5 | 6 | 6 | 6 | 6 | 5 | 5 |
| 5 | 5 | 6 | 6 | 6 | 6 | 5 | 5 |
| 5 | 5 | 6 | 6 | 6 | 6 | 5 | 5 |
| 5 | 5 | 5 | 5 | 5 | 5 | 5 | 5 |
| 4 | 5 | 5 | 5 | 5 | 5 | 5 | 4 |

*Figure 9.6* Weighting values for prediction with motion vector of current block, $H_0(i, j)$

where $(MV_x^0, MV_y^0)$ denotes the motion vector for the current block, $(MV_x^1, MV_y^1)$ denotes the motion vector of the block either above or below and $(MV_x^2, MV_y^2)$ denotes the motion vector of the block either to the left or right of the current block. The matrices $H_0(i, j)$, $H_1(i, j)$ and $H_2(i, j)$ are the current, top–bottom and left–right weighting matrices, respectively. Weighting matrices of $H_1(i, j)$ and $H_2(i, j)$ are shown in Figures 9.4 and 9.5, respectively, and the weighting matrix for prediction with the motion vector of the current block, $H_0(i, j)$, is shown in Figure 9.6.

If one of the surrounding blocks is not coded or is in intra mode, the corresponding remote motion vector is set to zero. However, in PB frames mode (see section 9.5), a candidate motion vector predictor is not set to zero if the corresponding macro block is intra mode.

If the current block is at the border of the picture and therefore a surrounding block is not present, the corresponding remote motion vector is replaced by the current motion vector. In addition, if the current block is at the bottom of the macroblock, the remote motion vector corresponding with an $8 \times 8$ luminance block in the macroblock below the current macroblock is replaced by the motion vector for the current block.

### 9.4.3 Importance of motion estimation

In order to demonstrate the importance of motion compensation and to some extent the compression superiority of H.263 over H.261 and MPEG-1, in an experiment the CIF test image sequence Claire was coded at 256 kbit/s (30 frames/s) with the following encoders:

- H.261
- MPEG-1, with a GOP length of 12 frames and two B-frames between the anchor pictures, i.e. $N = 12$ and $M = 3$ (MPEG-GOP)
- MPEG-1, with only P-pictures, i.e. $N = \infty$ and $M = 1$ (MPEG-IPPPP...)
- H.263 with advanced mode (H.263-ADV).

Figure 9.7 illustrates the peak-to-peak signal-to-noise ratio (PSNR) of the coded sequence. At this bit rate, the worst performance is that of MPEG-1, with a GOP structure of 12 frames per GOB, and two B-frames between the anchor pictures, (IBBPBBPBBPBBIBB...). The main reason for the poor performance of this codec at this bit rate is that I-pictures consume most of the bits and, compared with the other coding modes, relatively lower bits are assigned to the P and B-pictures.

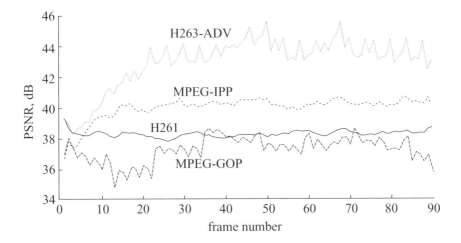

*Figure 9.7* PSNR of Claire sequence coded at 256 kbit/s, with MPEG-1, H.261 and H.263

The second poorest is the H.261, where all the consecutive pictures are interframe coded with an integer pixel precision motion compensation. The second best performance is the MPEG-1 with only P-pictures. It is interesting to note that this mode is similar to H.261 (every frame is predictively coded), except that motion compensation is carried out with half pixel precision. Hence this mode shows the advantage of using half pixel precision motion estimation. The amount of improvement for the used sequence at 256 kbit/s is almost 2 dB.

Finally, the best performance comes from the advanced mode of H.263, which results in an almost 4 dB improvement over the best of MPEG-1 and 6 dB over H.261. The following are some of the factors that may have contributed to such a good performance:

- motion compensation on smaller block sizes of 8 × 8 pixels results in smaller error signals than for the macroblock compensation used in the other codecs
- overlapped motion compensation; by removing the blocking artefacts on the block boundaries, the prediction picture has a better quality, so reducing the error signal, and hence the number of significant DCT coefficients
- efficient coding of DCT coefficients through three-dimensional (last, run, level)
- efficient representation of the combined macroblock type and block patterns.

Note that, in this experiment, other options such as PB frames mode, arithmetic coding etc. were not used. Had the arithmetic coding been used, it is expected that the picture quality would be further improved by 1–2 dB. Experimental results have confirmed that arithmetic coding has approximately 5–10 per cent better compression efficiency over the Huffman [10].

### 9.4.4 Deblocking filter

At very low bit rates, the blocks of pixels are mainly made of low frequency DCT coefficients. In these areas, when there is a significant difference between the DC levels of the adjacent blocks, they appear as block borders. At the extreme case pictures break into blocks, and the blocking artefacts can be very annoying.

The overlapped block matching motion compensation to some extent reduces these blocking artefacts. For further reduction in the blockiness, the H.263 specification recommends deblocking of the picture through the block edge filter [25-J]. The filtering is performed on 8 × 8 block edges and assumes that 8 × 8 DCT is used and the motion vectors may have either 8 × 8 or 16 × 16 resolution. Filtering is equally applied to both luminance and chrominance data and no filtering is permitted on the frame and slice edges.

Consider four pixels A, B, C and D on a line (horizontal or vertical) of the reconstructed picture, where A and B belong to block 1 and C and D belong to a neighbouring, block 2, which is either to the right of or below block 1, as shown in Figure 9.8.

In order to turn the filter on for a particular edge, either block 1 or block 2 should be an intra or a coded macroblock with the code COD = 0. In this case $B_1$ and $C_1$

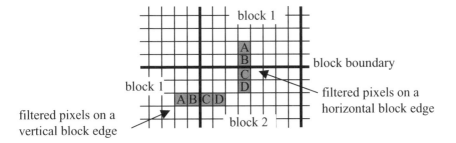

*Figure 9.8   Filtering of pixels at the block boundaries*

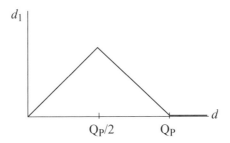

*Figure 9.9   $d_1$ as a function of d*

replace values of the boundary pixels $B$ and $C$, respectively, where:

$$B_1 = B + d_1$$
$$C_1 = C - d_1$$
$$d_1 = \text{sign}(d) \times \text{Max}(0, |d| - \text{Max}(0, 2 \times |d| - Q_P)) \qquad (9.4)$$
$$d = (3A - 8B + 8C - 3D)/16$$
$$Q_P = \text{quantisation parameter of block 2}$$

The amount of alteration of pixels, $\pm d_1$, is related to a function of pixel differences across the block boundary, $d$, and the quantiser parameter $Q_P$, as shown in eqn. 9.4. The sign of $d_1$ is the same as the sign of $d$.

Figure 9.9 shows how the value of $d_1$ changes with $d$ and the quantiser parameter $Q_P$, to make sure that only block edges which may suffer from blocking artefacts are filtered, and not the natural edges. As a result of this modification, only those pixels on the edge are filtered so that their luminance changes are less than the quantisation parameter, $Q_P$.

## 9.4.5 Motion estimation/compensation with spatial transforms

The motion estimation we have seen so far is based on matching a block of pixels in the current frame against a similar size block of pixels in the previous frame, the so-called block matching algorithm (BMA). It relies on the assumptions that the motion of objects is purely translational and the illumination is uniform, which of course are not realistic. In practice, motion has a complex nature that can be decomposed into translation, rotation, shear, expansion and other deformation components and the illumination changes are nonuniform. To compensate for these nonuniform changes between the frames a block of pixels in the current frame can be matched against a deformed block in the previous frame. The deformation should be such that all the components of the complex motion and illumination changes are included.

A practical method for deformation is to transform a square block of $N \times N$ pixels into a quadrilateral of irregular shape, as shown in Figure 9.10.

One of the methods for this purpose is the bilinear transform, defined as [11]:

$$x = \alpha_0 u + \alpha_1 v + \alpha_2 uv + \alpha_3$$
$$y = \alpha_4 u + \alpha_5 v + \alpha_6 uv + \alpha_7 \tag{9.5}$$

where a pixel at a spatial coordinate $(u, v)$ is mapped onto a pixel at coordinate $(x, y)$. To determine the eight unknown mapping parameters $\alpha_0$–$\alpha_7$, eight simultaneous equations relating the coordinates of vertices A, B, C and D into E, F, G and H of Figure 9.10 must be solved. To ease computational load, all the coordinates are offset to the position of coordinates at $u = 0$ and $v = 0$, as shown in the Figure. Referring to the Figure, the eight mapping parameters are derived as:

$$\alpha_0 = \frac{x_F - x_E}{k}; \quad \alpha_4 = \frac{y_F - y_E}{k}; \quad \alpha_1 = \frac{x_H - x_E}{k}; \quad \alpha_5 = \frac{y_H - y_E}{k}$$
$$\alpha_2 = \frac{x_E - x_F + x_G - x_H}{k}; \quad \alpha_6 = \frac{y_E - y_F + y_G - y_H}{k}; \quad \alpha_3 = x_E; \quad \alpha_7 = y_E \tag{9.6}$$

where $k = N - 1$ and $N$ is the block size, e.g. for $N = 16$, $k = 15$.

To use this kind of spatial transformation as a motion estimator, the four corners E, F, G and H in the previous frame are chosen among the pixels within a vertex search window. Using the offset coordinates of these pixels in eqn. 9.6, the motion

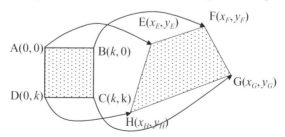

Figure 9.10  Mapping of a block to a quadrilateral

*Video coding for low bit rate communications (H.263)* 243

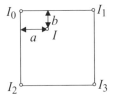

*Figure 9.11   Intensity interpolation of a nongrid pixel*

parameters that transform all the pixels in the current square macroblock of ABCD into a quadrilateral are derived. The positions of E, F, G and H that result in the lowest difference between the pixels in the quadrilateral and the pixels in the transformed block of ABCD are regarded as the best match. Then $\alpha_0$–$\alpha_7$ of the best match is taken as the parameters of the transform that define the motion estimation by spatial transformation.

It is obvious that in general the number of pixels in the quadrilateral is not equal to the $N^2$ pixels in the square block. To match these two unequal size blocks, the corresponding pixel locations of the square block in the quadrilateral must be determined. Since these in general do not coincide with the pixel grid, their values should be interpolated from the intensity of their four surrounding neighbours, $I_0$, $I_1$, $I_2$ and $I_3$, as shown in Figure 9.11.

The interpolated intensity of the mapped pixels, $I$, from its four immediate neighbours, is inversely proportional to their distances and is given by:

$$I = (1-a)(1-b)I_0 + a(1-b)I_1 + (1-a)bI_2 + abI_3 \qquad (9.7)$$

which is simplified to

$$I = (I_1 - I_0)a + (I_2 - I_0)b + (I_0 + I_3 - I_1 - I_2)ab + I_0$$

where $a$ and $b$ are the horizontal and vertical distances of the mapped pixel from the pixel with intensity $I_0$.

Note that this type of motion estimation/compensation is much more complex than the simple block matching. First, in block matching for a maximum motion speed of $\omega$ pixel/frame, there are $(2\omega + 1)^2$ matching operations, but in the spatial transforms, since each vertex is free to move in any direction, the number of matching operations becomes $(2\omega+1)^8$. Still each operation is more complex than BMA, since for each match the transformation parameters $\alpha_0$–$\alpha_7$ have to be calculated, and all the mapped pixels should be interpolated.

There are numerous methods of simplifying the operations [11]. For example, a fast matching algorithm, like the orthogonal search algorithm (OSA) introduced in section 3.3, can be used. Since for a motion speed of eight pixels/frame OSA needs only 13 operations, the total number of search operations becomes $13^4 = 28\,561$, which is practical and is much less than using the brute force method of

244  *Standard codecs: image compression to advanced video coding*

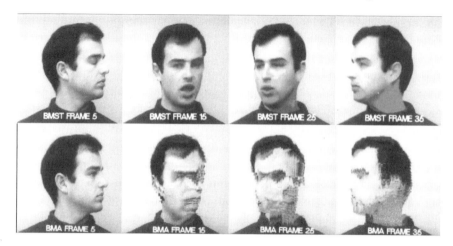

*Figure 9.12*  *Reconstructed pictures with the BMST and BMA motion vectors operating individually*

$(2 \times 8+1)^8 = 7 \times 10^9$, which is not practical! Also, the use of simplified interpolation reduces the interpolation complexity to some extent [12].

In order to appreciate the motion compensation capability of this method, a head-and-shoulders video sequence was recorded, at a speed of almost 12 frames per second, where the head moves from one side to another in three seconds. Assuming that the first frame is available at the decoder, the remaining 35 frames were reconstructed by the motion vectors only, and no motion compensated error was coded. Figure 9.12 shows frames 5, 15, 25 and 35 of the reconstructed pictures by the bilinear transform, called here block matching with spatial transform (BMST) and the conventional block matching algorithm (BMA). At frame 5, where the eye should be closed, the BMA cannot track it as this method only translates the initial eye position of frame one, where it was open, but BMST tracks it well. Also throughout the sequence, the BMST tracks all the eye's and head's movements (plus opening and closing the mouth) and produces almost good quality picture, but BMA produces a noisy image.

To explain why the BMST can produce such a remarkable performance, consider Figure 9.13, where the reconstructed pictures around frame 30, i.e. frames, 28–32, are shown. Looking at the back of the ear, we see that from frame to frame the hair grows, such that at frame 32 it looks quite natural. This is because if, for example, the quadrilateral is only made up of a single black dot, and it is then interpolated over the 16 × 16 pixels, to be matched to the current block of the same size in hair, then the current block can be made from a single pixel.

Note that in BMST, each block requires eight transformation parameters equal to four motion displacements at the four vertices of the quadrilateral. Either the eight parameters $\alpha_0$–$\alpha_7$ or four displacement vectors at the four vertices of the quadrilateral as four motion vectors should be sent. The second method is preferred, since $\alpha_0$–$\alpha_7$

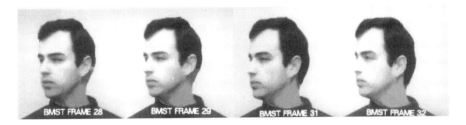

*Figure 9.13  Frame by frame reconstruction of the pictures by BMST*

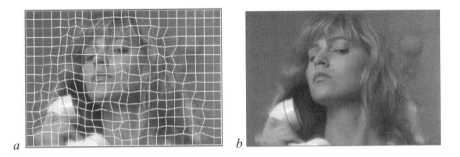

*Figure 9.14  Mesh-based motion compensation*
    *a*  mesh
    *b*  motion compensated picture

are in general noninteger values and need more bits than the four motion vectors. Hence, the motion vector overhead of this method is four times that of BMA for the same block size. However, if BMA of 8 × 8 pixels is used as in the advanced prediction mode [25-F], then we have the same overhead. Again this is irrespective of the block size, since BMA compensates for translational motion only, and cannot produce any better results than those above.

One way of reducing the motion vector overhead is to force the vertices of the four adjacent quadrilaterals to a common vertex. This generates a net-like structure or mesh, as shown in Figure 9.14*a*. As can be seen, the motion compensated picture (Figure 9.14*b*) is smooth and free from blocking artefacts.

To generate such a mesh, the three vertices of a quadrilateral are fixed to their immediate neighbours and only one (bottom right vertex) is free to move. This constrains the efficiency of the motion estimation and for better performance motion estimation for the whole frame has to be iterated several times. Thus it is not expected to perform as well as the unconstrained movement of the vertices applied to Figure 9.12. Despite this, since mesh-based motion estimation creates a smooth boundary between the quadrilaterals, the motion compensated picture will be free from blockiness. Also, it needs only one motion vector per quadrilateral, similar to BMA. Thus this mesh-based motion estimation is expected to be better than

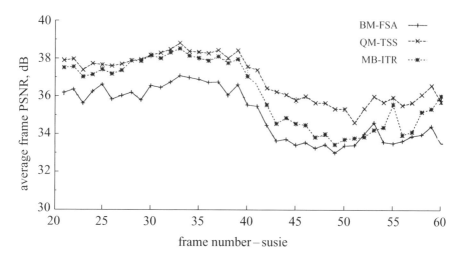

*Figure 9.15   Performance of spatial transform motion compensation*

the BMA with the same motion vector overhead, but of course with increased computational complexity. Figure 9.15 compares the motion compensation efficiency of the full search block matching algorithm (BM-FSA) with quadrilateral matching, using three-step search (QM-TSS) and the mesh-based iterative algorithm (MB-ITR). The motion compensation is applied between the incoming pictures, to eliminate accumulation of errors. Also, since QM requires four motion vectors per block, in order to reduce motion vector overhead, each macroblock is first tested with BMA, and if the MBA motion compensated error is larger than a given threshold, QM is used, otherwise BMA is used. Hence the MQ overhead is less than four times that of BMA. Our investigations show that for head-and-shoulders type pictures, about 20–30 per cent of the macroblocks need QM and the rest can be faithfully compensated by the BMA method. In this Figure, QM also used overlap motion compensation [13]. However, the mesh-based (MB) method, requires the same overhead as for BMA (slightly less, no need at the picture borders).

As the Figure shows, mesh-based motion compensation is superior to the conventional block matching technique, with the same motion vector overhead. Considering the smooth motion compensated picture of the mesh-based method (Figure 9.14) and its superiority over block matching, it is a good candidate to be used in standard codecs.

More information on motion estimation with spatial transforms is given in [11,14,15]. In these papers some other spatial transforms such as Affine and Perspective are also tested. Methods for their use in a video codec to generate equal overhead to those used in H.263 are also explained.

## 9.5 Treatment of B-pictures

B-pictures play an important role in low bit rate applications. If they are coded at lower quality, the quantisation distortion is not accumulated (since they are not used for prediction, see section 7.6). This is not the case for P-pictures, where any gain in reducing the bits in one frame may have to be returned at a higher cost later, when the distortion accumulates in a noise-like signal, which is difficult to code. For very low bit rate video, such as video for mobile networks, normally the frame rate is low (e.g. 5–10 frames/s), and hence the number of B-pictures between the anchor P and I-pictures cannot be large. Apparently only one B-picture is an ideal choice. Also, in these applications I-pictures are hardly used, or if they are used, the GOP length is normally very large. Hence it is plausible to assume, if there is any B-picture in a video, that it is accompanied by a neighbouring P-picture. Thus one can nearly always code B-pictures in relation to the P-picture counterpart, and interrelate their addressing. Two of these are used as annexes in the H.263 family, and will be discussed in the following.

### 9.5.1 PB frames mode

A PB frame consists of two P and B-pictures coded as one unit [25-G]. The P-picture is predicted from the last decoded P-picture and the B-picture is predicted both from the last decoded P-picture and the P-picture currently being decoded. The prediction process is illustrated in Figure 9.16.

#### 9.5.1.1 Macroblock type

Since in the PB frames mode a unit of coding is a combined macroblock from P and B-pictures, the composite macroblock comprises 12 blocks. First, the data for the six P-blocks is transmitted as the default H.263 mode then the data for the six B-blocks. The composite macroblock may have various combinations of coding status for the P- and B-blocks, which are dictated by the combined macroblock block pattern MCBPC.

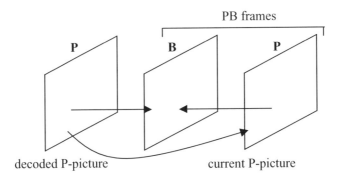

*Figure 9.16    Prediction in PB frames mode*

One of the modes of the MCBPC is the intra macroblock type that has the following meaning:

- the p-blocks are intra coded
- the B-blocks are inter coded with prediction as for an inter block.

The motion vector data (MVD) is also included for intra blocks in pictures for which the type information PTYPE indicates inter. In this case the vector is used for the B-block only. The codewords $MVD_{2-4}$ are never used for intra. The candidate motion vector predictor is not set to zero if the corresponding macroblock was coded in intra mode.

### 9.5.1.2 Motion vectors for B-pictures in PB frames

In the PB frames mode, the motion vectors for the B-pictures are calculated as follows. Assume we have a motion vector component $MV$ in half pixel units to be used in the P-pictures. This $MV$ represents a vector component for an $8 \times 8$ luminance block. If only one motion vector per macroblock is transmitted, then $MV$ has the same value for each of the $8 \times 8$ luminance blocks.

For prediction of the B-picture we need both forward and backward vector components $MV_F$ and $MV_B$. Assume also that $MV_D$ is the delta vector component given by the motion vector data of a B-picture (MVDB) and corresponds to the vector component $MV$. Now $MV_F$ and $MV_B$ are given in half pixel units by the following formulae:

$$MV_F = \frac{TR_B \times MV}{TR_D} + MV_D$$

$$MV_B = \frac{(TR_B - TR_D) \times MV}{TR_D} \quad \text{if } MV_D = 0 \qquad (9.8)$$

$$MV_B = MV_F - MV \quad \text{if } MV_D \neq 0$$

Here $TR_D$ is the increment of temporal reference $TR$ from the last picture header. In the optional PB frames mode, $TR$ only addresses P-pictures. $TR_B$ is the temporal reference for the B-pictures, which indicates the number of nontransmitted pictures since the last P or I-picture and before the B-picture.

Division is done by truncation and it is assumed that the scaling reflects the actual position in time of P and B-pictures. Care is also taken that the range of $MV_F$ should be constrained. Each variable length code for MVDB represents a pair of difference values. Only one of the pairs will yield a value for $MV_F$ falling within the permitted range of $-16$ to $+15.5$. The above relations between $MV_F$, $MV_B$ and $MV$ are also used in the case of intra blocks, where the vector is used for predicting B-blocks.

For chrominance blocks, the forward and backward motion vectors, $MV_F$ and $MV_B$, are derived by calculating the sum of the four corresponding luminance vectors and dividing this sum by 8. The resulting one-sixteenth pixel resolution vectors are modified towards the nearest half pixel position.

### 9.5.1.3 Prediction for a B-block in PB frames

In PB frames mode, predictions for the 8 × 8 pixel B-blocks are related to the blocks in the corresponding P macroblock. First, it is assumed that the forward and backward motion vectors $MV_F$ and $MV_B$ are calculated. Secondly, it is assumed that the luminance and chrominance blocks of the corresponding P-macroblock are decoded and reconstructed. This macroblock is called $P_{REC}$. Based on $P_{REC}$ and its prediction, the prediction for the B-block is calculated.

The prediction of the B-block has two modes that are used for different parts of the block:

(i) For pixels where the backward motion vector $MV_B$ points to inside $P_{REC}$, use bidirectional prediction. This is obtained as the average of the forward prediction using $MV_F$ relative to the previously decoded P-picture, and the backward prediction using $MV_B$ relative to $P_{REC}$. The average is calculated by dividing the sum of the two predictions by two with truncation.

(ii) For all other pixels, forward prediction using $MV_F$ relative to the previously decoded P-picture is used.

Figure 9.17 shows forward and bidirectionally predicted B-blocks. Part of the block that is predicted bidirectionally is shaded and the part that uses forward prediction only is shown unshaded.

### 9.5.2 Improved PB frames

This mode is an improved version of the optional PB frames mode of H.263 [25-M]. Most parts of this mode are similar to the PB frames mode, the main difference being that in the improved PB frames mode, the B part of the composite PB-macroblock, known as the $B_{PB}$-macroblock, may have a separate motion vector for forward and backward prediction. This is in addition to the bidirectional prediction mode that is also used in the normal PB frames mode.

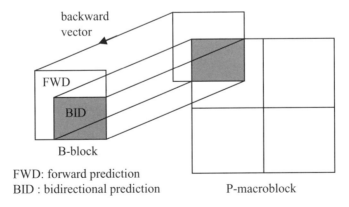

*Figure 9.17* Forward and bidirectional prediction for a B-block

Hence there are three different ways of coding a $B_{PB}$-macroblock and the coding type is signalled by the MVDB parameter. The $B_{PB}$-macroblock coding modes are:

1 *Bidirectional prediction*: in the bidirectional prediction mode, prediction uses the reference pictures before and after the $B_{PB}$-picture. These references are the P-picture part of the temporally previous improved PB frames and the P-picture part of the current improved PB frame. This prediction is equivalent to the prediction in normal PB frames mode when $MV_D = 0$. Note that in this mode the motion vector data (MVD) of the PB macroblock must be included if the P macroblock is intra coded.

2 *Forward prediction*: in the forward prediction mode the vector data contained in MVDB are used for forward prediction from the previous reference picture (an intra or inter picture, or the P-picture part of PB or improved PB frames). This means that there is always only one $16 \times 16$ vector for the $B_{PB}$-macroblock in this prediction mode. A simple prediction is used for coding of the forward motion vector. The rule for this predictor is that if the current macroblock is not at the far left edge of the current picture or slice and the macroblock to the left has a forward motion vector, then the predictor of the forward motion vector for the current macroblock is set to the value of the forward motion vector of the block to the left; otherwise the predictor is set to zero. The difference between the predictor and the desired motion vector is then VLC coded in the same way as vector data to be used for the P-picture (MVD).

3 *Backward prediction*: in the backward prediction mode the prediction of the $B_{PB}$-macroblock is identical to $B_{REC}$ of normal PB frames mode. No motion vector data is used for the backward prediction.

### 9.5.3 Quantisation of B-pictures

In normal mode the quantisation parameter quant is used for each macroblock of P and B-pictures. In PB frames mode, quant is used for P-blocks only, while for the B-blocks a different quantisation parameter bquant is used. In the header information a relative quantisation parameter known as dbquant is sent which indicates the relation between quant and bquant, as defined in Table 9.2.

Table 9.2 *Dbquant codes and relation between quant and bquant*

| dbquant | bquant |
|---------|--------|
| 00 | $(5 \times quant)/4$ |
| 01 | $(6 \times quant)/4$ |
| 10 | $(7 \times quant)/4$ |
| 11 | $(8 \times quant)/4$ |

Division is done by truncation, and bquant ranges from 1 to 31. If the range exceeds these values they are clipped to their limits. Note that since dbquant is a two-bit codeword whereas quantisation information, such as quant, is a five-bit word (indicating quantisation parameters in the range of 1 to 31), such a strategy significantly reduces the overhead information.

## 9.6 Advanced variable length coding

H.263 pays special attention to variable length coding (VLC) for two different reasons. First, since H.263 is a low bit rate codec, it uses any means as well as arithmetic coding as an efficient VLC to enhance the compression efficiency. On the other hand, since H.263 is intended for mobile applications, where the channel error can be very severe and VLC coded data is very prone to the effects of errors, it uses a less compression efficient VLC to localise the side effect of channel errors. These two contradictory requirements are of course for two different applications and both are optional. In the normal mode, H.263 like the other standard video codecs uses the conventional VLC.

### *9.6.1 Syntax-based arithmetic coding*

In the optional arithmetic coding mode of H.263 [25-E], all the corresponding variable length coding/decoding operations may be replaced by binary arithmetic coding/decoding. This mode is used to improve the compression efficiency. It is shown that use of arithmetic coding will improve the compression efficiency over the conventional Huffman coded VLC by approximately 5–10 per cent, depending on the type of data to be coded [10]. However, arithmetic coded data is as prone to channel errors as normal VLC data. Hence, care should be taken to protect data against channel errors.

The type of arithmetic coding is called syntax-based, since a symbol is VLC encoded using a specific table based on the syntax of the coder. This table typically stores lengths and values of the VLC codewords. The symbol is mapped to an entry of the table in a table look-up operation, and then the binary codeword specified by the entry is sent out normally to a buffer for transmitting to the receiver.

In variable length decoding (VLD), the received bit stream is matched entry by entry in a specific table based on the syntax of the coder. This table must be the same as the one used at the encoder for encoding the current symbol. The matched entry in the table is then mapped back to the corresponding symbol that is the end result of the VLD decoder and is then used for recovering the video pictures.

The use of this mode is indicated by the type information, ptype. Details of syntax-based binary arithmetic coding (SAC) were given in Chapter 3. The model for the arithmetic coding of each symbol is specified by the cumulative frequency of that symbol.

### 9.6.2 Reversible variable length coding

To decode VLC coded data, decoders need to find the beginning of the codeword that starts after the resynchronisation marker. The marker has a unique pattern which is known to the decoder. VLC coded data is decoded one bit at a time, and each time the found bits are compared against a set of codewords in a lookup table. If a valid codeword is found, the symbol is decoded, otherwise another bit from the bit stream is appended to the code and tested again. In the event of any error in the bit stream, either a wrong symbol is decoded or the result declared invalid. In the former, it is more likely that the decoded symbols that follow will all be wrong, and/or eventually an invalid codeword is detected. In the case of an invalid codeword, decoding is halted, and the decoder waits for the next resynchronisation marker to start decoding. Thus a single bit error may cause a large part of the picture, from the occurrence of the error to the next resynchronisation marker, to be corrupted.

One way of reducing the damaged area is to be able to decode the bit stream backward as well as forward. This is called reversible variable length coding (RVLC). The decoder normally decodes in the forward mode, but when an invalid codeword is detected it stops decoding and stores the remaining data, up to the next resynchronisation marker. It then decodes backward from the marker, to find an invalid codeword, as shown in Figure 9.18. The area between the forward and reverse nondecodable part becomes the erroneous part.

A variable length code able to work in both ways (RVLC) is required to be symmetric and its success very much depends on finding an invalid codeword arising from the error. If this is not found, then there is no way of identifying the erroneous area unless by postprocessing, which will be discussed in section 9.7.4. Fortunately, due to symmetry of RVLC codes, any error will most likely destroy the symmetry and will cause a nonvalid codeword.

### 9.6.3 Resynchronisation markers

The resynchronisation markers play an important role on the performance of H.263 video codec. If used too often, they limit the damaged area more tightly, so improving the error resilience of the codec. On the other hand, they incur some overheads, which can be costly for low bit rate video. In secure communication environments, they might be used only along with the picture header, where a single bit error can damage the whole picture. In this case, the overhead is minimised and the error-free picture quality is at its best. In normal transmission media, the resynchronisation markers are

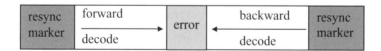

*Figure 9.18*  *A reversible VLC*

preferred to be used at each group of blocks (GOB), to give a balance between the compression efficiency (nonerroneous picture quality) and resilience to errors.

For an optimum balance between the resilience and the coding efficiency, the resynchronisation markers may be inserted where they are needed most. For example, if a GOB does not produce enough bits, or if it belongs to a B-picture, then markers may be inserted between several GOBs. Similarly, if a part of a picture is very active, such as the intraframe coded macroblocks, then within a GOB several markers can be inserted.

This optional mode of H.263 is defined under Annex K and is called the slice structure mode [25-K]. In this mode the slice header information within the bit stream acts as the resynchronisation marker. A slice header has more information than a GOB header (e.g. repeating picture header), such that out of order decoding of slices within a picture is possible. This is particularly useful for packetised transmission of H.263 coded data, where out of sequence decoding of packets reduces the decoding delay. Note that there is no complete independence between the slices, since some processing tools like deblocking filter mode interrelate adjacent slices [25-J].

In order to ensure that slice boundary locations can act as resynchronisation points, and ensure that slices can be sent out of order without causing additional decoding delays, the following rules are adopted in the slice structure mode:

1   The prediction of motion values is the same as if a GOB header was present (see section 9.1.2), preventing the use of motion vectors of blocks outside the current slice for the prediction of the values of motion vectors within the slice.
2   The advanced intra coding mode [25-I] treats the slice boundary as if it was a picture boundary with respect to the prediction of intra block DCT coefficient values.
3   The assignment of remote motion vectors for use in overlapped block motion compensation within the advanced prediction mode [25-F] also prevents the use of motion vectors of blocks outside of the current slice for use as remote motion vectors.

For complete independence between slices, the recommendation describes the optional independent segment decoding mode [25-R]. When this mode is used, the slice boundaries are treated like the picture boundaries, including the treatment of motion vectors which cross these boundaries. If need be, the boundary pixels are extrapolated to be able to use other optional modes such as: unrestricted motion vector, advanced prediction mode, deblocking filter and scalability [25-R].

### 9.6.4   Advanced intra/inter VLC

For further improvement to compression efficiency, H.263 specifies some optional modes that can use the normal Huffman designed VLC codes differently from the other standard video codecs. The following two optional modes describe situations where proper use of VLC improves the encoding efficiency.

### 9.6.4.1 Advanced intra coding

In this optional mode [25-I], intra blocks are predictively coded using nearby blocks in the image to predict values in each intra block. A separate VLC is used for the intra VLC coefficients, and also the quantisation of the DC coefficient for intra is different. This is all done to improve the coding efficiency of the intra macroblocks.

The prediction may be made from the block above or the block to the left of the current block being decoded. An exception occurs in the special case of an isolated intra coded macroblock in an inter coded frame with neither the macroblock above nor to the left being intra coded. In this case, no prediction is made. In prediction, DC coefficients are always predicted in some manner, although either the first row or column of AC coefficients may or may not be predicted as signalled on a macroblock-by-macroblock basis. Inverse quantisation of the intra DC coefficient is identical to the inverse quantisation of AC coefficients for predicted blocks, unlike the core H.263 or other standards that use a fixed quantiser of eight bits for intra DC coefficients.

Also, in addition to zigzag scanning, two more scans are employed, alternate horizontal and alternate vertical scans, as shown in Figure 9.19. Alternate vertical is similar to the alternate scan mode of MPEG-2. For intra predicted blocks, if the prediction mode is set to zero, a zigzag scan is selected for all blocks in a macroblock, otherwise the prediction direction is used to select a scan on a block basis. For instance, if the prediction refers to the horizontally adjacent block, an alternate vertical scan is selected for the current block, otherwise (for DC prediction referring to the vertically adjacent block), alternate horizontal scan is used for the current block.

For nonintra blocks, the $8 \times 8$ blocks of transform coefficients are always scanned with zigzag scanning, similar to all the other standard codecs. A separate VLC table is used for all intra DC and AC coefficients.

Depending on the value of intra_mode, either one or eight coefficients are the prediction residuals that must be added to a predictor. Figure 9.20 shows three $8 \times 8$ blocks of quantised DC levels and prediction residuals labelled $A(u, v)$, $B(u, v)$ and $E(u, v)$, where $u$ and $v$ are row and column indices, respectively.

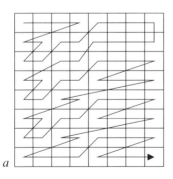 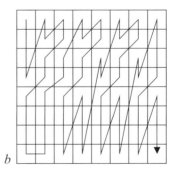

*Figure 9.19*    Alternate scans
         *a*    horizontal
         *b*    vertical

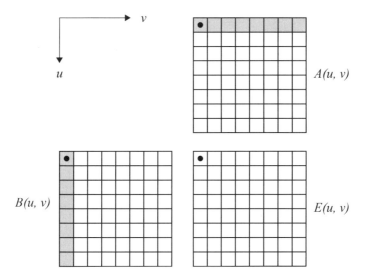

*Figure 9.20  Three neighbouring blocks in the DCT domain*

$E(u, v)$ denotes the current block that is being decoded. $A(u, v)$ denotes the block immediately above $E(u, v)$ and $B(u, v)$ denotes the block immediately to the left of $E(u, v)$. Define $C(u, v)$ to be the actual quantised DCT coefficient. The quantised level $C(u, v)$ is recovered by adding $E(u, v)$ to the appropriate prediction as signalled in the intra_mode field.

The reconstruction for each coding mode is given by:

Mode 0: DC prediction only

$$C(0, 0) = E(0, 0) + \frac{1}{2}\left(\frac{A(0, 0) \times QP_A}{QP_C} + \frac{B(0, 0) \times QP_B}{QP_C}\right) \quad (9.9)$$
$$C(u, v) = E(u, v) \quad u \neq 0, v \neq 0, \quad u = 0\ldots7, \quad v = 0\ldots7$$

Mode 1: DC and AC prediction from the block above

$$C(0, v) = E(0, v) + \frac{A(0, v) \times QP_A}{QP_C} \quad v = 0\ldots7 \quad (9.10)$$
$$C(u, v) = E(u, v) \quad u = 1\ldots7, \quad v = 0\ldots7$$

Mode 2: DC and AC prediction from the block to the left

$$C(u, 0) = E(u, 0) + \frac{B(u, 0) \times QP_B}{QP_C} \quad u = 0\ldots7 \quad (9.11)$$
$$C(u, v) = E(u, v) \quad u = 0\ldots7, \quad v = 1\ldots7$$

where $QP_A$, $QP_B$ and $QP_C$ denote the quantisation parameter (taking values between 1 and 31) used for $A(u, v)$, $B(u, v)$ and $C(u, v)$, respectively.

### 9.6.4.2 Advanced inter coding with switching between two VLC tables

At low frame rates (very common for low bit rate applications), the DCT coefficients of interframe coded macroblocks are normally large. Also, in general, the VLC tables designed for intraframe coded macroblocks suits larger value coefficients better. Hence to improve the compression efficiency of the H.263 codec, the inter coded macroblocks are allowed to use the VLC tables that primarily designed for intra macroblocks, but with a different interpretation of level and run. This is made optional and is called alternative inter VLC mode [25-S]. It is activated when significant changes are evident in the picture.

The intra VLC is constructed so that codewords have the same value for last (0 or 1) in both the inter and intra tables. The intra table is therefore produced by reshuffling the meaning of the codewords with the same value of last. Furthermore, for events with large level, the intra table uses a codeword which in the inter table has large run.

*Encoder action*

The encoder uses the intra VLC table for coding an inter block if the following two criteria are satisfied:

- the intra VLC results in fewer bits than Inter VLC
- if the coefficients are coded with the intra VLC table, but the decoder assumes that the inter VLC is used, coefficients outside the 64 coefficients of a $8 \times 8$ block are addressed.

With many large coefficients, this will easily happen due to the way the intra VLC is used.

*Decoder action*

At the decoder the following actions are taken:

- the decoder first receives all coefficient codes of a block
- the codewords are then interpreted assuming that inter VLC is used; if the addressing of coefficients stays inside the 64 coefficients of a block, the decoding is ended
- if coefficients outside the block are addressed, the codewords are interpreted according to the intra VLC.

## 9.7 Protection against error

H.263 provides error protection, robustness and resilience to allow accessing of video information over a wide range of transmission media. In particular, due to the rapid growth of mobile communications, it is extremely important that access is available to video information *via* wireless networks. This implies a need for useful operation of video compression algorithms in a very error prone environment at low bit rates (i.e. less than 64 kbit/s).

In the previous sections we studied the two important coding tools of VLC and resynchronisation markers in the H.263 codec. The former spreads the errors, and the latter tries to confine them into a small area. In this section, we introduce some more useful tools that can enhance the video quality beyond what we have seen so far. Some of these are recommended as options (or annexes) and some as postprocessing tools that can be implemented at the decoder without the help of the encoder. They can be used either together or individually, to improve video quality.

## 9.7.1 Forward error correction

Forward error correction is the simplest and most effective means of improving video quality in the event of channel errors. It is based on adding some redundancy bits, known as parity bits, to a group of data bits, according to some rules. At the receiver, the decoder, invoking the same rule, can detect if any error has occurred, and in certain cases even correct it. However, for video data, error correction is not as important as is error detection.

The forward error correction for H.263 is the same as for H.261, and is optional [25-H]. However, since the main usage of H.263 will be in a mobile environment with poor error characteristics, forward error correction is particularly important. In most cases (e.g. the GSM system), the error correction will be an integral part of the transmission channel. If it is not, or if additional protection is required, then it should be built into the H.263 system.

To allow the video data and error correction parity information to be identified by the decoder, an error correction framing pattern is included. This pattern consists of multiframes of eight frames, each frame comprising 1 bit framing, 1 bit fill indicator (FI), 492 bits of coded data and 18 bits parity. One bit from each one of the eight frames provide the frame alignment pattern of $(S_1 S_2 S_3 S_4 S_5 S_6 S_7 S_8) = (00011011)$ that will help the decoder to resynchronise itself after the occurrence of errors.

The error detection/correction code is a BCH (511, 493) [16]. The parity is calculated against a code of 493 bits, comprising a bit fill indicator (FI) and 492 bits of coded video data. The generator polynomial is given by:

$$g(x) = (x^9 + x^4 + 1)(x^9 + x^6 + x^4 + x^3 + 1) \tag{9.12}$$

The parity bits are calculated by dividing the 493 bits (left shifted by 18 bits) of the video data (including the fill bit) to this generating function. Since the generating function is a 19-bit polynomial, the remainder will be an 18-bit binary number (that is why the data bits had to be shifted by 18 bits to the left), to be used as the parity bits. For example, for the input data of 01111...11 (493 bits), the resulting correction parity bits are 011011010100011011 (18 bits). The encoder appends these 18 bits to the 493 data bits, and the whole 511 bits are sent to the receiver as a block of data. Now this 511-bit data is exactly divisible by the generating function, and the remainder will be zero. Thus, the receiver can perform a similar division, and if there is any remainder, it is an indication of channel error. This is a very robust form of error detection, since burst of errors can also be detected.

## 9.7.2 Back channel

The impact of error on interframe coded pictures becomes objectionable when error propagates through the picture sequence. Errors affecting only one video frame are easily tolerated by the viewers, especially at high frame rates. To improve quality of video services, propagation of errors through the picture frames must be prevented. A simple method for this task is: when the decoder detects errors in the bit stream (e.g. section 9.7.1), it may ask the encoder to code that part of the picture in the next frame in intra mode. This is called forced updating, and of course requires a back channel from the decoder to the encoder.

Since intraframe coded macroblocks (MB) generate more bits than interframe coded ones, forced updating may not be too impressive. In particular, in normal interframe coding, only a small number of MBs in a GOB are coded. Forced updating will encode all the MBs in the GOB (including the noncoded MBs) in intramode, which increases the bit rate significantly. This can have a side effect of impairing video quality in the subsequent frames. Moreover, if errors occur in more than one GOB, the situation becomes much worse, since the encoder can exceed its bit rate budget, dropping some picture frames. This results in picture jerkiness, which is equally annoying.

A better way of preventing propagation of errors is to ask the encoder to change its prediction to an error free picture. For example, if error occurs in frame $N$, then in coding of the next frame (frame $N + 1$) the encoder uses frame $N - 1$, which is free of error at the decoder. This of course requires some additional picture buffers at both the encoder and the decoder.

The optional reference picture selection mode of H.263 uses additional picture memory at the encoder to perform such a task [25-N]. The amount of additional picture memory accommodated in the decoder may be signalled by external means to help memory management at the encoder. The source encoder for this mode is similar to the generic interframe coder, but several picture memories are provided in order that the encoder may keep a copy of several past pictures, as shown in Figure 9.21.

The source encoder selects one of the picture memories according to the backward channel message GOB-by-GOB to suppress the temporal error propagation due to the interframe coding. The information to signal which picture is selected for prediction is included in the encoded bit stream. The decoder of this mode also has an additional plural number of picture memories, to store the correctly decoded video signals with its temporal reference (TR) information. The decoder uses the stored picture whose TR is TRP as the reference picture for interframe decoding, instead of the last decoded picture, if the TRP field exists in the forward message. When the picture whose TR is TRP is not available at the decoder, the decoder may send the forced intra update signal to the encoder.

A positive acknowledgment (ACK) or a negative acknowledgment (NACK) is returned depending on whether the decoder successfully decodes a GOB.

Both forced intra updating and multiple reference picture modes require a back channel from the decoder to the encoder. In case the back channel cannot be provided,

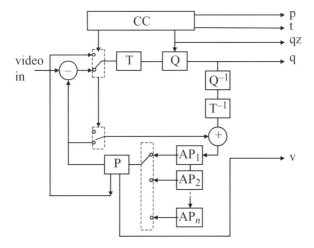

T: Transform    Q: quantiser    CC: coding control
P: picture memory with motion compensated variable delay
AP: additional picture memory    v: motion vector
p: flag for intra/inter    t: flag for transmitted or not.
qz: quantisation indication    q: quantisation index for DCT coefficients

Figure 9.21    An encoder with multiple reference pictures

the multiple reference picture can still be used to alleviate error propagation. For example, the encoder may always use an average of two previous frames for prediction. Of course, in an error-free environment, the compression efficiency is not as good as for the single-frame prediction, but has a good robustness against the channel errors.

Figure 9.22 illustrates the efficiency of the multiple reference picture in preventing error propagation. In the Figure errors occur at frame 30, where the picture quality drops by 3 dB (graph E) compared with the nonerror case (NE). With the back channel (BC), the encoder in coding of frame 31 uses prediction from frame 29 instead of frame 30, and the picture quality is not very different from the nonerror case. Without the back channel, with a prediction from the average of two previous frames $(2F + E)$, the picture quality improves. However, the improvement is not very significant, since the quality of the picture without error (2F) due to nonoptimum prediction is not as good as with the nonerror mode (NE).

### 9.7.3 Data partitioning

Although the individual bits of the VLC coded symbols in a bit stream are equally susceptible to channel errors, the impact of the error on the symbols is unequal. Between the two resynchronisation markers, symbols that appear earlier in the bit stream suffer less from the errors than those which come later. This is due to the

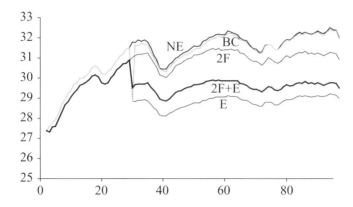

Figure 9.22    Use of multiple reference pictures with and without back channel

cumulative impact of VLC on decoding of the subsequent data. To illustrate the extent of the difference on the unequal susceptibility to errors, consider a segment of VLC coded video data between two resynchronisation markers. Also assume that the segment has $N$ symbols with an average VLC length of $L$ bits/symbol and a channel with a bit error rate of $P$. If any of the first $L$ bits of the bit stream (those immediately after the first marker) are in error, then the symbol would be in error with a probability of $LP$. The probability that the second symbol in the bit stream is in error now becomes $2LP$, since any error in the first $L$ bits also affects the second symbol. Hence the probability that the last symbol in the bit stream is in error will be $NLP$, since every error ahead of this symbol can change the value of this symbol. Thus the last symbol is $N$ times more likely to be in error than the first symbol in the bit stream.

In applications where some video data is more important than the others, like the macroblock addresses (as distinct from interframe DCT coefficients), by bringing the important data ahead of the nonimportant data, one can significantly reduce the channel error side effects. This form of partitioning the VLC coded data into segments of various importance is called data partitioning, which is one of the optional modes of H.263 [25-V]. Note that this form of data partitioning is different from the data partitioning used as a layering technique, described in section 8.5.2. There, through the priority break point, the DCT coefficients were divided into two parts and the lower frequency coefficients along with the other data comprised the base layer and the high frequency DCT coefficients were the second layer. Inclusion of the priority break points and other overheads increased the bit rate by about 3–4 per cent (see Figure 8.25). But here, the entire set of data in a GOB is partitioned and are ordered according to the importance of their contributions in video quality, without any additional overhead. For example, within a GOB, the order of importance of data can be: coding status of MBs, motion vectors, block pattern, quantiser parameter, DC coefficients, AC coefficients. Thus it is also possible to extract all the DC coefficients of the blocks in a GOB, and send them ahead of all the AC coefficients.

*Figure 9.23    Effects of errors*
        *a*   with data partitioning
        *b*   without data partitioning

To appreciate the importance of data partitioning in protecting video against channel errors, Figure 9.23 shows two snap shots of a video sequence with and without data partitioning. It was assumed that in data partitioning, only the DCT coefficients were subjected to errors, but for the normal mode, the bit error could affect any bit of the data. This is a plausible assumption, since normally the important data comprises a small fraction of the bit stream and it can be heavily protected against error. The important data can also use a reversible variable length code (RVLC), such that some of the corrupted data can be retrieved. In fact Annex V of data partitioning recommends RVLC for the slice header (including the macroblock type) and motion vectors [25-V]. The DCT coefficients according to this recommendation use normal VLC. The good picture quality under data partitioning over the normal as shown in Figure 9.23 justifies such a decision. This also shows the insignificance of the DCT coefficients, as their loss hardly affects the picture quality. It should be noted that, in this picture, all the macroblocks were interframe coded. Had there been any intraframe coded macroblock, then its loss would have been noticeable.

Table 9.3 compares the normal VLC and RVLC for the combined macroblock type and block pattern (MCBPC). Note that RVLC is symmetric and it has more bits than the normal VLC. Hence its use should be avoided, unless it is vital to prevent drastic image degradation.

Table 9.4 shows the average number of bits used in an experiment for each slice of a QSIF size salesman image test sequence (picture in Figure 9.23). The last column is the average bit/slice in normal coding of the sequence, for the whole nine slices. For data partitioning, the second column is the slice overhead (including the macroblock type, resynchronisation markers), the third column is the motion vector overhead and the fourth column is the number of bits used for the DCT coefficients. The sum of all the bits in data partitioning is given in the fifth column.

First, since the sum of the slice header and motion vectors is only 8–28 per cent of the data, less for more active slices, they can be easily protected without significantly

*Table 9.3  VLC and RVLC bits of MCBPC*

| Index | MB type | CBPC | Normal VLC | RVLC |
|---|---|---|---|---|
| 0 | 3 (intra) | 00 | 1 | 1 |
| 1 | 3 | 01 | 001 | 010 |
| 2 | 3 | 10 | 010 | 0110 |
| 3 | 3 | 11 | 011 | 01110 |
| 4 | 4 (intra + Q) | 00 | 0001 | 00100 |
| 5 | 4 | 01 | 000001 | 011110 |
| 6 | 4 | 10 | 000010 | 001100 |
| 7 | 4 | 11 | 000011 | 0111110 |

*Table 9.4  Number of bits per slice for data partitioning*

| Slice No | Slice header | MV | Coeff | SUM | Normal |
|---|---|---|---|---|---|
| 1 | 52 | 30 | 211 | 293 | 269 |
| 2 | 63 | 34 | 506 | 603 | 571 |
| 3 | 45 | 42 | 748 | 835 | 803 |
| 4 | 48 | 42 | 1025 | 1115 | 1083 |
| 5 | 45 | 71 | 959 | 1075 | 1043 |
| 6 | 41 | 46 | 844 | 931 | 899 |
| 7 | 48 | 34 | 425 | 507 | 475 |
| 8 | 51 | 32 | 408 | 491 | 459 |
| 9 | 38 | 24 | 221 | 283 | 251 |

increasing the total bit rate. Secondly, comparing the total number of bits in data partitioning with normal coding (columns 5 and 6), we see that data partitioning uses about 3–12 per cent more bits than does normal coding. Considering that this increase is due to the use of RVLC for only the header and the motion vectors and some more resynchronisation markers at the end of the important data, then had we used RVLC for the entire bits, the increase in bit rate would have been much higher. Hence the fact that DCT coefficients do not contribute too much to image quality and RVLC needs more bits than VLC; it is very wise not to use RVLC for the DCT coefficients, as Annex V recommends [25-V]. It should be noted that the main cause for the unpleasant appearance of the picture without data partitioning (Figure 9.23b) is the error on the important data of the bit stream, such as MB address and motion vectors. When the coding status of an MB is wrongly addressed to the decoder, visual information is misplaced. Also in the nondata partitioning mode, since the data, MB and motion vectors VLC coded are mixed, any bit error easily causes nonvalid codewords and a large area of the picture will be in error, as shown in Figure 9.23b.

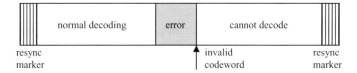

*Figure 9.24    Error in a bit stream*

Note that data partitioning is only used for P and B-pictures, because for I-pictures, DCT coefficients are all important and their absence degrades picture quality significantly.

### 9.7.4    Error detection by postprocessing

In the error correction/detection section of 9.7.1 we saw that with the help of parity bits the decoder can detect an erroneous bit stream. In data communications, the decoder normally ignores the entire segment of the bits and requests for retransmission. Due to the delay sensitive nature of visual services, in video communication retransmission is never used. Moreover, the decoder can decode a part of the bit stream, up to the point where it finds an invalid code word. Hence a part of the corrupted bit stream can be recovered limiting the damaged area.

However, the decoder still cannot identify the exact location of the error (if this were possible, it could have corrected it!). What is certain is that the bits after the invalid codeword up to the next resynchronisation marker are not decodable, as shown in Figure 9.24.

It is to be expected that several symbols are wrongly decoded before the decoder finds an invalid codeword. In some cases, the entire data may be decodable without encountering an invalid codeword, although this rarely happens. For example, the grey parts of the slices in Figure 9.23*b* are due to the invalid codewords that the decoder has given up decoding. Figure 9.23*b* also shows wrongly decoded blocks of pixels, where the decoder can still carry on decoding beyond these blocks. Hence in those parts that are decodable, the correctly decoded data cannot be separated from the wrongly decoded ones, unless some form of processing on the decoded pixels is carried out.

A simple and efficient method of separating correctly decoded blocks from the wrongly decoded ones is to test for pixel continuity at the macroblock (MB) boundaries. For nonerroneous pictures, due to high interpixel correlation pixel differences at the MB borders are normally small, and those due to errors create large differences. As shown in Figure 9.25*a*, for every decoded MB, the average of upper and lower pixel differences at the MB boundaries is calculated as:

$$BD = \frac{1}{N} \sum_{i=1}^{N} |P_i^{in} - P_i^{out}| \qquad (9.13)$$

where $N$ is the total number of pixels at the upper and lower borders of the MB.

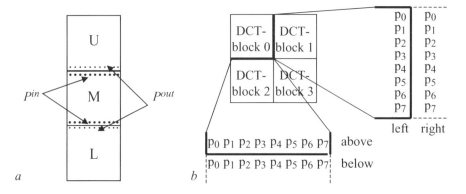

Figure 9.25  Pixels at the boundary of
*a*  a macroblock
*b*  four blocks

The boundary difference (BD) of each MB is then compared against a threshold, and for those MBs which are larger, the implication is that they are most likely to be erroneously decoded. Since due to texture or edges in the image there might be some inherent discontinuity at the MB boundaries, the boundary threshold can be made dependent on the local image statistics. For example, the mean value of the boundary differences of all the MB in the slice, or the slice above, with some tolerance (a few times the standard deviation of the mean differences) can be used as the threshold. Experiments show that mean plus four times the standard deviation is a good value for the threshold [17]. The boundary difference, BD, can be calculated separately for luminance and each of the colour differences. A macroblock might have been erroneously decoded if any of these boundary differences so indicated.

Another method is to calculate the boundary differences around the $8 \times 8$ pixel block boundaries, as shown in Figure 9.25*b*. In $4:2:0$ image format, each MB has four luminance blocks and one of each chrominance block, and hence the boundary difference is applied only to the luminance blocks.

In a similar fashion to the boundary difference of eqn. 9.13, the block boundary is calculated on the inner and outer pixels of the blocks, as shown in Figure 9.25*b*. Again, if any of the four block boundary values, BD, indicates a discontinuity, the macroblock is most likely to be erroneously decoded. Combining the boundary differences of the macroblock (Figure 9.25*a*) and the block (Figure 9.25*b*) increases the reliability of detection.

Assuming that these methods can detect an erroneously decoded MB, then if the first erroneous MB in a slice is found, and provided that the error had only occurred in the bits of this MB, in general it is possible to retrieve the remaining data. Here, after identifying the first erroneous MB, some of the bits are skipped and decoding is performed on the remaining bits. The process is continued such that the remaining bits up to the next resynchronisation marker are completely decodable (no invalid codeword is encountered). In doing so, even parts of the slice/GOB that were not

Figure 9.26   Step-by-step decoding and skipping of bits in the bit stream

decodable before are now decoded and the erroneous part of the GOB can be confined to one MB.

If errors occur in more than one MB, then it may not be possible to have perfect decoding (no invalid codeword) up to the next resynchronisation marker. Thus in general, when decoding proceeds up to the next resynchronisation marker, the number of erroneous MBs are counted. This number should be less than the number of erroneous MBs in the previous run. The process ends when any further skips in bits and decoding does not further reduce the number of erroneous macroblocks in a GOB.

Figure 9.26 shows the decoded pictures at each stage of this step-by-step skipping and decoding of the bits. For the purpose of demonstration, only one bit was introduced in the bit stream between the resynchronisation markers of some of the slices. The first picture shows the erroneous picture without any postprocessing. The second picture shows the reconstructed picture after the first round of bit skipping in each slice, and so on. As we see, in each stage the erroneous area (number of erroneous MBs) is reduced and further processing does not reduce the number of erroneous MBs (not much differences between pictures $d$ and $e$). There is only one erroneous MB in each slice of the final picture (Figure 9.26$e$), which can easily be concealed.

In the above example it was assumed that a single bit error was affecting one MB, or a burst of errors affected only one MB. If errors affect more than one MB, then at the end more than one MB will be in error and of course it will take more time to find these erroneous MBs in the decoding. This is because, after finding the first erroneous MB, since there are some erroneous MBs to follow, perfect decoding (not finding a valid codeword) is not possible. Experiments show that in most cases, all the macroblocks between the first and the last erroneous MB in a slice will be in error. However, it is still possible to recover some of the macroblocks, which without this sort of processing was not possible.

## 9.7.5 Error concealment

If any of the error resilience methods mentioned so far or their combinations is not sufficient to produce satisfactory picture quality, then one may try to hide the image degradation from the viewer. This is called error concealment.

The main idea behind error concealment is to replace the damaged pixels with pixels from parts of the video that have maximum resemblance. In general, pixel substitution may come from the same frame or from the previous frame. These are called intraframe and interframe error concealment, respectively [18].

### 9.7.5.1 Intraframe error concealment

In intraframe error concealment, pixels of an erroneous MB are replaced by those of a neighbouring MB with some form of interpolation. For example, pixels at the macroblock boundary may be directly replaced by the pixels from the other side of the border, and for the other pixels, the average of the neighbouring pixels inversely weighted by their distances may be substituted.

An efficient method of intraframe error concealment is shown in the block diagram of Figure 9.27. A block of pixels, larger than the size of a macroblock (preferably $48 \times 48$ pixels, equivalent to $3 \times 3$ MBs) encompassing the MB to be concealed, is fast Fourier transformed (FFT). Pixels of the MB to be concealed initially are filled with grey level values. The FFT coefficients are two-dimensionally lowpass filtered (LPF) to remove the discontinuity due to these inserted pixels. The resultant lowpass filtered coefficients are then inverse fast Fourier transformed (IFFT), to reconstruct a replica of the input pixels. Due to lowpass filtering, the reconstructed pixels are similar but not exactly the same as the input pixels. The extent of dissimilarity depends on the cutoff frequency of the lowpass filter. The lower the cutoff frequency, the stronger is the influence of the neighbouring pixels into the concealed MB. The centre MB at the output now replaces the centre MB at the input, and the whole process of FFT, LPF and IFFT repeats again. The process is repeated several times, and at each time the cutoff frequency of the LPF is gradually increased. To improve the quality of error concealment, the lowpass filter can be made directional, based on the characteristics of the surrounding pixels. The process is terminated when the difference between the pixels of the concealed MB at the input and output is less than a threshold.

This form of error concealment assumes an isolated erroneous MB surrounded by eight immediate nonerroneous neighbours. This is suitable for JPEG or motion JPEG coded pictures, where error is localised (see Figure 5.17), or for interframe coded pictures, if by means of postprocessing error is confined to an MB (e.g. Figure 9.26*e*). For video, where there is a danger of error at the same slice/GOB, then pixels of the

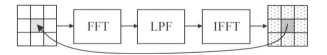

*Figure 9.27* An example of intraframe error concealment

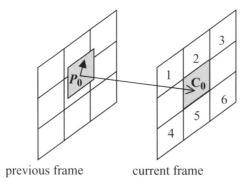

*Figure 9.28  A grid of 3 × 3 macroblocks in the current and previous frame*

top and bottom slices should be used, and the two right and left MBs are treated as if they are in error. This impairs the performance of the concealment, and may not be suitable. For video, a more suitable error concealment is interframe error concealment, which is explained in the following.

### 9.7.5.2  Interframe error concealment

In interframe error concealment, pixels from the previous frame are substituted for the pixels of the MB to be concealed, as shown in Figure 9.28. This could be either by direct substitution, or by their motion compensated version, using an estimated motion vector. Obviously, due to movement, motion compensated substitution is better. The performance of this method depends on how accurately the motion vector for concealment is estimated. In the following, several methods of estimating this motion vector are explained and their error concealment fidelity are compared against each other.

*Zero mv*

Direct substitution of pixels from the MB of the previous frame at the same spatial position of the MB to be concealed (zero motion vector). This is the simplest method of substitution, and is effective in the picture background, or in the foreground with slow motion.

*Previous mv*

The estimated motion vector is the same as the motion vector of the spatially similar MB in the previous frame. This method, which assumes a uniform motion of objects, performs well most of the time, but, however, it is eventually bound to fail.

*Top mv*

The estimated motion vector is the same as the motion vector of MB at the top of the wanted MB (e.g. MB number 2 of Figure 9.28). Similarly, the motion vector of the bottom MB (e.g. MB number 5) may be used. Since these two MBs are closest to the current MB, it is expected that their MVs will have the highest similarity. However, this method is as simple as the direct substitution (zero mv) and previous mv.

### Mean mv

The average of the motion vectors of the six immediate neighbours represents the estimated mv. The mean value for horizontal displacement, $x_0$, and vertical displacement, $y_0$, are taken separately:

$$x_0 = \frac{1}{6} \sum_{i=1}^{6} x_i; \quad y_0 = \frac{1}{6} \sum_{i=1}^{6} y_i \quad (9.14)$$

where $x_i$ and $y_i$ are the horizontal and vertical components of the motion vector $i$; $mv_i(x_i, y_i)$. Note that, due to averaging, any small perturbations in the neighbouring motion vector components will cancel each other. Thus the estimated motion vector will be different from the motion vector of the neighbouring MB. This method of error concealment may not produces a smooth picture. The discontinuity at the MB boundaries produces a blocking artefact that appears very annoying. Hence this method is not good for parts of the picture with motion in various directions, such as the movement of lips and eyes of a talking head.

### Majority mv

The majority of the motion vectors are grouped together, and their mean or other representative value is taken as the estimated motion vector.

$$x_0 = \frac{1}{N} \sum_{i=1}^{N} x_i; \quad y_0 = \frac{1}{N} \sum_{i=1}^{N} y_i \quad N \leq 6 \quad (9.15)$$

where $N$ out of six motion vectors are almost at the same direction. Since in general all motion vectors can differ from each other, then to find the majority, the motion vectors should be vector quantised, and the majority is found among their original values. This method works well for rigid body movement, where the neighbouring motion vectors normally move in the same direction. However, since there are only six neighbouring motion vectors, a definite majority among them cannot be found reliably. Hence for nonrigid movement, such as lips and eyes, this method may not work well.

### Vector median mv

The median of a group of vectors is one of the vectors in the group that has the smallest Euclidean distance from all. Thus among the six neighbouring motion vectors $mv_1$–$mv_6$ of Figure 9.28, the $j$th motion vector, $mv_j$, is the median if:

$$dist_j = \frac{1}{5} \sum_{i=1}^{6} \sqrt{(x_i - x_j)^2 + (y_i - y_j)^2} \quad i \neq j \quad (9.16)$$

such that for all motion vectors $mv_k$, $1 \leq k \leq 6$, the distance of vector $j$, $dist_j$ is less than the distance of vector $k$, $dist_k$ [19].

This method is expected to produce a good result, because since the median of vectors has the least distance from all, it then has the largest correlation with them. Also, since the macroblock to be concealed is at the centre of all and has the highest correlation with them, then it has the same property as the median vector. This good performance is achieved at a higher computational cost. Here, a Euclidean distance of each vector from all the five other vectors should be calculated first, which requires $\frac{1}{2}(6 \times 5) = 15$ vector distance calculations. Then for each, five distances to be averaged to represent the average distance of a vector from the others. Finally, they should be rank ordered to find the minimum distance.

To compare the relative error concealment performance of each method, four sets of head-and-shoulders type image sequences at QSIF resolutions were subjected to channel errors. In the event of error, the whole GOB was concealed by the above mentioned methods. This is because, due to VLC coding, a single bit error may cause the remaining bits up to the next GOB nondecodable, as shown on the erroneous picture of Figure 9.23. Tables 9.5 and 9.6 summarise the quality of these error concealment methods for QCIF video at 5 and 12.5 frames/s, respectively. To show just the impact of error concealment, measurements were carried out only on the concealed areas.

As the Tables show, the vector median method gives the best result at both high and low frame rates. That of the majority method is the second best. In all cases, the performance of the average method is as poor as the simple method of top and, in some cases, it is even poorer (seq-1 and seq-4 of Table 9.5). The poor performance of the previous mv means that motion is not uniform. This is particularly evident at the low frame rate of five frames/s. However, all the methods are superior to zero motion, implying that loss concealment by an estimated motion vector improves picture quality.

Also, note that since quality measurements were carried out at the error concealed areas, then the performance at a lower frame rate is poorer than at the higher frame rate. That is, as the frame rate is reduced, the estimated motion vector is less similar to the actual motion vector. Despite this, estimating motion vectors by all the methods gives better performance than not estimating them (zero motion vector). Figure 9.29 shows an accumulated erroneous picture of seq-3 at five frames/s and its concealed one with the median vector method [19].

*Bidirectional mv*

If B-pictures are present in the group of pictures (GOP), then due to stronger relation between the motion vector of a B and its anchor P or I-picture, a better estimation of the motion vector can be made. As an example consider a GOP of $N = \alpha$ and $M = 2$, that is, the image sequence is made of alternate P and B-pictures, as shown in Figure 9.30.

To estimate a missing motion vector for a P-picture, say $P_{31}$, the available motion vectors of the same spatial coordinates of the B-pictures can be used, with the

*Table 9.5   PSNR [dB] of various error concealment methods at 5 frames/s*

| Type | 64 Kbps, 5 fps, QCIF | | | |
|---|---|---|---|---|
| Sequence | Seq-1 | Seq-2 | Seq-3 | Seq-4 |
| Zero | 17.04 | 18.15 | 14.54 | 13.08 |
| Previous | 17.28 | 18.34 | 14.53 | 13.48 |
| Top | 19.27 | 21.08 | 17.04 | 16.25 |
| Average | 19.18 | 21.74 | 17.51 | 16.18 |
| Majority | 19.35 | 21.83 | 17.89 | 16.61 |
| Median | 19.87 | 22.52 | 18.29 | 16.89 |
| No errors | 22.57 | 26.94 | 20.85 | 19.88 |

*Table 9.6   PSNR [dB] of the various error concealment methods at 12.5 frames/s*

| Type | 64 Kbps, 12.5 fps, QCIF | | | |
|---|---|---|---|---|
| Sequence | Seq-1 | Seq-2 | Seq-3 | Seq-4 |
| Zero | 20.62 | 22.11 | 18.64 | 16.62 |
| Previous | 22.49 | 22.19 | 18.19 | 16.53 |
| Top | 22.97 | 25.16 | 20.97 | 20.04 |
| Average | 22.92 | 25.99 | 21.24 | 20.08 |
| Majority | 23.33 | 26.32 | 21.57 | 20.36 |
| Median | 24.36 | 26.72 | 22.16 | 20.81 |
| No errors | 26.12 | 29.69 | 23.93 | 23.04 |

*Figure 9.29   An erroneous picture along with its error concealed version*

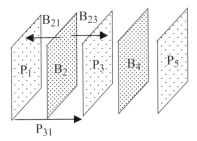

Figure 9.30    A group of alternate P and B-pictures

following substitutions:

if only $B_{23}$ is available, then $P_{31} = 2 \times B_{23}$
if only $B_{21}$ is available, then $P_{31} = -2 \times B_{21}$
if both $B_{23}$ and $B_{21}$ are available, then $P_{31} = B_{23} - B_{21}$
if none of them are available, then set $P_{31} = 0$

To estimate a missing motion vector of a B-picture, simply divide that of the P-picture by two: $B_{23} = \frac{1}{2}P_{31}$ or $B_{21} = -\frac{1}{2}P_{31}$.

Here we have used a simple previous mv estimation method, explained earlier. Although in the tests of Tables 9.5 and 9.6 (images sequences made of P-pictures only) this method did not perform well, since the relation here between P and B-pictures is strong, the method does work well. For example, using MPEG-1 video we have achieved about 3–4 dB improvement over the majority method [20]. The amount of improvement is picture dependent, and it appears for QCIF images coded with H.263; at least 1 dB improvement over the majority can be achieved. Interested readers should consult [20] for further detailed information.

### 9.7.5.3  Loss concealment

In transmission of video over packet networks such as IP, ATM or wireless packet networks, the video data is packed into the payload of the packets. In this transmission mode two types of distortion may occur. One is the error in the payload, which results in erroneous reception of the bit stream, similar to the effect of channel errors. The second one is either error in the packet header, which results in a packet loss, or if the packet is queued in a congested network. Excessively delayed packets will be of no use and hence they will be discarded either by the switching nodes (routers) or by the receiver itself.

Detection, correction and concealment of the error in the packet payload is similar to that for the previous methods mentioned. For packet loss the methods can be slightly different. First, the decoder by examining the packet sequence number discovers that a packet is missing. Second, when a packet is lost, unlike channel errors, no part of the video data is decodable. Hence, loss concealment is more vital to video over packet networks than is error concealment in a nonpacket transporting environment.

Considering that in coding of video, in particular at low bit rates, not all parts of the picture are coded, then the best concealment for noncoded macroblocks is the direct copy of the previous macroblock without any motion compensation (i.e. zero mv). For those which are coded, as Tables 9.5 and 9.6 show, a motion compensated macroblock gives a better result. However, the information as to which macroblock was or was not coded is not available at the decoder. It is obvious that any attempt to replace the noncoded area by the motion compensated macroblock will degrade the image quality rather than improve it. Our simulations show that replacing a noncode MB with an estimated motion compensated MB would degrade the quality of the pixels in that MB by 7–10 dB [21].

Therefore, for proper loss concealment, the coded and noncoded maroblocks should be discriminated from each other. A noncoded MB should be directly copied from the previous frame, but for the coded one, it should be motion compensated by an estimated motion vector (any of the estimation methods of section 9.7.5.2). Decision on the coding status of a missing MB can be made on the coding status of the MB at the same spatial location in the previous frame. Investigations show that if an MB is coded, it is about 70 per cent certain that it will be coded in the next frame [21]. Also, if an MB is not coded, it is 90 per cent certain that it will not be coded in the next frame. Thus a decision on whether a lost MB should be replaced by direct substitution, or its motion compensated version, can be made based on the coding status of that MB in the previous frame. We call this method of loss concealment, selective loss concealment.

To demonstrate the image enhancement due to loss concealment, the Salesman test image sequence coded at 144 kbit/s, 10 Hz, was exposed to channel errors at a rate of $10^{-2}$ bit rate, using the channel error model given in Appendix E [22].

Figure 9.31 shows the objective quality of the entire decoded picture sequence with loss and loss concealment. As the Figure shows, while the quality of the decoded video due to loss is impaired by more than 10 dB, loss concealment enhances degraded image quality by around 7 dB. Figure 9.31 also shows the improvement due to selective concealment *versus* the full concealment, where a lost macroblock is always replaced by the motion compensated previous macroblock, irrespective of whether a macroblock was coded or not.

### 9.7.5.4 Selection of best estimated motion vector

Although Tables 9.5 and 9.6 show that one method of estimating a lost motion vector is better than the other, nevertheless they represent the average quality over the entire video sequence. Had we compared these methods on a macroblock by macroblock basis, there can be situations in which an overall best method will not perform well. The reason is that the quality of such error/loss concealment depends on the directions and values of the surrounding motion vectors of that macroblock. What makes poor error/loss concealment is that the motion compensated replacement macroblock shows some pixel discontinuity. This makes the reconstructed picture look blocky, which is very disturbing.

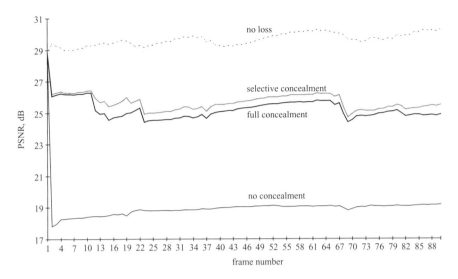

*Figure 9.31* Quality of decoded video with and without loss concealment with a bit error ratio of $10^{-2}$

To improve the error/loss concealed image quality, one may apply all the above motion estimation methods, and test for image discontinuity around the reconstructed macroblock. The method that gives the least discontinuity is then chosen. Methods introduced in section 9.7.4 can be used as a discontinuity measure.

## 9.8 Scalability

Although we have extensively described the scalability under JPEG2000 and MPEG-2, in H.263 it is used with different terminology and we visit this subject again. It might be useful to know that scalability in H.263 is not used for distribution purposes, but more as a layering technique. Hence, by unequal error protection on the base layer, this method in conjunction with the other error resilience methods, explained in section 9.7, further improves the robustness of this codec.

Extensions of H.263 also support temporal, SNR and spatial scalability as optional modes [25-O]. This mode is normally used in conjunction with the error control scheme. The capability of this mode and the extent to which its features are supported is signalled by external means such as H.245 [9].

There are three types of enhancement picture in the H.263+ codec that are known as B, EI and EP-pictures [5]. Each of these has an enhancement layer number, ELNUM, which indicates to which layer it belongs, and a reference layer number, RLNUM, which indicates which layer is used for its prediction. The encoder may use any of its basic scalability nodes of temporal, SNR, spatial or their combinations in a multilayer scalability mode. Details of the basic and multilayer scalabilities were

274  Standard codecs: image compression to advanced video coding

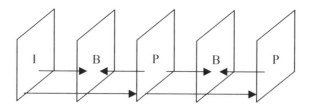

Figure 9.32  B-picture prediction dependency in the temporal scalability

given in section 8.5. However, due to the different nature and application of H.263 to MPEG-2, there are some differences.

## 9.8.1  Temporal scalability

Temporal scalability is achieved using bidirectionally predicted pictures or B-pictures. As usual, B-pictures use prediction from either or both of a previous and subsequent reconstructed picture in the reference layer. These B-pictures differ from the B-picture part of a PB or improved PB frames, in that they are separate entities in the bit stream. They are not syntactically intermixed with a subsequent P or its enhancement part EP.

B-pictures and the B part of PB or improved PB frames are not used as reference pictures for the prediction of any other pictures. This property allows for B-pictures to be discarded if necessary without adversely affecting any subsequent pictures, thus providing temporal scalability. There is no limit to the number of B-pictures that might be inserted between the pairs of the reference pictures in the base layer. A maximum number of such pictures may be signalled by external means (e.g. H.245). However, since H.263 is normally used for low frame rate applications (low bit rates, e.g. mobile), then due to larger separation between the base layer I and P-pictures, there is normally one B-picture between them. Figure 9.32 shows the position of base layer I and P-pictures and the B-pictures of the enhancement layer for most applications.

## 9.8.2  SNR scalability

In SNR scalability, the difference between the input picture and lower quality base layer picture is coded. The picture in the base layer which is used for the prediction of the enhancement layer pictures may be an I-picture, a P-picture, or the P part of a PB or improved PB frame, but should not be a B-picture or the B part of a PB or its improved version.

In the enhancement layer two types of picture are identified, EI and EP. If prediction is only formed from the base layer, then the enhancement layer picture is referred to as an EI-picture. In this case the base layer picture can be an I or a P-picture (or the P part of a PB frame). It is possible, however, to create a modified bidirectionally predicted picture using both a prior enhancement layer picture and temporally simultaneous base layer reference picture. This type of picture is referred to as an

# Video coding for low bit rate communications (H.263) 275

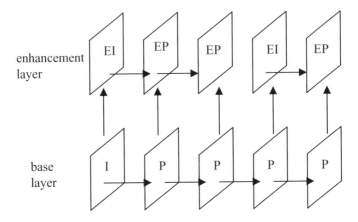

*Figure 9.33    Prediction flow in SNR scalability*

EP-picture or enhancement P-picture. Figure 9.33 shows the positions of the base and enhancement layer pictures in an SNR scalable coder. The Figure also shows the prediction flow for the EI and EP enhancement pictures.

For both EI and EP-pictures, prediction from the reference layer uses no motion vectors. However, as with normal P-pictures, EP pictures use motion vectors when predicting from their temporally-prior reference picture in the same frame.

## 9.8.3  Spatial scalability

The arrangement of the enhancement layer pictures in the spatial scalability is similar to that of SNR scalability. The only difference is that before the picture in the reference layer is used to predict the picture in the spatial enhancement layer, it is downsampled by a factor of two either horizontally or vertically (one-dimensional spatial scalability), or both horizontally and vertically (two-dimensional spatial scalability). Figure 9.34 shows the flow of the prediction in the base and enhancement layer pictures of a spatial scalable encoder.

## 9.8.4  Multilayer scalability

Undoubtedly multilayer scalability will increase the robustness of H.263 against the channel errors. In the multilayer scalable mode, it is possible not only for B-pictures to be temporally inserted between the base layer pictures of type I, P, PB and improved PB, but also between the enhancement picture types of EI and EP, whether these consist of SNR or spatial enhancement pictures. It is also possible to have more than one SNR or spatial enhancement layer in conjunction with the base layer. Thus a multilayer scalable bit stream can be a combination of SNR layers, spatial layers and B-pictures. With increasing the layer number, the size of a picture cannot decrease. Figure 9.35 illustrates the prediction flow in a multilayer scalable encoder.

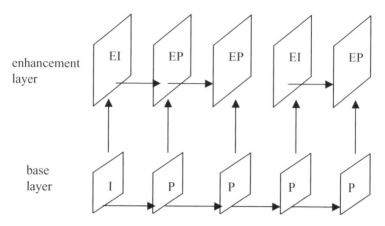

Figure 9.34   Prediction flow in spatial scalability

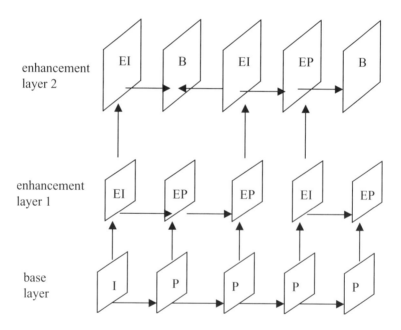

Figure 9.35   Positions of the base and enhancement layer pictures in a multilayer scalable bit stream

As with the two-layer case, B-pictures may occur in any layer. However, any picture in an enhancement layer which is temporally simultaneous with a B-picture in its reference layer must be a B-picture or the B-picture part of a PB or improved PB frame. This is to preserve the disposable nature of B-pictures. Note, however, that B-pictures may occur in any layers that have no corresponding picture in the lower layers. This allows an encoder to send enhancement video with a higher picture rate than for the lower layers.

The enhancement layer number and the reference layer number of each enhancement picture (B, EI, or EP) are indicated in the ELNUM and RLNUM fields, respectively, of the picture header (when present). If a B-picture appears in an enhancement layer in which temporally surrounding SNR or spatial pictures also appear, the reference layer number (RLNUM) of the B-picture shall be the same as the enhancement layer number (ELNUM). The picture height, width and pixel aspect ratio of a B-picture shall always be equal to those of its temporally subsequent reference layer picture.

## 9.8.5 *Transmission order of pictures*

Pictures, which are dependent on other pictures, shall be located in the bit stream after the pictures on which they depend. The bit stream syntax order is specified such that for reference pictures (i.e. pictures having types I, P, EI, EP, or the P part of PB or improved PB), the following two rules shall be obeyed:

1  All reference pictures with the same temporal reference shall appear in the bit stream in increasing enhancement layer order. This is because each lower layer reference picture is needed to decode the next higher layer reference picture.
2  All temporally simultaneous reference pictures as discussed in item 1 above shall appear in the bit stream prior to any B-pictures for which any of these reference pictures is the first temporally subsequent reference picture in the reference layer of the B-picture. This is done to reduce the delay of decoding all reference pictures, which may be needed as references for B-pictures.

Then, the B-pictures with earlier temporal references shall follow (temporally ordered within each enhancement layer). The bit stream location of each B-picture shall comply with the following rules:

1  Be after that of its first temporally subsequent reference pictures in the reference layer. This is because the decoding of the B-pictures generally depends on the prior decoding of that reference picture.
2  Be after that of all reference pictures that are temporally simultaneous with the first temporally subsequent reference picture in the reference layer. This is to reduce the delay of decoding all reference pictures, which may be needed as references for B-pictures.
3  Precede the location of any additional temporally subsequent pictures other than B-pictures in its reference layer. Otherwise, it would increase picture storage memory requirement for the reference layer pictures.
4  Be after that of all EI and EP pictures that are temporally simultaneous with the first temporally subsequent reference picture.
5  Precede the location of all temporally subsequent pictures within the same enhancement layer. Otherwise, it would introduce needless delay and increase picture storage memory requirements for the enhancement layer.

278  Standard codecs: image compression to advanced video coding

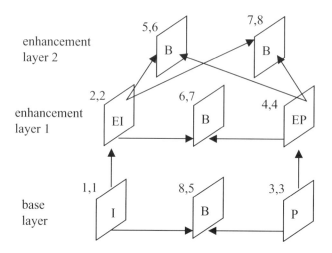

*Figure 9.36  Example of picture transmission order*

Figure 9.36 shows two allowable picture transmission orders given by the rules above for the layering structure shown as an example. Numbers next to each picture indicate the bit stream order, separated by commas for the two alternatives.

## 9.9 Buffer regulation

Regulation of output bit rates for better distribution of the target bit rate among the encoding parameters is an important part of any video encoder. This is particularly vital in the H.263 encoder, at least for the following reasons:

- better bit rate regulation requires larger buffer sizes, hence longer delays
- H.263 is intended for visual telephony, and the encoding delay should be limited, hence smaller buffer sizes are preferred
- the target bit rate is in the order of 24 kbit/s, and even small size buffers can introduce long delays.

There is no best known method for buffer regulation, and Recommendation H.263 does not standardise any method (neither do other standard encoders). However, at least for the laboratory simulations, one can use those methods designed for the test models. The following is a method that can be used in the simulations [5]. The bit rate is controlled at a macroblock level, by changing the quantiser parameter, $Q_P$, depending on the bit rate, the source and target frame rates.

For the first picture, which is intraframe coded, the quantisation parameter is set to its mid range $Q_P = 16$ ($Q_P$ varies from 1 to 31). After the first picture, the buffer content is set to:

$$\frac{R}{f_{target}} + 3 \times \frac{R}{FR} \quad \text{and} \quad B_{i-1} = \bar{B} \qquad (9.17)$$

For the following pictures the quantiser parameter is updated at the beginning of each new macroblock line. The formula for calculating the new quantiser parameter is:

$$QP_{new} = \overline{QP_{i-1}} \left(1 + \frac{\Delta_1 B}{2\bar{B}} + \frac{12\Delta_2 B}{R}\right)$$

$$\Delta_1 B = B_{i-1} - \bar{B} \tag{9.18}$$

$$\Delta_2 B = B_{i,mb} - \frac{mb}{MB} \times \bar{B}$$

where:

$\overline{QP_{i-1}}$ = mean quantiser parameter for the previous picture
$B_{i-1}$ = number of bits spent for the previous picture
$\bar{B}$ = target number of bits per picture
$mb$ = present macroblock number
$MB$ = number of macroblocks in a picture
$B_{i,mb}$ = number of bits spent until now for the picture
$R$ = bit rate
$FR$ = frame rate of the source picture (typically 25 or 30 Hz)
$f_{target}$ = target frame rate

The first two terms of the above formula are fixed for macroblocks within a picture. The third term adjusts the quantiser parameter during coding of the picture.

The calculated new quantisation parameter, $QP_{new}$, must be adjusted so that the difference fits within the definition of dquant. The buffer content is updated after each complete picture by using the following C function:

```
buffer_content=buffer_content+Bi,99 ;
while(buffer_content>(3R/FR)){
buffer_content=buffer_content - (R/FR);
frame_incr++;
}
```

The variable `frame_incr` indicates how many times the last coded picture must be displayed. It also indicates which picture from the source is coded next.

To regulate the frame rate, $f_{target}$, a new $\bar{B}$ is calculated at the start of each frame:

$$f_{target} = 10 - \frac{\overline{QP_{i-1}}}{4} \quad 4 < f_{target} < 10$$

$$\bar{B} = \frac{R}{f_{target}} \tag{9.19}$$

For this buffer regulation, it is assumed that the process of encoding is temporarily stopped when the physical transmission buffer is nearly full, preventing buffer overflow. However, this means that no minimum frame rate and delay can be guaranteed.

## 9.10 Advanced video coding (H.26L)

The long-term objective of the ITU-T video coding experts group under the advanced video coding project is to provide a video coding recommendation to perform substantially better than the existing standards (e.g. H.263+) at very low bit rates. The group worked closely with the MPEG-4 experts group of ISO/IEC for more than six years (from 1997 to 2002). The joint work of the ITU-T and the ISO/IEC is currently called H.26L, with L standing for long-term objectives. The final recommendations of the codec will be made available in 2003, but here we report only on what are known so far. These may change over time, but we report on those parts that not only look fundamental to the new codec, but are also more certain to be adopted as recommendations. The codec is expected to be approved as H.264 by the ITU-T and as MPEG-4 part 10 (IS14496-10) by ISO/IEC [7].

Simulation results show that H.26L has achieved substantial superiority of video quality over that achieved by the existing most optimised H.263 and MPEG-4 codecs [23]. Most notable features of the H.26L are:

- *Up to 50 per cent in bit rate saving*: compared with the H.263+ (H.263V2) or MPEG-4 simple profile (see Chapter 10), H.26L permits an average reduction in bit rate of up to 50 per cent for a similar degree of encoder optimisation at most bit rates. This means that H.26L offers consistently higher quality at all bit rates including low bit rates.
- *Adaptation to delay constraints*: H.26L can operate in a low delay mode to adapt telecommunication applications (e.g. H.264 for videoconferencing), while allowing higher processing delay in applications with no delay constraints such as video storage and server-based video streaming applications (MPEG-4V10).
- *Error resilience*: H.26L provides the tools necessary to deal with packet loss in packet networks and bit errors in error-prone wireless networks.
- *Network friendliness*: the codec has a feature that conceptually separates the video coding layer (VCL) from the network adaptation layer (NAL). The former provides the core high compression representation of the video picture content and the latter supports delivery over various types of network. This facilitates easier packetisation and better information priority control.

### 9.10.1 How does H.26L (H.264) differ from H.263

Despite the above mentioned features, the underlying approach of H.26L (H.264) is similar to that adopted by the H.263 standard [7]. That is, it follows the generic transform-based interframe encoder of Figure 3.18, which employs block matching motion compensation, transform coding of the residual errors, scalar quantisation with an adaptive quantiser step size, zigzag scanning and run length coding of the transform coefficients. However, some specific changes are made to the H.26L codec to make it not only more compression efficient, but also more resilient against channel errors. Some of the most notable differences between the core H.26L and the H.263

codecs are:

- H.26L employs 4 × 4 integer transform block sizes, as opposed to 8 × 8 floating points DCT transform used in H.263
- H.26L employs a much larger number of different motion compensation block sizes per 16 × 16 pixel macroblock; H.263 only supports two such block sizes in its optional mode
- higher precision of spatial accuracy for motion estimation with quarter pixel accuracy as the default mode for lower complexity mode and eighth pixel accuracy for the higher complexity mode; H.263 uses only half pixel and MPEG-4 uses quarter pixel accuracy
- the core H.26L uses multiple previous reference pictures for prediction, whereas the H.263 standard uses this feature under the optional mode
- in addition to I, P and B-pictures, H.26L uses a new type of interstream transitional picture, called an SP-picture
- the deblocking filter in the motion compensation loop is a part of the core H.26L, whereas H.263 uses it as an option.

In the following sections some details of these new changes known so far are explained.

### 9.10.2 Integer transform

H.26L is a unique standard coder that employs a purely integer transform as opposed to the DCT transform with noninteger elements used in the other standard codecs. The core H.26L specifies a 4 × 4 integer transform which is an approximation to the 4 × 4 DCT (compare matrices in eqn. 9.20), hence it has a similar coding gain to the DCT transform. However, since the integer transform, $T_{int}$, has an exact inverse transform, there is no mismatch between the encoder and the decoder. Note that in DCT, due to the approximation of cosine values, the forward and inverse transformation matrices cannot be exactly the inverse of each other and hence encoder/decoder mismatch is a common problem in all standard DCT-based codecs that has to be rectified. Transformation matrices in eqn. 9.20 compare the 4 × 4 $T_{int}$ integer transform against the 4 × 4 DCT transform of the same dimensions.

$$T_{int} = \begin{bmatrix} 1 & 1 & 1 & 1 \\ 2 & 1 & -1 & -2 \\ 1 & -1 & -1 & 1 \\ 1 & -2 & 2 & -1 \end{bmatrix}$$

$$T_{cos} = \begin{bmatrix} 1 & 1 & 1 & 1 \\ 1.3 & 0.54 & -0.54 & -1.3 \\ 1 & -1 & -1 & 1 \\ 0.54 & -1.3 & 1.3 & -0.54 \end{bmatrix} \quad (9.20)$$

It may be beneficial to note that applying the transform and then the inverse transform does not return the original data. This is due to scaling factors built into the transform definition, where the scaling factors are different for different frequency coefficients. This scaling effect is removed partly by using different quantiser step sizes for the different frequency coefficients, and by a bit shift applied after the inverse transform. Moreover, use of the smaller block size of 4 × 4 reduces the blocking artefacts. Even more interesting, use of integer numbers makes transformation fast, as multiplication by 2 is simply a shift of data by one bit to the left. This can significantly reduce the processing power at the encoder, which can be very useful in power constrained processing applications, such as video over mobile networks, and allow increased parallelism.

Note that, as we saw in Chapter 3, for the two-dimensional transform the second stage transform is applied in the vertical direction of the first stage transform coefficients. This means transposing the transform elements for the second stage. Hence, for integer transform of a block of 4 × 4 pixels, $x_{ij}$, the 16 transform coefficients, $y_{ij}$ are calculated as:

$$\begin{bmatrix} y_{00} & y_{01} & y_{02} & y_{03} \\ y_{10} & y_{11} & y_{12} & y_{13} \\ y_{20} & y_{21} & y_{22} & y_{23} \\ y_{30} & y_{31} & y_{32} & y_{33} \end{bmatrix} = \begin{bmatrix} 1 & 1 & 1 & 1 \\ 2 & 1 & -1 & -2 \\ 1 & -1 & -1 & 1 \\ 1 & -2 & 2 & -1 \end{bmatrix} \times \begin{bmatrix} x_{00} & x_{01} & x_{02} & x_{03} \\ x_{10} & x_{11} & x_{12} & x_{13} \\ x_{20} & x_{21} & x_{22} & x_{23} \\ x_{30} & x_{31} & x_{32} & x_{33} \end{bmatrix} \times \begin{bmatrix} 1 & 2 & 1 & 1 \\ 1 & 1 & -1 & -2 \\ 1 & -1 & -1 & 2 \\ 1 & -2 & 1 & -1 \end{bmatrix}$$

(9.21)

As an optional mode, H.26L also specifies an integer transform of length 8. In this mode, the integer transform, which again is an approximation to the DCT of length 8, is defined as:

$$T_{\text{int}} = \begin{bmatrix} 13 & 13 & 13 & 13 & 13 & 13 & 13 & 13 \\ 19 & 15 & 9 & 3 & -3 & -9 & -15 & -19 \\ 17 & 7 & -7 & -17 & -17 & -7 & 7 & 17 \\ 9 & 3 & -19 & -15 & 15 & 19 & -3 & -9 \\ 13 & -13 & -13 & 13 & 13 & -13 & -13 & 13 \\ 15 & -19 & -3 & 9 & -9 & 3 & 19 & -15 \\ 7 & -17 & 17 & -7 & -7 & 17 & -17 & 7 \\ 3 & -9 & 15 & -19 & 19 & -15 & 9 & -3 \end{bmatrix}$$

(9.22)

Note that the elements of this integer transform are not powers of two, and hence some multiplications are required. Also, in this optional mode, known as the adaptive block transform (ABT), the two-dimensional blocks could be either: 4 × 4, 4 × 8, 8 × 4 or 8 × 8, depending on the texture of the image. Hence, a macroblock may be coded by a combination of these blocks, as required.

### 9.10.3 Intra coding

Intra coded blocks generate a large amount of data, which could be undesirable for very low bit rate applications. Observations indicate that there are significant

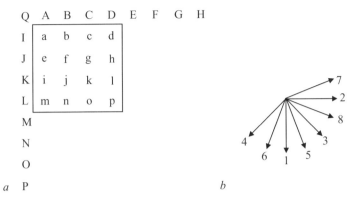

Figure 9.37  A 4 × 4 luminance pixel block and its eight directional prediction modes

correlations among the pixels of adjacent blocks, particularly when the block size is as small as 4 × 4 pixels. In core H.26L, for an intra coded macroblock of 16 × 16 pixels, the difference between the 4 × 4 pixel blocks and their predictions are coded.

In order to perform prediction, the core H.26L offers eight directional prediction modes plus a DC prediction (mode 0) for coding of the luminance blocks, as shown in Figure 9.37 [7]. The arrows of Figure 9.37b indicate the directions of the predictions used for each pixel of the 4 × 4 block in Figure 9.37a.

For example, when DC prediction (mode 0) is chosen, then the difference between every pixel of the block and a DC predictor defined as:

$$DC_{prd} = (A + B + C + D + I + J + K + L)//8 \quad (9.23)$$

is coded, where $A$, $B$, $C$ etc. are the pixels at the top and left sides of the block to be coded and // indicates division by rounding to the nearest integer. As another example, in mode 1, which is a vertical direction prediction mode, as shown in Figure 9.37b, every column of pixels uses the top pixel at the border as the prediction. In this mode, the prediction for all four pixels a, e, i and m would be A, and the prediction for, say, pixels d, h, l and p would be pixel D. Some modes can have a complex form of prediction. For example, mode 6, which is called the vertical right prediction mode, indicates that the prediction for, say, pixel c is:

$$prd = (C + D)//2$$

and the prediction for pixel e in this mode would be

$$prd = (A + 2B + C + K + 2L + M)//8 \quad (9.24)$$

For the complete set of prediction modes, the H.26L Recommendation should be consulted [7].

In the plain areas of the picture there is no need to have nine different prediction modes, as they unnecessarily increase the overhead. Instead, H.26L recommends only four prediction modes: DC, horizontal, vertical and plane [7]. Moreover, in the intra macroblocks of plain areas, normally AC coefficients of each $4 \times 4$ are small and may not be needed for transmission. However, the DC coefficients can be large, but they are highly correlated. For this reason, the H.26L standard suggests that the 16 DC coefficients in a macroblock should be decorrelated by the Hadamard transform. Hence if the DC coefficient of the $4 \times 4$ integer transform coefficient of block $(i, j)$ is called $x_{ij}$ then the two-dimensional Hadamard transform of the 16 DCT coefficients gives 16 new coefficients $y_{ij}$ as:

$$\begin{bmatrix} y_{00} & y_{01} & y_{02} & y_{03} \\ y_{10} & y_{11} & y_{12} & y_{13} \\ y_{20} & y_{21} & y_{22} & y_{23} \\ y_{30} & y_{31} & y_{32} & y_{33} \end{bmatrix} = \begin{bmatrix} 1 & 1 & 1 & 1 \\ 1 & 1 & -1 & -1 \\ 1 & -1 & -1 & 1 \\ 1 & -1 & 1 & -1 \end{bmatrix} \times \begin{bmatrix} x_{00} & x_{01} & x_{02} & x_{03} \\ x_{10} & x_{11} & x_{12} & x_{13} \\ x_{20} & x_{21} & x_{22} & x_{23} \\ x_{30} & x_{31} & x_{32} & x_{33} \end{bmatrix} \times \begin{bmatrix} 1 & 1 & 1 & 1 \\ 1 & 1 & -1 & -1 \\ 1 & -1 & -1 & 1 \\ 1 & -1 & 1 & -1 \end{bmatrix}$$
(9.25)

where $y_{00}$ is the DC value of all the DC coefficients of the $4 \times 4$ integer transform blocks. The overall DC coefficient $y_{00}$ is normally large, but the remaining $y_{ij}$ coefficients are small. This mode of intra coding is called the intra $16 \times 16$ mode [7].

For prediction of chrominance blocks (both intra and inter), since there are only four chrominance blocks of each type in a macroblock, then first of all the recommendation suggests using four prediction modes for a chrominance macroblock. Second, the four DC coefficients of the chrominance $4 \times 4$ blocks are now Hadamard transformed with transformation matrix of

$$\begin{bmatrix} 1 & 1 \\ 1 & -1 \end{bmatrix}$$

Note all these transformation (as well as the inverse transformation) matrices should be properly weighted such that the reconstructed pixel values are positive with eight bits resolution (see problem 5 in section 9.11).

### 9.10.4 Inter coding

Interframe predictive coding is where H.26L makes most of its gain in coding efficiency. Motion compensation on each $16 \times 16$ macroblock can be performed with different block sizes and shapes, as shown in Figure 9.38. In mode $16 \times 16$, one motion vector per macroblock is used, and in modes $16 \times 8$ and $8 \times 16$ there are two motion vectors. In addition, there is a mode known as $8 \times 8$ split, in which each of the $8 \times 8$ blocks can be further subdivided independently into $8 \times 8$, $8 \times 4$, $4 \times 8$, $4 \times 4$ or $8 \times 8$ intra blocks.

Experimental results indicate that using all block sizes and shapes can lead to bit rate savings of more than 15 per cent compared with the use of only one motion vector per $16 \times 16$ macroblock. Another 20 per cent saving in bit rate is achieved by

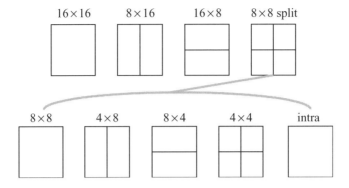

*Figure 9.38 Various motion compensation modes*

representing the motion vectors with quarter pixel spatial accuracy as compared with integer pixel spatial accuracy. It is expected that eighth pixel spatial accuracy will increase the saving rate even further.

*9.10.5 Multiple reference prediction*

The H.26L standard offers the option of using many previous pictures for prediction. This will increase the coding efficiency as well as producing a better subjective image quality. Experimental results indicate that using five reference frames for prediction results in a bit rate saving of about 5–10 per cent. Moreover, using multiple reference frames improves the resilience of H.26L to errors, as we have seen from Figure 9.22.

In addition to allowing immediately previous pictures to be used for prediction, H.26L also allows pictures to be stored for as long as desired and used at any later time for prediction. This is beneficial in a range of applications, such as surveillance, when switching between a number of cameras in fixed positions can be encoded much more efficiently if one picture from each camera position is stored at the encoder and the decoder.

*9.10.6 Deblocking filter*

The H.26L standard has adopted some of the optional features of H.263 that have been proven to significantly improve the coding performance. For example, while data partitioning was an option in H.263, it is now in the core H.26L. The other important element of the core H.26L is the use of an adaptive deblocking filter that operates on the horizontal and vertical block edges within the prediction loop to remove the blocking artefacts. The filtering is based on $4 \times 4$ block boundaries, in which pixels on either side of the boundary may be filtered, depending on the pixel differences across the block boundary, the relative motion vectors of the two blocks, whether the blocks have coefficients and whether one or other block is intra coded [7].

*9.10.7 Quantisation and scanning*

Similar to the H.263 standard, the H.26L standard also uses a dead band zone quantiser and zigzag scanning of the quantised coefficients. However, the number of quantiser step sizes and the adaptation of the step size is different. While in H.263 there are thirty-two different quantisation levels, H.26L recommends fifty-two quantisation levels. Hence, H.26L coefficients can be coded much more coarsely than those of H.263 for higher compression and more finely to produce better quality, if the bit rate budget permits (see problem 8, section 9.11). In H.26L the quantiser step size is changed at a compound rate of 12.5 per cent. The fidelity of the chrominance components is improved by using finer quantisation step sizes as compared with those used for the luminance component, particularly when the luminance quantiser step size is large.

## 9.10.8 Entropy coding

Before transmission generated data of all types is variable length (entropy) coded. As we discussed in Chapter 3, entropy coding is the third element of the redundancy reduction technique that all the standard video encoders try to employ as best they can. However, the compression efficiency of entropy coding is not at the same degree of spatial and temporal redundancy reduction techniques. Despite this, H.26L has put a lot of emphasis on entropy coding. The reason is that, although with sophisticated entropy coding the entropy of symbols may be reduced by a bit or a fraction of a bit, for very low bit rate applications (e.g. 20 kbit/s), when the symbols are aggregated over the frame or video sequence, the reduction can be quite significant.

The H.26L standard, like its predecessor, specifies two types of entropy coding: Huffman and arithmetic coding. To make Huffman encoding more efficient, it adaptively employs a set of VLC tables based on the context of the symbols. For example, Huffman coding of zigzag scanned transform coefficients is performed as follows. Firstly, a symbol that indicates the number of nonzero coefficients, and the number of these at the end of the scan that have magnitude of one, up to a maximum of three, is encoded. This is followed by one bit for each of the indicated trailing ones to indicate the sign of the coefficient. Then any remaining nonzero coefficient levels are encoded, in reverse zigzag scan order, finishing with the DC coefficient, if nonzero. Finally, the total number of zero coefficients before the last nonzero coefficient is encoded, followed by the number of zero coefficients before each nonzero coefficient, again in reverse zigzag scan order, until all zero coefficients have been accounted for. Multiple code tables are defined for each of these symbol types, and the particular table used adapts as information is encoded. For example, the code table used to encode the total number of zero coefficients before the last nonzero coefficient depends on the actual number of nonzero coefficients, as this limits the maximum value of zero coefficients to be encoded. The processes are complex and for more information the H.26L recommendation [7] should be consulted. However, this complexity is justified by the additional compression advantage that it offers.

For arithmetic coding, H.26L uses context-based adaptive binary arithmetic coding (CABAC). In this mode the intersymbol redundancy of already coded symbols in the neighbourhood of the current symbol to be coded is used to derive a context model. Different models are used for each syntax element (e.g. motion vectors and transform coefficients use different models) and, like Huffman, the models are adapted based on the context of the neighbouring blocks. When the symbol is not binary, it is mapped onto a sequence of binary decisions, called bins. The actual binarisation is done according to a given binary tree. Each binary decision is then encoded with the arithmetic encoder using the new probability estimates, which have been updated during the previous context modelling stage. After encoding of each bin, the probability estimate for the binary symbol that was just encoded is incremented. Hence the model keeps track of the actual statistics. Experiments with test images indicate that context-based adaptive binary arithmetic coding in H.26L improves bit saving by up to 10 per cent over the Huffman code, where the Huffman used in H.26L itself is more efficient than the conventional nonadaptive methods used in the other codecs.

### 9.10.9 Switching pictures

One of the applications for the H.26L standard is video streaming. In this operation the decoder is expected to decode a compressed video bit stream generated by a standard encoder.

One of the key requirements in video streaming is to adapt the transmission bit rate of the compressed video according to the network congestion condition. If the video stream is generated by an online encoder (real time), then according to the network feedback, rate adaptation can be achieved on the fly by adjusting the encoder parameters such as the quantiser step size, or in the extreme case by dropping frames. In typical streaming, where preencoded video sequences are to be streamed from a video server to a client, the above solution cannot be used. This is because any change in the bit stream makes the decoded picture different from the locally decoded picture of the encoder. Hence the quality of the decoded picture gradually drifts from that of the locally decoded picture of the encoder, and the visual artefacts will further propagate in time so that the quality eventually becomes very poor.

The simplest way of achieving scalability of preencoded bit streams is by producing multiple copies of the same video sequence at several bit rates, and hence qualities. The server then dynamically switches between the bit streams, according to the network congestion or the bandwidth available to the client. However, in switching on the P-pictures, since the prediction frame in one bit stream is different from the other, the problem of picture drift remains unsolved.

To rectify picture drift in bit stream switching, the H.26L Recommendation introduces a new type of picture, called the switching picture or the secondary picture, for short: SP-picture. SP-pictures are generated by the server at the time of switching from one bit stream to another and are transmitted as the first picture after the switching. To see how SP-pictures are generated and how they can prevent picture drift, consider switching between two bit streams as an example, shown in Figure 9.39.

Consider a video sequence encoded at two different bit rates, generating bit streams A and B. Also, consider a client who up to time $N$ has been receiving bit stream A, and at time $N$ wishes to switch to bit stream B. The bit streams per frame at this time are identified as $S_A$ and $S_B$, respectively.

To analyse the picture drift, let us assume in both cases that the pictures are entirely predictively coded (P-picture). Thus at time $N$, in bit stream A the difference

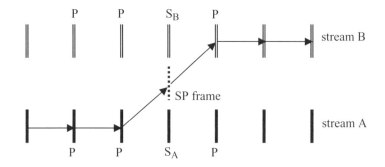

*Figure 9.39    Use of S frame in bit stream switching*

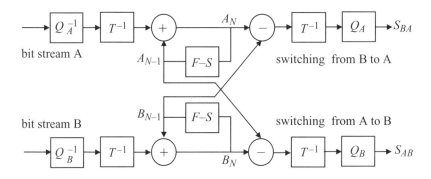

*Figure 9.40    Creation of the switching pictures from the bit streams*

between frame $N$ and $N-1$, $S_A = A_N - A_{N-1}$, is transmitted and for bit stream B, $S_B = B_N - B_{N-1}$. The decoded picture at the receiver prediction loop before switching would be frame $A_{N-1}$. Hence, for drift-free pictures, at the switching time the decoder expects to receive $S_{AB} = B_N - A_{N-1}$, from bit stream B, which is different from $S_B = B_N - B_{N-1}$. Therefore, at the switching time, a new bit stream for the duration of one frame, SP, should be generated, as shown in Figure 9.39. SP now replaces $S_B$ at the switching time, and if switching was done from bit stream B to A, we needed $SP = S_{BA} = A_N - B_{N-1}$, to replace $S_A$.

The switching pictures $S_{AB}$ and $S_{BA}$ can be generated in two ways, each for a specific application. If it is desired to switch between two bit streams at any time, then at the switching point we need a video transcoder to generate the switching pictures where they are wanted. Figure 9.40 shows such a transcoder, where each bit stream is decoded and then reencoded with the required prediction picture. If a macroblock is intra coded, the intra coded parts of the S frame of the switched bit stream are directly copied in the SP picture.

Although the above method makes it possible to switch between two bit streams at any time, it is expensive to use it along with the bit stream. A cheaper solution would

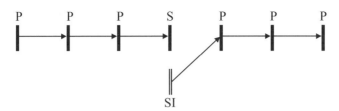

*Figure 9.41*   *The position of the secondary picture in random accessing of the bit stream*

be to generate these switching pictures ($S_{AB}$ and $S_{BA}$) offline and store them along the bit streams A and B. Normally, these pictures are created periodically, with reasonable intervals. Here there is a trade-off between the frequency of switching between the bit streams and the storage required to store these pictures. It is in this application that these pictures are also called secondary pictures, as well as the switching pictures.

### 9.10.10   Random access

Another vital requirement in video streaming is to be able to access the video stream and decode the pictures almost at any time. In accessing preencoded video, we have seen that the approach taken by the MPEG codecs was to insert I-pictures at regular intervals (e.g. every 12 frames). Due to the high bit rates of these pictures, their use in very low bit rate codecs, such as H.26L, would be very expensive. The approach taken by the H.26L standard is to create secondary pictures at the time access is required. These pictures are intraframe coded I-pictures and are encoded in such a way as to generate a decoded picture identical to the frame to be accessed. Figure 9.41 shows the position of the secondary picture for accessing the bit stream at the switching frame, S.

In the Figure the bit stream is made of P-pictures for high compression efficiency. One of these pictures, identified as picture S, is the picture where the client can access the bit stream if he wishes to. From the original picture an SI-picture is encoded to provide an exact match for picture S. Note that the first part of the accessed bit stream is now the data generated by the SI-picture. Thus the amount of bits at the start is high. However, since in video streaming at each session accessing occurs only once, these extra bits compared with the duration of the session are negligible. The average bit rate hardly changes, and is much less than when regular I-pictures are present.

In the above method, similar to the offline creation of secondary pictures for bit stream switching, the frequency of SI pictures is traded against the storage capacity required to store them along with the bit stream.

### 9.10.11   Network adaptation layer

An important feature of H.26L is the separation of the video coding layer (VCL) from the network adaptation layer (NAL). The VCL is responsible for high compression representation of the video picture content. In the previous sections some of the important methods used in the video coding layer were presented. In this section we describe the NAL which is responsible for efficient delivery of compressed video over the underlying transport network. Specifications for the NAL depend on the nature of the transport network. For example, for fixed bit rate channels, the MPEG-2 transport layer can be used. In this mode, 184 bytes of compressed video bits along with a four-byte header are packed into a packet. However, even if the channel is variable in rate, provided that some form of guaranteeing quality of service can be provided, such as ATM networks, the MPEG-2 transport protocol can still be used. In this case each MPEG-2 packet is segmented into four ATM cells, each cell carrying 47 bytes in its payload.

Perhaps the most interesting type of NAL is the one for the transport of a compressed video stream over a nonreliable network such as the best effort Internet protocol (IP) network. In this application NAL takes the compressed bit stream and converts it into packets that can be conveyed directly over the real time transport protocol (RTP) [24].

RTP has emerged as the *de facto* specification for the transport of time sensitive streamed media over IP networks. RTP provides a wrapper for media data that identifies the media type and provides synchronisation information in the form of a time stamp. These features enable individual packets to be reconstructed into a media stream by a recipient. To enable flow control and management of the media stream, additional information is carried between the sender and the receiver using the associated RTP control protocol (RTCP). Whenever an RTP stream is required, associated out of band RTCP flows between the sender and receiver are established. This enables the sender and receiver to exchange information concerning the performance of the media transmission and may be of use to higher level application control functions. Typically, RTP is mapped onto and carried in user data packets (UDP). In such cases, an RTP session is usually associated with an even numbered port and its associated RTCP flows with the next highest odd numbered port.

Encapsulation of video bit stream into RTP packets should be done with the following procedures:

- arrange partitions in an intelligent way into packets
- eventually split/merge partitions to match the packet size constraints
- avoid the redundancy of (mandatory) RTP header information and information in the video stream
- define receiver/decoder reactions to packet losses.

In packetised transmission, the packetisation scheme has to make some assumptions on typical network conditions and constraints. The following set of assumptions has been made on packetised transmission of H.26L video streams:

- *maximum packet size*: around 1500 bytes for any transmission media except the dial-up link; for this link 500 byte packets to be used
- *packet loss characteristics*: nonbursty, due to drop tail router implementations and assuming reasonable pacing algorithm (e.g. no bursting occurs at the sender)
- *packet loss rate*: up to 20 per cent.

In packetised transmission, the quality of video can be improved by stronger protection of the important packets against errors or losses. This is very much in line with the data partitioning feature of the H.26L coded bit stream. In the NAL it is expected that the bit stream generated per slice will be data partitioned into three parts, each being carried in a different packet. The first packet contains the important part of the compressed data, such as header type, macroblock type and address, motion vectors etc. The second packet contains intra information, and the third packet is assembled with the rest of the video bit stream, such as transform coefficients.

This configuration allows decoding of the first packet independently from the second and third packets, but not *vice versa*. As the first packet is more important than the second packet, which is more important than the third (it carries the most vital video information and is also necessary to decode the second and third packets), it is more heavily protected with forward error correction than the second and third packets.

It should be noted that slices should be structured to ensure that all packets meet the maximum packet size to avoid a network splitting/recombination process. This means inserting the slice header at appropriate points.

## 9.11 Problems

1 Assume the DCT coefficients of problem 3 of Chapter 8 are generated by an H.263 encoder. After zigzag scanning, and quantisation with $th = q = 8$, they are converted into three-dimensional events of (last, run, index). Identify these events.

2 The neighbouring motion vectors of the motion vector MV are shown in Figure 9.42. Find:

  a  the median of the neighbouring motion vectors
  b  the motion vector data (MVD), if the motion vector MV is (2,1).

| 5, –2 | 4, 3 | –1, 1 |
|---|---|---|
| 3, –3 | MV | |

*Figure 9.42*

3 The intensities of four pixels A, B, C and D of the borders of two macroblocks are given in Figure 9.43.

Using the deblocking filter of eqn. 9.4, find the interpolated pixels, $B_1$ and $C_1$ at the macroblock boundary for each of:

  (i)  $A = 100, B = 150, C = 115$ and $D = 50$
  (ii) $A = B = 150$ and $C = D = 50$

Assume the quantiser parameter of macroblock 2 is $Q_P = 16$.

$$\begin{array}{c|c} A\ B & C\ D \\ MB1 & MB2 \end{array}$$

*Figure 9.43*

4 Figure 9.44 shows the six neighbouring macroblocks of a lost motion vector. The value of these motion vectors are also given. Calculate the estimated motion vector for this macroblock for each of the following loss concealment methods:

   (i) top
   (ii) bottom
   (iii) mean
   (iv) majority
   (v) vector median

| 1 | 2 | 3 |
|---|---|---|
|   | mv |   |
| 4 | 5 | 6 |

$mv_1 = (2, 3);\ mv_2 = (3, 4);\ mv_3 = (-2, -1)$
$mv_4 = (4, 1);\ mv_5 = (0, -3);\ mv_6 = (-1, -1)$

*Figure 9.44*

5 Show that for the integer transforms of lengths four to be orthonormal, the DC and the second AC coefficients should be divided by 2, but the first and the third AC coefficients should be divided by $\sqrt{10}$. Determine the inverse transformation matrix and show it is an orthonomal matrix.

6 A block of 4 × 4 pixels given by:

$$[x] = \begin{bmatrix} 100 & 120 & 85 & 10 \\ 80 & 70 & 60 & 50 \\ 110 & 90 & 100 & 120 \\ 180 & 200 & 150 & 200 \end{bmatrix}$$

is two-dimensionally transformed by a 4 × 4 DCT and the integer transform of problems 5. The coefficients are zigzag scanned and $N$ out of 16 coefficients in the scanning order are retained. For each transform determine the reconstructed block, and its PSNR value, given that the number of retained coefficients are:

   (i) $N = 10$
   (ii) $N = 6$
   (iii) $N = 3$

7 The fifty-two quantiser levels of an H.26L encoder may be indexed from 0 to 51. If at the lowest index index_0, the quantiser step size is $Q$, find the quantiser step sizes at the following indices:

  (i) index_8
  (ii) index_16
  (iii) index_32
  (iv) index_48
  (v) index_51

8 If the lowest index of H.26L has the same quantisation step size as the lowest index of H.263, show that the H.26L quantiser is finer at lower step sizes but is coarser at larger step sizes than that of H.263.

## 9.12 References

1 Draft ITU-T Recommendation, H.263: 'Video coding for low bit rate communication'. July 1995
2 H.261: 'ITU-T Recommendation, H.261, video codec for audiovisual services at p × 64 kbit/s'. Geneva, 1990
3 MPEG-1: 'Coding of moving pictures and associated audio for digital storage media at up to about 1.5 Mbit/s'. ISO/IEC 1117–2: video, November 1991
4 MPEG-2: 'Generic coding of moving pictures and associated audio information'. ISO/IEC 13818–2 video, draft international standard, November 1994
5 Draft ITU-T Recommendation H.263+: 'Video coding for very low bit rate communication'. September 1997
6 ITU-T recommendation, H.263++: 'Video coding for low bit rate communication'. ITU-T SG16, February 2000
7 Joint video team (JVT) of ISO/IEC MPEG and ITU-T VCEG: text of committee draft of joint video specification. ITU-T Rec, H.264 | ISO/IEC 14496–10 AVC, May 2002
8 MPEG-4: 'Testing and evaluation procedures document'. ISO/IEC JTC1/SC29/WG11, N999, July 1995
9 ITU-T Recommendation, H.245 'Control protocol for multimedia communication'. September 1998
10 WALLACE, G.K.: 'The JPEG still picture compression standard', *Commun. ACM*, 1991, **34**, pp.30–44
11 GHANBARI, M., DE FARIA, S. GOH, I.N., and TAN, K.T.: 'Motion compensation for very low bit rate video', *Signal Process., Image Commun.*, 1994 **7**, pp.567–580
12 NETRAVALI, A.N., and ROBBINS, J.B.: 'Motion-compensated television coding: Part I', *Bell Syst. Tech. J.*, 1979, **58**, pp.631–670
13 LOPES, F.J.P., and GHANBARI, M.: 'Analysis of spatial transform motion estimation with overlapped compensation and fractional-pixel accuracy', *IEE Proc., Vis. Image Signal Process.*, 1999, **146**, pp.339–344

14 SEFERIDIS, V., and GHANBARI, M.: 'General approach to block matching motion estimation', *Opt. Eng.*, 1993, **32**, pp.1464–1474
15 NAKAYA, Y., and HARASHIMA, H.: 'Motion compensation based on spatial transformation', *IEEE Trans. Circuits Syst. Vide. Technol.*, 1994, **4**, pp.339–356
16 BLAHUT, R.E.: 'Theory and practice of error control codes' (Addison-Wesley, 1983)
17 KHAN, E., GUNJI, H., LEHMANN, S., and GHANBARI, M.: 'Error detection and correction in H.263 coded video over wireless networks'. International workshop on packet video, PVW2002, Pittsburgh, USA
18 GHANBARI, M., and SEFERIDIS, V.: 'Cell loss concealment in ATM video codecs', *IEEE Trans. Circuits Syst. Video Technol.*, Special issue on packet video, 1993, **3:3**, pp.238–247
19 GHANBARI, S., and BOBER, M.Z.: 'A cluster-based method for the recovery of lost motion vectors in video coding'. The 4th IEEE conference *on Mobile and wireless communications networks*, MWCN'2002, September 9–11, 2002, Stockholm, Sweden
20 SHANABLEH, T., and GHANBARI, M.: 'Loss concealment using B-pictures motion information', *IEEE Trans. Multimedia* (to be published in 2003)
21 LIM, C.P., TAN, E.A.W., GHANBARI, M., and GHANBARI, S.: 'Cell loss concealment and packetisation in packet video', *Int. J. Imaging Syst. Technol.*, 1999, **10**, pp.54–58
22 ITU SGXV working party XV/I, Experts Group for ATM video coding, working document AVC-205, January 1992
23 JOCH, A., KOSSENTINI, F., and NASIOPOULOS, P., 'A performance analysis of the ITU-T draft H.26L video coding standard'. Proceedings of 12th international *Packet video workshop*, 24–26 April 2002, Pittsburgh, USA
24 SCHULZRINE, H, CASNER, S., FREDRICK, R., and JACOBSON, V.: 'RFC 1889: RTP: A transport protocol for real-time applications'. Audio-video transport working group, January 1996
25 Some of the H.263 annexes used in this chapter:
    Annex C: 'Considerations for multipoint'
    Annex D: 'Unrestricted motion vector mode'
    Annex E: 'Syntax-based arithmetic coding mode'
    Annex F: 'Advanced prediction mode'
    Annex G: 'PB frames mode'
    Annex H: 'Forward error correction for coded video signal'
    Annex I: 'Advanced intra coding mode'
    Annex J: 'Deblocking filter mode'
    Annex K: 'Slice structured mode'
    Annex M: 'Improved PB frames mode'
    Annex N: 'Reference picture selection mode'
    Annex O: 'Temporal, SNR, and spatial scalability mode'
    Annex R: 'Independent segment decoding mode'
    Annex S: 'Alternative inter VLC mode'
    Annex V: 'Data partitioning'

*Chapter 10*
# Content-based video coding (MPEG-4)

MPEG-4 is another ISO/IEC standard, developed by MPEG (Moving Picture Experts Group), the committee that also developed the Emmy Award winning standards of MPEG-1 and MPEG-2. While MPEG-1 and MPEG-2 video aimed at devising coding tools for CD-ROM and digital television respectively, MPEG-4 video aims at providing tools and algorithms for efficient storage, transmission and manipulation of video data in multimedia environments [1,2]. The main motivations behind such a task are the proven success of digital video in three fields of digital television, interactive graphics applications (synthetic image content) and the interactive multimedia (worldwide web, distribution and access to image content). The MPEG-4 group believe these can be achieved by emphasising the functionalities of the proposed codec, which include efficient compression, object scalability, spatial and temporal scalability, error resilience etc.

The approach taken by the experts group in coding of video for multimedia applications relies on a content-based visual data representation of scenes. In content-based coding, in contrast to conventional video coding techniques, a scene is viewed as a composition of video objects (VO) with intrinsic properties such as shape, motion and texture. It is believed that such a content-based representation is a key to facilitating interactivity with objects for a variety of multimedia applications. In such applications, the user can access arbitrarily shaped objects in the scene and manipulate these objects.

The MPEG-4 group has defined the specifications of their intended video codec in the form of verification models (VM) [3]. The verification model in MPEG-4 has the same role as the reference and test models defined for H.261 and MPEG-2, respectively. The verification model has evolved over time by means of core experiments in various laboratories round the world. It is regarded as a common platform with a precise definition of the encoding and decoding algorithms that can be represented as tools addressing specific functionalities of MPEG-4. New algorithms/tools are added to the VM and old algorithms/tools are replaced in the VM by successful core experiments.

So far the verification model has been gradually evolved from version 1.0 to version 11.0, and during each evolution new functionalities have been added. In this Chapter we do not intend to review all of them, but instead to address those functionalities that have made MPEG-4 video coding radically different from its predecessors. Hence, it is intended to look at the fundamentals of new coding algorithms that have been introduced in MPEG-4. Before going into details of this coding technique, let us examine the functionalities of this codec.

## 10.1 Levels and profiles

MPEG-4, as with MPEG-2, has so many functionalities that users may only be interested in a subset of them. These are defined as profiles. For each profile, a number of resolution states such as bit rate and frame rate and pixel resolutions can be defined as levels. Since MPEG-4 is a very versatile video coding tool it has several levels and profiles, and every now and then many more are added to them. Profile in MPEG-4 is also synonymous with the support of a set of annexes in H.263. Some well known profiles with the associated commonly used levels are as follows:

- *Simple profile*: this profile provides the simplest tool for low cost applications, such as video over mobile and Internet. It supports up to four rectangular objects in a scene within QCIF pictures. There are three levels in this profile to define bit rates from 64–384 kbit/s (64, 128 and 384 kbit/s for level-1, level-2 and level-3, respectively). The simple profile also supports most of the optionalities (annexes in H.263) that are mainly useful for error resilience transmission. In addition to I and P-VOPs (video object planes to be defined shortly) they include: AC/DC prediction, four motion vectors, unrestricted motion vectors, quarter-pixel spatial accuracy, slice synchronisation, data partitioning and reversible VLC. This profile can decode a bit stream generated by the core H.263.
- *Simple scalable profile*: this adds the support for B-VOPs and temporal and spatial scalability to the simple profile. It provides services to the receivers requiring more than one level of quality of service, such as video over Internet.
- *Advanced real time simple*: this adds error protection to the simple profile, through the introduction of the back channel. In response to a negative acknowledgement from the decoder, the encoder encodes the affected parts of the picture in intra mode. This profile improves the robustness of real time visual services over error prone channels such as videophone.
- *Advanced simple profile*: this improves the compression efficiency of the simple profile, by supporting quarter-pixel resolution and global motion estimation in addition to B-VOPs.
- *Fine granular scalability profile*: this is similar to SNR scalability, but the enhancement layers are represented in bit planes, to offer up to eight scalable layers. It is mainly used with the simple or advanced simple profile as the base layer.
- *Core profile*: this adds scalability to still textures, B-VOPs, binary shape coding and temporal scalability of rectangular as well as binary shape objects to the

simple profile. Its maximum bit rate for level-1 is 384 kbit/s and for level-2 is 1 Mbit/s. This profile is useful for high quality interactive services, as well as mobile broadcast services.
- *Core scalable visual profile*: this adds object-based SNR, spatial and temporal scalability to the core profile.
- *Main profile*: this supports for interlaced video, greyscale alpha maps and sprites. This profile is intended for broadcast services that can handle both progressive and interlaced video. It can handle up to 32 objects with a maximum bit rate of 38 Mbit/s.
- *Advanced coding efficiency*: this profile is an extension of the main profile, but for bit rates of less than 1 Mbit/s. It adds quarter pixel and global motion estimation to the main profile to improve the encoding efficiency. However, it does not support sprites.
- *Simple studio profile*: this only supports I-VOP pictures coded at very high quality up to 1200 Mbit/s. As the name implies it is designed for studio applications, and can support high resolution video for HDTV and digital cinema. Each I-VOP may have an arbitrary shape and have several alpha planes.
- *Core studio profile*: this adds P-VOPs to the simple studio profile, to reduce the bit rate for very high quality video.

## 10.2 Video object plane (VOP)

In object-based coding the video frames are defined in terms of layers of video object planes (VOP). Each video object plane is then a video frame of a specific object of interest to be coded, or to be interacted with. Figure 10.1*a* shows a video frame that is made of three VOPs. In this Figure, the two objects of interest are the balloon and the aeroplane. They are represented by their video object planes of $VOP_1$ and $VOP_2$. The remaining part of the video frame is regarded as a background, represented by $VOP_0$. For coding applications, the background is coded only once, and the other object planes are encoded through time. At the receiver the reconstructed background is repeatedly added to the other decoded object planes. Since in each frame the encoder only codes the objects of interest (e.g. $VOP_1$ and/or $VOP_2$), and usually these objects represent a small portion of the video frame, then the bit rate of the encoded video stream can be extremely low. Note that, had the video frame of Figure 10.1*a* been coded with a conventional codec such as H.263, since clouds in the background move, the H.263 encoder would have inevitably encoded most parts of the picture with a much higher bit rate than that generated from the two objects.

The VOP can be a semantic object in the scene, such as the balloon and aeroplane in Figure 10.1. It is made of $Y$, $U$ and $V$ components plus their shapes. The shapes are used to mask the background, and help to identify object boarders.

In MPEG-4 video, the VOPs are either known by construction of the video sequence (hybrid sequence based on blue screen composition or synthetic sequences) or are defined by semiautomatic segmentation. In the former, the shape information

298  *Standard codecs: image compression to advanced video coding*

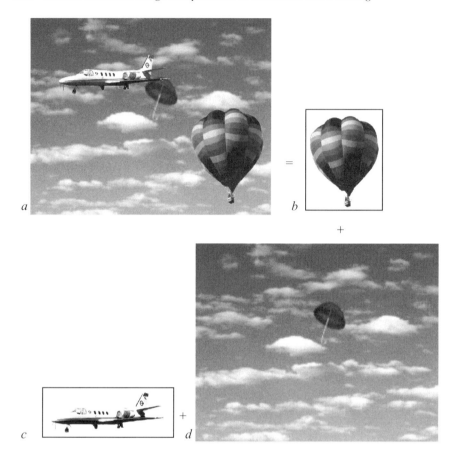

*Figure 10.1*  *a*  A video frame composed of
  *b*  balloon $VOP_1$,
  *c*  aeroplane $VOP_2$ and
  *d*  the background $VOP_0$

is represented by eight bits, known as the greyscale alpha plane. This plane is used to blend several video object planes to form the video frame of interest. Thus with eight bits, up to 256 objects can be identified within a video frame. In the second case, the shape is a binary mask, to identify individual object borders and their positions in the video frames.

Figure 10.2 shows the binary shapes of the balloon and aeroplane in the above example. Both cases are currently considered in the encoding process. The VOP can have an arbitrary shape. When the sequence has only one rectangular VOP of fixed size displayed at a fixed interval, it corresponds to frame-based coding. Frame-based coding is similar to H.263.

## Content-based video coding (MPEG-4) 299

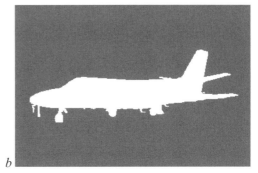

*Figure 10.2  Shape of objects*
*a*  balloon
*b*  aeroplane

### 10.2.1  Coding of objects

Each video object plane corresponds to an entity that after being coded is added to the bit stream. The encoder sends, together with the VOP, composition information to indicate where and when each VOP is to be displayed. Users are allowed to trace objects of interest from the bit stream. They are also allowed to change the composition of the entire scene displayed by interacting with the composition information.

Figure 10.3 illustrates a block diagram of an object-based coding verification model (VM). After defining the video object planes, each VOP is encoded and the encoded bit streams are multiplexed to a single bit stream. At the decoder the chosen object planes are extracted from the bit stream and then are composed into an output video to be displayed.

### 10.2.2  Encoding of VOPs

Figure 10.4 shows a general overview of the encoder structure for each of the video object planes (VOPs). The encoder is mainly composed of two parts: the shape encoder and the traditional motion and texture encoder (e.g. H.263) applied to the same VOP.

Before explaining how the shape and the texture of the objects are coded, in the following we first explain how a VOP should be represented for efficient coding.

300  *Standard codecs: image compression to advanced video coding*

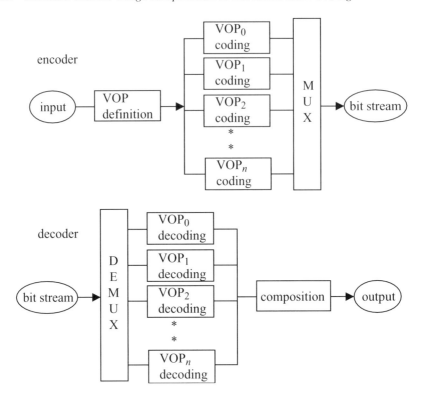

*Figure 10.3  An object-based video encoder/decoder*

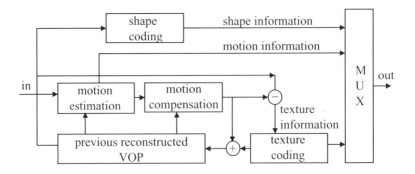

*Figure 10.4  VOP encoder structure*

### 10.2.3  Formation of VOP

The shape information is used to form a VOP. For maximum coding efficiency, the arbitrary shape VOP is encapsulated in a bounding rectangle such that the object

## Content-based video coding (MPEG-4)    301

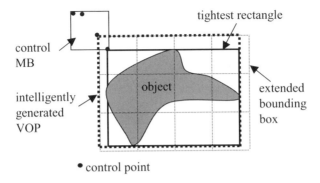

Figure 10.5    Intelligent VOP formation

contains the minimum number of macroblocks. To generate the bounding rectangle, the following steps are followed:

1. Generate the tightest rectangle around the object, as shown in Figure 10.5. Since the dimensions of the chrominance VOP are half of the luminance VOP (4 : 2 : 0), then the top left position of the rectangle should be an even numbered pixel.
2. If the top left position of this rectangle is the origin of the frame, skip the formation procedure.
3. Form a control macroblock at the top left corner of the tightest rectangle, as shown in Figure 10.5.
4. Count the number of macroblocks that completely contain the object, starting at each even numbered point of the control macroblock. Details are as follows:
   (i) Generate a bounding rectangle from the control point to the right bottom side of the object that consists of multiples of 16 × 16 pixel macroblocks.
   (ii) Count the number of macroblocks in this rectangle that contain at least one object pixel.
5. Select that control point which results in the smallest number of macroblocks for the given object.
6. Extend the top left coordinate of the tightest rectangle to the selected control coordinate.

This will create a rectangle that completely contains the object but with the minimum number of macroblocks in it. The VOP horizontal and vertical spatial references are taken directly from the modified top left coordinate.

## 10.3    Image segmentation

If VOPs are not available, then video frames need to be segmented into objects and a VOP derived for each one. In general, segmentation consists of extracting image regions of similar properties such as brightness, colour or texture. These regions are then used as masks to extract the objects of interest from the image [4]. However, video segmentation is by no means a trivial task and requires the use of a combination of techniques that were developed for image processing, machine vision and video coding. Segmentation can be performed either automatically or semiautomatically.

### 10.3.1 Semiautomatic segmentation

In semiautomatic video segmentation, the first frame is segmented by manually tracing a contour around the object of interest. This contour is then tracked through the video sequence by dynamically adapting it to the object's movements. The use of this deformable contour for extracting regions of interest was first introduced by Kass et al. [5], and it is commonly known as active contour or active snakes.

In the active snakes method, the contour defined in the first frame is modelled as a polynomial of an arbitrary degree, the most common being a cubic polynomial. The coefficients of the polynomial are adapted to fit the object boundary by minimising a set of constraints. This process is often referred to as the minimisation of the contour energy. The contour stretches or shrinks dynamically while trying to seek a minimal energy state that best fits the boundary of the object. The total energy of a snake is defined as the weighted sum of its internal and external energy given by:

$$E_{total} = \alpha E_{internal} + \beta E_{external} \tag{10.1}$$

In this equation, the constants $\alpha$ and $\beta$ give the relative weightings of the internal and the external energy. The internal energy ($E_{internal}$) of the snake constrains the shape of the contour and the amount it is allowed to grow or shrink. External energy ($E_{external}$) can be defined from the image property, such as the local image gradient. By suitable use of the gradient-based external constraint, the snake can be made to wrap itself around the object boundary. Once the boundary of the object is located in the first frame, the shape and positions of the contours can be adapted automatically in the subsequent frames by repeated applications of the energy minimisation process.

### 10.3.2 Automatic segmentation

Automatic video segmentation aims to minimise user intervention during the segmentation process and is a significantly harder problem than semiautomatic segmentation. Foreground objects or regions of interest in the sequence have to be determined automatically with minimal input from the user. To achieve this, spatial segmentation is first performed on the first frame of the sequence by partitioning it into regions of homogeneous colour or grey level. Following this, motion estimation is used to identify moving regions in the image and to classify them as foreground objects.

Motion information is estimated for each of the homogenous regions using a future reference frame. Regions that are nonstationary are classified as belonging to the foreground object and are merged to form a single object mask. This works well in simple sequences that contain a single object with coherent motion. For more complex scenes with multiple objects of motion disparity, a number of object masks have to be generated. Once the contours of the foreground objects have been identified, they can be tracked through the video sequence using techniques such as active snakes. In the following, each element of video segmentation is explained.

## 10.3.3 Image gradient

The first step of the segmentation process is the partitioning of the first video frame into regions of homogenous colour or grey level. A technique that is commonly used for this purpose is the watershed transform algorithm [6]. This algorithm partitions the image into regions which correspond closely to the actual object boundaries, but direct application of the algorithm on the image often leads to over segmentation. This is because, in addition to the object borders in the image, texture and noise may also create artificial contours. To alleviate this problem, the image gradient is used as input to the watershed transform instead of the actual image itself.

### 10.3.3.1 Nonlinear diffusion

To eliminate the false contours the image has to be smoothed, in such a way that edges are not affected. This can be done by diffusing the image iteratively without allowing the diffusion process to cross the object borders. One of the best known filters for this purpose is nonlinear diffusion, where the diffusion is prohibited across edges but unconstrained in regions of almost uniform texture.

Nonlinear diffusion can be realised by space variant filtering, which adapts the Gaussian kernel dynamically to the local gradient estimate [7]. The Gaussian kernel is centred at each pixel and its coefficients are weighted by a function of the gradient magnitude between the centre pixel and its neighbours. The weighted coefficients $a_{ij}$ are given by:

$$a_{ij} = \frac{w_{ij}c_{ij}}{\sum_{i,j=1} w_{ij}c_{ij}} \qquad w_{ij} = g(\| \nabla I_{ij} \|) \tag{10.2}$$

where $c_{ij}$ is the initial filter coefficient, $w_{ij}$ is the weight assigned to each coefficient based on the local gradient magnitude $\nabla I_{ij}$ and $g$ is the diffusion function [8]. If $w_{ij} = 1$, then the diffusion is linear. The local gradient is computed for all pixels within the kernel by taking the difference between the centre and the corresponding pixels. More precisely, $\nabla I_{ij} = I_{ij} - I_c$, where $I_c$ is the value of the centre pixel and $I_{ij}$ are the pixels within the filter kernel.

Figure 10.6 shows the diffusion function and the corresponding adapted Gaussian kernel to inhibit diffusion across a step edge. The Figure shows that for large gradient magnitudes, the filter coefficient is reduced in order not to diffuse across edges. Similarly, at the lower gradient magnitudes (well away from the edges) within the area of almost uniform texture, the diffusion is unconstrained, so as to have maximum smoothness. The gradient image generated in this form is suitable to be segmented at its object borders by the watershed transform.

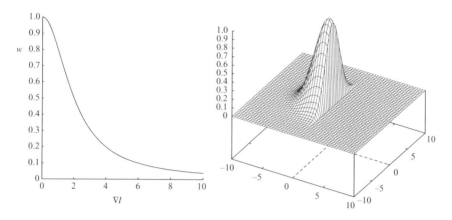

Figure 10.6  Weighting function (left), adapted Gaussian kernel (right)

### 10.3.3.2  Colour edge detection

Colour often gives more information about the object boundaries than the luminance. However, although gradient detection in greyscale images is to some extent easy, deriving the gradient of colour images is less straightforward due to separation of the image into several colour components. Each pixel in the RGB colour space can be represented by a position vector in the three-dimensional RGB colour cube:

$$\vec{p} = R\vec{i} + G\vec{j} + B\vec{k} \tag{10.3}$$

Likewise, pixels in other colour spaces can be represented in a similar manner. Hence, the problem of detecting edges in a colour image can be redefined as the problem of computing the gradient between vectors in the three-dimensional colour space. Various methods have been proposed for solving the problem of colour image gradient detection and they include component-wise gradient detection and vector gradient.

*Component-wise gradient*

In this approach, gradient detection is performed separately for each colour channel, followed by the merging of the results. The simplest way is to apply the Prewitt or the Sobel operator separately on each channel and then combine the gradient magnitude using some linear or nonlinear function [10]. For example:

$$\nabla I = \frac{\nabla I_R + \nabla I_G + \nabla I_B}{3} \quad \text{or} \quad \nabla I = \max(\nabla I_R; \nabla I_G; \nabla I_B) \tag{10.4}$$

where $\nabla I_R, \nabla I_G, \nabla I_B$ represent the gradient magnitude computed for each of the colour channels.

*Vector gradient*

This approach requires the definition of a suitable colour difference metric in the three-dimensional colour space. This metric should be based on human perception of colour difference, such that a large value indicates significant perceived difference, and a small value indicates colour similarity. A metric that is commonly used is the Euclidean distance between two colour vectors. For the $n$-dimensional colour space, the Euclidean distance between two colour pixels, $A$ and $B$, is defined as follows:

$$euclid(A, B) = \sqrt{\sum_{i=1}^{n}(A_i - B_i)^2} \qquad (10.5)$$

The gradient of the image is then derived, based on the sum of the colour differences between each pixel and its neighbours. This approximates the Prewitt kernel and is given by the following equations.

$$\frac{\partial I}{\partial x} = \sum_{a=y-1}^{y+1} euclid[I(x+1, a), I(x-1, a)]$$

$$\frac{\partial I}{\partial y} = \sum_{a=x-1}^{x+1} euclid[I(a, y+1), I(a, y-1)] \qquad (10.6)$$

The gradient magnitude is then defined as:

$$\|\nabla I\| = \sqrt{\left(\frac{\partial I}{\partial x}\right)^2 + \left(\frac{\partial I}{\partial y}\right)^2}$$

The use of Euclidean distance in the RGB colour space does not give a good indication of the perceived colour differences. Hence, in [9], it was proposed that the vector gradient should be computed in the CIELUV colour space. CIELUV is an approximation of a perceptually uniform colour space, which is designed so that the Euclidean distance could be used to quantify colour similarity or difference [10].

## 10.3.4 Watershed transform

The watershed transform algorithm can be visualised by adopting a topological view of the image, where high intensity values represent peaks and low intensity values represent valleys. Water is flooded into the valleys and the lines where the waters from different valleys meet are called the watershed lines, as shown in Figure 10.7. The watershed lines separate one region (catchment basin) from another. In actual implementation, all pixels in an image are initially classified as unlabelled and they are examined starting from the lowest intensity value. If the pixel that is currently being examined has no labelled neighbours, it is classified as a new region or a new catchment basin. On the other hand, if the pixel was a neighbour to exactly one

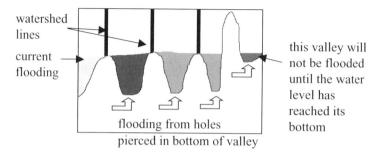

*Figure 10.7   Immersion-based watershed flooding*

labelled region, it would be classified as belonging to that region. However, if the pixel separates two or more regions, it would be labelled as part of the watershed line.

The algorithm can be seeded with a series of markers (seed regions), which represent the lowest point in the image from where water is allowed to start flooding. In other words, these markers form the basic catchment basins for the flooding. In some implementations, new basins are not allowed to form and, hence, the number of regions in the final segmented image will be equal to the number of markers. In other implementations, new regions are allowed to be formed if they are at a higher (intensity) level as compared with the markers. In the absence of markers, the lowest point (lowest intensity value) in the valley is usually chosen as the default marker.

There are two basic approaches to the implementation of the watershed transform and they are the immersion method [11] and the topological distance method [5]. Results obtained using different implementations are usually not identical.

#### 10.3.4.1   Immersion watershed flooding

Efficient implementations of immersion-based watershed flooding are largely based on Vincent and Soille's algorithm [11]. Immersion flooding can be visualised by imagining that holes are pierced at the bottom of the lowest point in each valley. The image is then immersed in water, which flows into the valley from the holes. Obviously, the lowest intensity valley will get flooded first before valleys that have a higher minimum point, as shown in Figure 10.7.

Vincent and Soille's algorithm uses a sorted queue in which pixels in the image are sorted based on their intensity level. In this way, pixels of low intensity (lowest points in valleys) could be examined first before the higher intensity pixels. This algorithm achieves fast performance since the image is scanned in its entirety only once during the queue setup phase.

#### 10.3.4.2   Topological distance watershed

In the topological distance approach, a drop of water that falls on the topology will flow to the lowest point *via* the slope of the steepest descent. The slope of the steepest descent is loosely defined as the shortest distance from the current position to the local minima. If a certain point in the image has more than one possible steepest descent, that point would be a potential watershed line that separates two or more regions. A more rigorous mathematical definition of this approach can be found in [5].

### 10.3.5 Colour similarity merging

Despite the use of nonlinear diffusion and the colour gradient, the watershed transformation algorithm still produces an excessive number of regions. For this reason, the regions must be merged based on colour similarity in a suitable colour space. The CIELUV uniform colour space is a plausible candidate due to the simplicity of quantifying colour similarity based on Euclidean distances. The mean luminance $L$ and colour differences $u^*$ and $v^*$ values are computed for each region of the watershed transformed image. The regions are examined in the increasing order of size and merged with their neighbours if the colour difference between them is less than a threshold ($T_h$). More precisely, the regions $A$ and $B$ are merged if:

$$euclid(A_{L,u^*,v^*}, B_{L,u^*,v^*}) < T_h \tag{10.7}$$

where $A_{L,u^*,v^*}$ and $B_{L,u^*,v^*}$ are the mean CIELUV values of the respective regions. Since it is likely that each region would have multiple neighbours, the neighbour that minimises the Euclidean distance is chosen as the target for merging. After each merging, the mean $L$, $u^*$ and $v^*$ values of the merged region are updated accordingly.

### 10.3.6 Region motion estimation

Spatial segmentation alone cannot discriminate the foreground objects in a video sequence. Motion information between the current frame and a future reference frame must be utilised to identify the moving regions of the image. These moving regions can then be classified as foreground objects or as regions of interest.

Motion estimation is performed for each of the regions after the merging process. Since regions can have any arbitrary shapes, the boundary of each region is extended to its tightest fitting rectangle. Using the conventional block matching motion estimation algorithm (BMA), the motion of the fitted rectangular is estimated.

Figure 10.8 gives a pictorial summary of the whole video segmentation process. The first frame of the video after filtering and gradient operator generates the gradient of the diffused image. This image is then watershed transformed into various regions, some of which might be similar in colour and intensity. In the above example, the watershed transformed image has about 2660 regions, but many regions are identical in colour. Through the colour similarity process, identical colour regions are merged into about 60 regions (in this picture). Finally, homogeneous neighbouring regions are grouped together, and are separated from the static background, resulting in two objects: the ball and the hand and the table tennis racket.

### 10.3.7 Object mask creation

The final step of the segmentation process is the creation of the object mask. The object mask is synonymous with object segmentation, since it is used to extract the

308  *Standard codecs: image compression to advanced video coding*

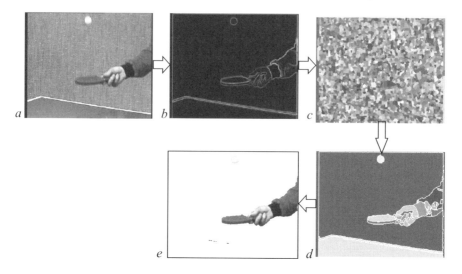

*Figure 10.8*   A pictorial representation of video segmentation
     *a*   original
     *b*   gradient
     *c*   watershed transformed
     *d*   colour merged
     *e*   segmented image

object from the image. The mask can be created from the union of all regions that have nonzero motion vectors. This works well for sequences that have a stationary background with a single moving object. For more complex scenes with nonstationary background and multiple moving objects, nine object masks are created for the image. Each mask is the union of all regions with coherent motion. Mask 0 corresponds to regions of zero motion, and the other eight masks are regions with motion in the N, NE, E, SE, S, SW, W and NW directions, respectively. These object masks are then used to extract the foreground objects from the video frame.

Figure 10.9 shows two examples of creating the object mask and extracting the object, for two video sequences. The mother and daughter sequence has a fairly stationary background and hence the motion information, in particular motion over several frames, can easily separate foreground from the background.

For the BBC car, due to the camera following the car, the background is not static. Here before estimating the motion of the car, the global motion of the background needs to be compensated. In section 10.7 we will discuss global motion estimation.

The contour or shape information for each object can also be derived from the object masks and used for tracking the object through the video sequence. Once the object is extracted and its shape is defined, the shape information has to be coded and sent along with the other visual information to the decoder. In the following section several methods for coding of shapes are presented.

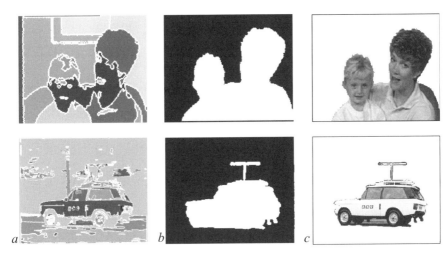

*Figure 10.9*  *a*  gradient image
  *b*  object mask
  *c*  segmented object

## 10.4 Shape coding

The binary and greyscale shapes are normally referred to as binary and greyscale alpha planes. Binary alpha planes are encoded with one of the binary shape coding methods (to be explained later), and the greyscale alpha planes are encoded by motion compensated DCT similar to texture coding (e.g. H.263). An alpha plane is bounded by a rectangle that includes the shape of the VOP, as described in the formation of VOP, in section 10.2.3. The bounding rectangle of the VOP is then extended on the right bottom side to multiples of $16 \times 16$ pixel macroblocks. The extended alpha samples are set to zero. Now the extended alpha plane can be partitioned into exact multiples of $16 \times 16$ pixel macroblocks. Hereafter, these macroblocks are referred to as alpha blocks, and the encoding and decoding process for block-based shape coding is carried out per alpha block.

If the pixels in an alpha block are all transparent (all zero), the block is skipped before motion and/or texture coding. No overhead is required to indicate this mode since this transparency information can be obtained from shape coding. This skipping applies to all I, P and B-VOPs. Since shape coding is unique to MPEG-4 (no other standard codecs use it), then in the following sections we pay special attention to various shape coding methods.

### 10.4.1 Coding of binary alpha planes

A binary alpha plane is encoded in the intra mode for I-VOPs and the inter mode for P-VOPs and B-VOPs. During the development of MPEG-4 several methods for coding of the binary alpha planes have been considered. These include chain coding of the

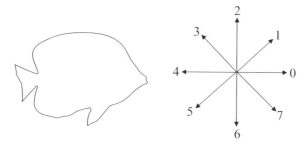

*Figure 10.10  Object boundaries for chain coding and the eight directions of the chain around the object*

object contours, quad tree coding, modified modified Reed (MMR) and context-based arithmetic encoding (CAE) [3,12]. It appears that CAE, recommended in the latest version of the verification model (VM-11) [3], is the best. Hence in the introduction of these methods, more details are given on the CAE.

### 10.4.2 Chain code

In the chain code method, the object boundaries are represented by a closed contour, as shown in Figure 10.10, and the chain codes are then applied to the contour. Derivation of the contour from the binary alpha plane is similar to detection of the edge, as discussed in section 10.3. For coding, it involves moving on the contour in one of eight directions, as shown in the Figure, and coding the direction. The chain code terminates when the starting point is revisited.

Each chain code contains a start point data followed by the first chain code, and the subsequent differential chain codes. If VOP contains several closed contours, then plural chain codes are coded following the data for the number of regions.

Since a chain code has a cyclic property, a differential chain code in eight directions can be expressed in the range from $-3$ to 4 by the following definition:

$$d = \begin{cases} c_n - c_{n-1} + 8, & \text{if } c_n - c_{n-1} < -3 \\ c_n - c_{n-1} - 8, & \text{if } c_n - c_{n-1} > 4 \\ c_n - c_{n-1}, & \text{otherwise} \end{cases} \quad (10.8)$$

where $d$ is the differential chain code, $c_n$ is the current chain code and $c_{n-1}$ is the previous chain code. Huffman code is used to encode the differential chain code $d$. The Huffman table is shown in Table 10.1.

At the receiver, after the variable length decoding of $d$, the current chain code, $c_n$, is then reconstructed as follows:

$$c_n = (c_{n-1} + d + 8) \text{ mode } 8 \quad (10.9)$$

*Content-based video coding (MPEG-4)* 311

*Table 10.1 Huffman table for the differential chain code*

| d  | code    |
|----|---------|
| 0  | 1       |
| 1  | 00      |
| −1 | 011     |
| 2  | 0100    |
| −2 | 01011   |
| 3  | 010100  |
| −3 | 0101011 |
| 4  | 0101010 |

### 10.4.3 Quad tree coding

Each binary alpha block (BAB) of 16 × 16 pixels, represented by binary data (white 255 and black 0), is first quad tree segmented. The indices of the segments, according to the rules to be explained, are calculated and then Huffman coded. Figure 10.11 shows the quad tree structure employed for coding of a binary alpha block.

At the bottom level (level-3) of the quad tree, a 16 × 16 alpha block is partitioned into 64 subblocks of 2 × 2 samples. Each higher level as shown also contains 16 × 16 pixels, but in groups of 4 × 4, 8 × 8 and 16 × 16 subblocks.

The calculation of the indices is as follows:

- The indexing of subblocks starts at level-3, where an index is assigned to each 2 × 2 subblock pixel.
- For the four pixels of the subblock $b[0]$ to $b[3]$ of Figure 10.12, the index is calculated as:

$$index = (27 \times b[0]) + (9 \times b[1]) + (3 \times b[2]) + b[3] \qquad (10.10)$$

where $b[i] = 2$ if the sample value is 255 (white) and $b[i] = 0$ if it is black. Hence there are 16 different index values with a minimum of 0 and a maximum of 80.

**Step 1: indexing of subblocks at level 3**

These indices then become inputs to level-2 for the generation of a new set of indices at this level. However, to increase inter block correlation, the subblocks are swapped in decreasing order of indices. The swapping also causes the distribution of index values to be predominantly low and hence this nonuniform distribution is more efficiently variable length coded. Arrangement for the swap of the four subblocks is carried out according to the relationship of the neighbouring indices in the following order. The upper and left neighbouring indices are shown in Figure 10.13.

312  *Standard codecs: image compression to advanced video coding*

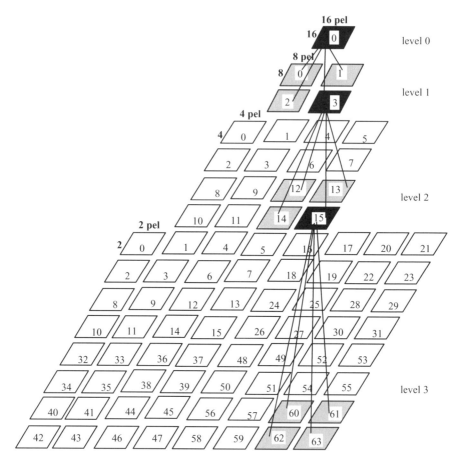

*Figure 10.11   Quad tree representation of a shape block*

| b[0] | b[1] |
|------|------|
| b[2] | b[3] |

*Figure 10.12   A subblock of 2 × 2 pixels*

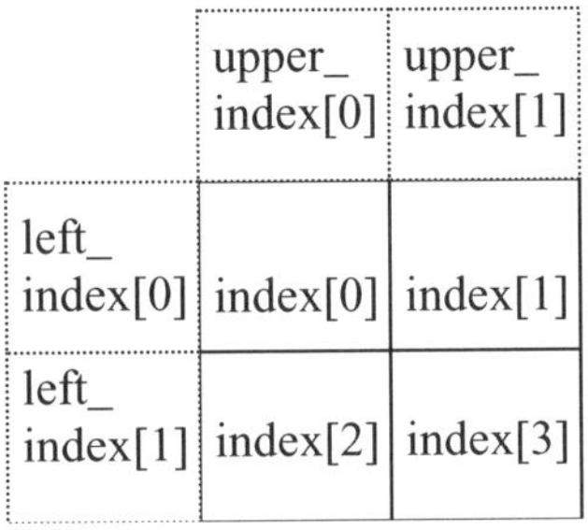

|  | upper_index[0] | upper_index[1] |
|---|---|---|
| left_index[0] | index[0] | index[1] |
| left_index[1] | index[2] | index[3] |

*Figure 10.13   Upper and left level indices of a subblock*

a  If upper_index[0] is less than upper_index[1], then swap index[0] with index[1] and index[2] with index[3], except for subblocks numbered 0,1,4,5,16,17,20 and 21.
b  If left_index[0] is less than left_index[1], then swap index[0] with index[2] and index[1] with index[3] except for subblocks numbered 0,2,8,10,32,34,40 and 42.
c  If upper_index[0]+upper_index[1] is less than left_index[0]+left_index[1], then swap index[1] with index[2] except for subblocks numbered 0,1,2,4,5,8,10,16,17, 20,21,32,34,40 and 42,
d  The index of level-2 is computed from index[0], index[1], index[2] and index[3] after swapping according to:

$$index\_level\_2 = 27 \times f(index[0]) + 9 \times f(index[1]) + 3 \times f(index[2]) + f(index[3])$$

where

$$f(x) = 0 \quad \text{if } x = 0$$
$$f(x) = 2 \quad \text{if } x = 80 \quad \quad (10.11)$$
$$f(x) = 1 \quad \text{otherwise}$$

The current subblock is then reconstructed and used as a reference when processing subsequent subblocks.

**Step 2: grouping process for higher levels**

The grouping process of blocks at the higher level first starts at level-2 where four subblocks from level-3 are grouped to form a new subblock. The grouping process involves swapping and indexing similarly to that discussed for level-3, except that in this level a 4 × 4 pixel block is represented by a 2 × 2 subblock whose elements are indices rather than pixels. The current subblock is then reconstructed and used as a reference when processing subsequent subblocks. At the decoder swapping is done following a reverse sequence of steps as at the encoder.

314  *Standard codecs: image compression to advanced video coding*

The grouping process is also performed similarly for level-1 where four sub-blocks from level-2 are grouped to form a new subblock. The swapping, indexing and reconstruction of a subblock follows grouping, the same as that for other levels.

Now arrangement of subblocks in decreasing order of their indices at level-2, to utilise interblock correlation, is done as follows:

a  If $f(\text{upper\_index}[0])$ is less than $f(\text{upper\_index}[1])$, then swap index[0] with index[1] and index[2] with index[3], except for subblocks numbered 0,1,4 and 5.
b  If $f(\text{left\_index}[0])$ is less than $f(\text{left\_index}[1])$, then swap index[0] with index[2] and index[1] with index[3] except for subblocks numbered 0,2,8 and 10.
c  If $f(\text{upper\_index}[0]) + f(\text{upper\_index}[1])$ is less than $f(\text{left\_index}[0]) + f(\text{left\_index}[1])$, then swap index[1] with index[2] except for subblocks numbered 0,1,2,4,5,8 and 10.
d  The index of level-1 is computed from index[0], index[1], index[2] and index[3] after swapping, according to eqn. 10.11.

At level-1 no swapping is required.

**Step 3: encoding process**

The encoding process involves use of results from the grouping process which produces a total of 85 ($=1 + 4 + 16 + 64$) indices for a $16 \times 16$ alpha block. Each index is encoded from the topmost level (level 0). At each level, the order for encoding and transmission of indices is shown by numbers in Figure 10.11. Indices are Huffman coded. Note that indices at level-3 can take only 16 different values, but at the other levels they take 80 different values. Hence for efficient variable length coding, two different Huffman tables are used, one with 16 symbols at level-3 and the other with 80 symbols at levels 0 to 2. These tables are shown in Appendix C.

*10.4.4  Modified modified Reed (MMR)*

During the course of MPEG-4 development, another shape coding method named modified modified Reed (MMR) was also investigated [12]. The basic idea in this method is to detect the pixel intensity changes (from opaque to transparent and *vice versa*) within a block, and to code the positions of the changing pixels. The changing pixels are defined by pixels whose colour changes while scanning an alpha block in raster scan order. Figure 10.14 illustrates the changing pixels in an intra and motion compensated inter alpha block. Also in the Figure are the top and left reference row and column of pixels, respectively.

To code the position of the changing pixels, the following parameters, as shown in Figure 10.14, are defined.

For each length, the starting pixel is denoted by $a_0$. Initially $a_0$ is positioned at the right end in the top reference and addressed as abs_$a_0 = -1$. The first changing pixel appearing after $a_0$ is called $a_1$. For intra blocks, the pixel located above $a_0$ is called $b_0$. The area from the next pixel to $b_0$ down to $a_0$ is called the reference area in intraMMR. In this case the reference changing pixel is called $b_1$. This pixel is found in reference area $r_1$ or $r_2$, shown in Figure 10.15a. Searching in $r_1$, the first changing

Content-based video coding (MPEG-4) 315

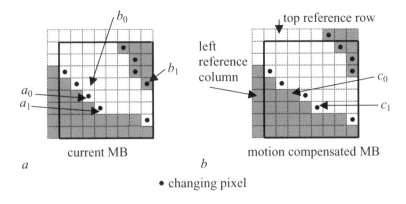

*Figure 10.14* Changing pixels
 a intra alpha block
 b inter alpha block

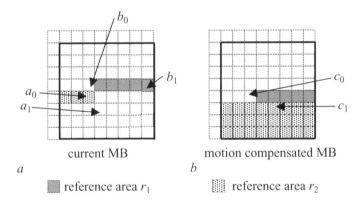

*Figure 10.15* Reference area for detecting reference changing pixel
 a $b_1$
 b $c_1$

pixel whose colour is the opposite of $a_0$ is named $b_1$; if not found, the first changing pixel in $r_2$ is named $b_1$. Thus $b_1$ might become identical to $a_0$. Exceptionally $b_1$ may not be found when abs_$a_0 = -1$.

For interMMR, the pixel at the corresponding position to $a_0$ is called $c_0$. The area from the next pixel to $c_0$ down to the end of alpha block is called the reference area in interMMR. The reference changing pixel in this mode is called $c_1$. This pixel is found in the reference area $r_1$ or $r_2$, shown in Figure 10.15b. Searching in $r_1$, the first changing pixel whose colour is the opposite of $a_0$ is named $c_1$; if not found, the first changing pixel in $r_2$ is named $c_1$.

Given $a_0$, the eventual task is to identify $a_1$ by referring to either $b_1$ or $c_1$, depending on the intraMMR or interMMR mode, respectively. This is achieved in three different modes, the vertical, horizontal and vertical pass modes. The vertical mode is the first option that is considered and if it is decided not to be applicable, the horizontal mode or the vertical pass mode is employed in turn.

In the vertical mode the position of $a_1$ is identified relative to that of $b_1$. It is invoked only when the absolute value of the relative distance defined by the relative address $DIST = r\_a_1 - r\_b_1$, is equal to or less than a predefined threshold [12]. The DIST is then variable length coded using the appropriate intraMMR and interMMR VLC tables.

If the vertical mode is not used (DIST > threshold), then the horizontal mode is considered for coding. In this mode the position of $a_1$ is identified on the basis of the absolute distance from $a_0$. If this distance is less than the width of the alpha block, then it is used; otherwise the vertical pass mode is used, which implies that one row of the pixels in the alpha block is passed (not coded).

Finally, the decision to use intraMMR or interMMR is to first scan the alpha block in the horizontal and vertical scanning directions. The one requiring the least number of bits is chosen. In the case of a tie, horizontal scanning is chosen. For the final decision between intraMMR and interMMR, again the one is selected that gives the least coding bits.

### 10.4.5 Context-based arithmetic coding

Each intracoded binary alpha block (BAB), and the intercoded one after being motion compensated by block-based motion compensation, is context-based arithmetic encoded (CAE). In general each binary alpha block is coded according to one of the following seven modes (in C terminology):

1 MVDs == 0 && No_Update
2 MVDs != 0 && No_Update
3 All_0
4 All_255
5 IntraCAE
6 MVDs == 0 && InterCAE
7 MVDs != 0 && InterCAE

The first and second modes indicate that the shape will not be updated, and All_0 and All_255 indicate that the BAB contains only black and white pixels, respectively. None of these modes is required to be arithmetic coded. Also, in the quad tree and MMR methods, All_0 and All_255 are not coded further.

IntraCAE is the mode for context-based arithmetic coding of BABs that contain a mixture of black and white pixels. In modes 6 and 7, the interframe BABs (mixed black and white pixels) with and without motion compensation, respectively, are arithmetic coded.

The motion vector data of shape, represented by MVDS, is the difference between the shape motion vector and its predictor, MVP. The prediction procedure is similar to

## Content-based video coding (MPEG-4)

that of the motion vector data for texture, described in section 9.1.2. However, there are differences, such as:

- the prediction motion vectors can be derived from the candidate motion vectors of shape MVS1, MVS2, MVS3 or the candidate motion vectors of the texture, MV1, MV2 and MV3, similar to those of H.263 illustrated in Figure 9.2; the prediction motion vector is determined by taking the first encountered motion vector that is valid; if no candidate is valid, the prediction is set to zero
- overlapped, half pixel precision and $8 \times 8$ motion compensation is not carried out
- in the case that the region outside the VOP is referred to, the value for that is set to zero
- for B-VOPs, only forward motion compensation is used and neither backward nor interpolated motion compensation is allowed.

It should be noted that when the shape prediction motion vector MVPS is determined, the difference between the motion compensated BAB indicated with MVPS and the current BAB is calculated. If the motion compensated error is less than a certain threshold (AlphaTH) for any $4 \times 4$ subblock of the BAB, the MVPS is directly employed as the best prediction. If this condition is not met, MV is searched around the prediction vector MVPS, by comparing the BAB indicated by the MV and the current BAB. The MV that minimises the error is taken as the best motion vector for shape, MVS, and the motion vector data for shape (MVDS) is given by MVDS = MVS − MVPS.

### 10.4.5.1 Size conversion

Rate control and rate reduction in MPEG-4 is realised through size conversion of the binary alpha information. This method is also applicable to quad tree and MRR. It is implemented in two successive steps.

In the first step, if required, the size of the VOP can be reduced by half in each of the horizontal and vertical directions. This is indicated in the VOP header, as the video object plane conversion ratio, VOP_CR, which takes a value of either 1 or $\frac{1}{2}$. When VOP_CR is $\frac{1}{2}$, size conversion is carried out on the original bounding box of Figure 10.5.

In the case that the value of VOP_CR is $\frac{1}{2}$, the locally decoded shape, which is size converted at the VOP level, is stored in the frame memory of the shape frame. For shape motion estimation and compensation, if VOP_CR of the reference shape VOP is not equal to that of the current shape VOP, the reference shape frame (not VOP) is size converted corresponding to the current shape VOP.

For P-VOPs, if the VOP_CR is $\frac{1}{2}$, the components of the shape motion information vector are measured on the downsampled shape frame. The predicted motion vector for the shape, MVPS, is calculated only using the shape motion vectors MVS1, MVS2 and MVS3.

In the second step, when required, the size conversion is carried out for every binary alpha block, BAB, except for All_0, All_255 and No_Update. At the block level the conversion ratio (CR) can be one of $\frac{1}{4}, \frac{1}{2}$ and 1 (the original size).

For CR = $\frac{1}{2}$, if the average of pixel values in a 2 × 2 pixel block is equal to or larger than 128, the pixel value of the downsampled block is set to 255, otherwise it is set to zero. For CR = $\frac{1}{4}$, if the average of pixels in a 4 × 4 pixel block is equal to or larger than 128, the pixel value of the downsampled block is set to 255, otherwise it is set to zero. In either of these cases, upsampling is carried out for the BAB. The values of the interpolated pixels are calculated from their neighbouring pixels, according to their Euclidean distances.

Selection of a suitable value for the conversion ratio (CR) is done based on the conversion error between the original BAB and the BAB which is once downsampled and then reconstructed by upsampling. The conversion error is computed for each 4×4 subblock, by taking the absolute difference of pixels in the corresponding subblocks of the original and reconstructed BABs. If this difference is greater than a certain threshold, this subblock is called an error pixel block (Error_PB). Size conversion at a certain conversion ratio is accepted if there is no Error_PB at that ratio. Figure 10.16 summarises determination of the conversion ratio, CR.

If a downsampled BAB turns out to be all transparent or all opaque and the conversion error in any 4 × 4 subblocks in the BAB is equal to or lower than the threshold, the shape information is coded as shape_mode = All_0 or All_255. Unless this is the case, the BAB is coded with a context-based arithmetic coding at the size determined by the algorithm for the rate control.

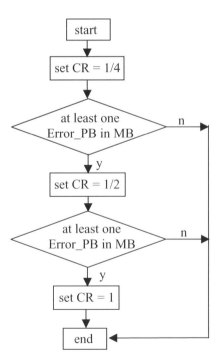

*Figure 10.16   CR determination algorithm*

### 10.4.5.2 Generation of context index

The pixels in the binary alpha blocks (BABs) are context-based arithmetic coded for both intra and inter modes. The number of pixels in the BAB is determined by the conversion ratio (CR), which is either 16 × 16, 8 × 8 or 4 × 4 pixels for CR values of 1, $\frac{1}{2}$ and $\frac{1}{4}$, respectively.

The context-based arithmetic encoding (CAE) is a binary arithmetic coding, where the symbol probability is determined from the context of the neighbouring pixels. Such a coding is applied to each pixel of the BAB in the following manner. First, prior to encoding of each BAB, the arithmetic encoder is initialised. Each binary pixel is then encoded in the raster scan order. The process for coding a given pixel is carried out using the following steps:

1. compute the context number
2. index a probability table using the context number
3. use the indexed probability to derive an arithmetic encoder.

When the final pixel has been encoded the arithmetic code is terminated.

Figure 10.17 shows the template of the neighbouring pixels that contribute to the creation of the context number for intra and inter shape pixels.

For intra coded BABs, a ten-bit context $C = \sum_{k=0}^{9} c_k \cdot 2^k$ is calculated for each pixel, as shown in Figure 10.17a. In this Figure the pixel to be encoded is represented by ? and the ten neighbouring pixels are ordered as shown. For inter coded BABs, in addition to spatial redundancy, temporal redundancy is exploited by using pixels from the bordered motion compensated BAB, to make up part of the context, as shown in Figure 10.17b. In this mode, only nine bits are required to calculate the context number, e.g. $C = \sum_{k=0}^{8} c_k 2^k$.

In both modes there are some special cases to note:

- In building contexts, any pixel outside the bounding box of the current VOP to the left and above are assumed to be zero.

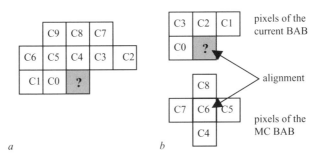

*Figure 10.17* Template for the construction of the pixels of
    *a*   the intra and
    *b*   inter BABs. The pixel to be coded is marked with '?'

- The template may cover pixels from BABs which are not known at the decoding time. The values of these unknown pixels are estimated by template padding in the following manner:

  a When constructing the intra context, the following steps are taken in sequence:

  (1) if (C7 is unknown) C7 = C8
  (2) if (C3 is unknown) C3 = C4
  (3) if (C2 is unknown) C2 = C3.

  b When constructing the inter context, the following conditional assignment is performed:

  (1) if (C1 is unknown) C1 = C2.

Once the context number is calculated, it is used to derive a probability table for binary arithmetic coding. Two probability tables, a 10-bit for intra and a 9-bit for inter BABs, are given in Appendix D. These tables contain the probabilities for a binary alpha pixel being equal to 0 for intra and inter shape coding using the context-based arithmetic coding. All probabilities are normalised to the range of [1, 65535].

As an example let us assume that the neighbouring pixels for an intra BAB template have a black and white pattern as shown in Figure 10.18.

In this Figure, C0 = C1 = C2 = C3 = C4 = C7 = 1, and C5 = C6 = C8 = C9 = 0. Hence the context number for coding of pixel ? is $C = 2^0 + 2^1 + 2^2 + 2^3 + 2^4 + 2^7 = 159$.

If pixel ? was a black pixel it would have been coded with an Intra_prob [159]. This value according to Appendix D is 74 out of 65535. If it was a white pixel, its probability would have been $65535 - 74 = 65461$ out of 65535. Such a high probability for a white pixel in this example is expected, since this pixel is surrounded by many white pixels. Note also, although the given probability table is fixed, as the pattern of neighbouring pixels changes, the calculated context number changes such that the assigned probability to the pixel is better suited for that pattern. This is a form of adaptive arithmetic coding which only looks at the limited number of past coded symbols. It has been shown that adaptive arithmetic coding with limited past history has more efficient compression over fixed rate arithmetic coding [13].

|    |    | C9 | C8 | C7 |    |
|----|----|----|----|----|----|
|    | C6 | C5 | C4 | C3 | C2 |
|    | C1 | C0 | ?  |    |    |

*Figure 10.18  An example of an intra BAB template*

Content-based video coding (MPEG-4) 321

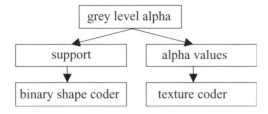

Figure 10.19    Greyscale shape coding

## 10.4.6   Greyscale shape coding

The grey level alpha plane is encoded as its support function and the alpha values on the support. The support is obtained by thresholding the grey level alpha plane by 0, and the remaining parts constitute the alpha values, as shown in Figure 10.19.

The support function is encoded by binary shape coding, as described in section 10.4.5. The alpha values are partitioned into $16 \times 16$ blocks and encoded the same way as the luminance of texture is coded.

## 10.5   Motion estimation and compensation

The texture of each VOP is motion compensated prior to coding. The motion estimation and compensation is similar to that of H.263 with the exception that the blocks on the VOP borders have to be modified to cater for the arbitrary shapes of the VOPs. These modified macroblocks are referred to as polygons, and the motion estimation is called polygon-based matching. Furthermore, since shapes change from time to time, some conversion is necessary to ensure the consistency of the motion compensation.

A macroblock that lies on the VOP boundary, called a boundary macroblock, is padded by replicating the boundary samples of the VOP towards the exterior. This process is carried out by repetitive padding in the horizontal and vertical directions. In case there are macroblocks completely outside the VOP, they are padded by extended padding.

In horizontal padding, each sample at the boundary of a VOP is replicated horizontally in the left or right direction in order to fill the transparent region outside the VOP of a boundary macroblock. If there are two boundary sample values for filling a sample outside a VOP, the two boundary samples are averaged. A similar method is used for vertical padding of the boundary macroblocks in the vertical direction.

Exterior macroblocks immediately next to boundary macroblocks are filled by replicating the samples at the border of the boundary macroblocks. The boundary macroblocks are numbered in a prioritised order according to Figure 10.20.

The exterior macroblock is then padded by replicating upwards, downwards, lefwards or rightwards the rows of sampling from the horizontal, vertial border of

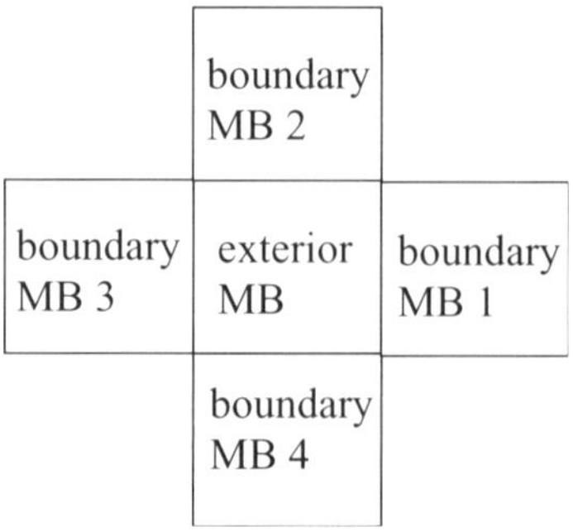

|              | boundary MB 2 |              |
|--------------|---------------|--------------|
| boundary MB 3 | exterior MB   | boundary MB 1 |
|              | boundary MB 4 |              |

*Figure 10.20  Priority of boundary macroblocks surrounding an exterior macroblock*

the boundary macroblock having the largest priority number. Note that the boundary macroblocks have already been padded by horizontal and vertical repetitive padding. The remaining macroblocks (not located next to any boundary macroblock) are filled with 128. The original alpha plane for the VOP is used to exclude the pixels of the macroblocks that are outside the VOP.

The reference VOP is padded based on its own shape information. For example, when the reference VOP is smaller than the current VOP, the reference is not padded up to the size of the current VOP.

The motion estimation and compensation with the padded VOPs can be carried out in several different forms, such as integer pixel motion estimation, half and quarter sample search, unrestricted motion estimation/compensation, overlapped motion compensation and advanced mode prediction. Motion vectors are then differentially encoded, similar to H.263.

## 10.6  Texture coding

The intra VOPs and motion compensated inter VOPs are coded with $8 \times 8$ block DCT. The DCT is performed separately for each of the luminance and chrominance planes.

For an arbitrarily shaped VOP, the macroblocks that completely reside inside the VOP shape are coded with a technique identical to H.263. Boundary macroblocks that are of intra type are padded with horizontal and vertical repetition. For inter macroblocks, not only is the macroblock repeatedly padded, but also the region outside the VOP within the block is padded with zeros. Transparent blocks are skipped and therefore are not coded. These blocks are then coded in a manner identical to the interior blocks. Blocks that lie outside the original shape are padded with 128, 128 and 128 for the luminance and the two chrominances in the case of intra and 0, 128 and 128 for inter macroblocks. Blocks that belong neither to the original nor the coded arbitrary shape but to the inside of the bounding box of the VOP are not coded at all.

## 10.6.1 Shape adaptive DCT

At the boundary macroblocks, the horizontally/vertically padded blocks can be coded with a standard 8 × 8 block DCT. This padding removes any abrupt transitions within a block, and hence reduces the number of significant DCT coefficients. At the decoder the added pixels are removed by the help of shape parameters from the decoded BABs.

Since the number of opaque pixels in the 8 × 8 blocks of some of the boundary macroblocks is usually less than 64 pixels, it would have been more efficient if these opaque pixels could have been DCT coded without padding. This method of DCT coding is called shape adaptive DCT (SA-DCT). The internal processing of SA-DCT is controlled by the shape information that has to be derived from the decoded BAB. Hence only opaque pixels within the boundary blocks are actually coded. As a consequence, in contrast to standard DCT, the number of DCT coefficients in an SA-DCT is equal to the number of opaque pixels.

There are two types of shape adaptive DCT, one used for inter blocks, known as SA-DCT and the other for intra blocks, known as $\Delta$SA-DCT, which is an extension of SA-DCT. For both cases, a two-dimensional separable DCT with a varying length of basis vector is used.

The basic concept of SA-DCT is shown in Figure 10.21. Segments of the opaque pixels are encapsulated in the picture grids of 8 × 8 pixel blocks as shown in Figure 10.21a. Each row of the pixels is shifted and aligned to the left, as shown in Figure 10.21b. The aligned pixels are then one-dimensionally DCT coded in the horizontal direction with variable basis functions, where the lengths of the basis functions are determined by the number of pixels in each line. For example, the first pixel of the segment is represented by itself, as a DC coefficient.

The second line of the pixels is DCT coded with a three-point transform, and the third line with a five-point transform, and so on. The coefficients of the $N$-point DCT, $c_j$, are defined by:

$$c_j = \sqrt{\frac{2}{N}} \, c_0 \cos\left[p(k + 0.5)\frac{\pi}{N}\right] \tag{10.12}$$

and

$$c_0 = \frac{\sqrt{2}}{2} \quad \text{if } p = 0;$$

$$c_0 = 1 \text{ otherwise}$$

Figure 10.21c illustrates the horizontal DCT coefficients, where the DC values are represented with a dot. These coefficients now become the input to the second stage one-dimensional DCT in the vertical direction. Again they are shifted upwards and aligned to the upper border, and the $N$-point DCT is applied to each vertical column, as shown in Figure 10.21e. The final two-dimensional DCT coefficients are shown in Figure 10.21f. Note that, since the shape information from the decoded BABs is known, these processes of shifting and alignments in the horizontal and vertical directions are reversible.

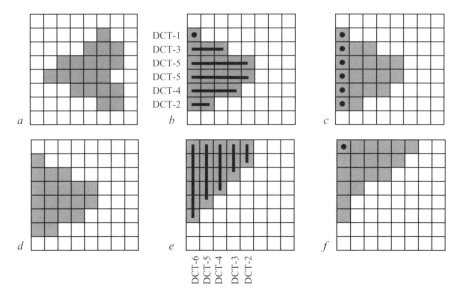

Figure 10.21  An example of SA-DCT
  a  original segment
  b  ordering of pixels and horizontal SA-DCT
  c  location 1D coefficients
  d  location of samples prior to vertical SA-DCT
  e  ordering of 1D samples and vertical SA-DCT
  f  location of 2D SA-DCT coefficients

The $\Delta$SA-DCT algorithm that is used for intra coded macroblocks is similar to SA-DCT but with extra processing. This additional processing is simply calculating the mean of the opaque pixels in the block, and subtracting the mean from each individual pixel. The resultant zero mean opaque pixels are then SA-DCT coded. The mean value is separately transmitted as the DC coefficient.

Zigzag scanning, quantisation and the variable length coding of the SA-DCT coefficients operations are similar to those of standard DCT used in H.263.

## 10.7  Coding of the background

An important advantage of the content-based coding approach in MPEG-4 is that the compression efficiency can be significantly improved by not coding the video background. If the background is completely static, all the other noncontent-based codecs would not code the background either. However, due to noise, camera shaking, or deliberate camera movement, such as panning, zooming, tilt etc., the background cannot be entirely static. One of the attractive features of content-based or object-based coding is that the background movement does not need to be represented in its exact form. Viewers normally do not pay attention to the accuracy of the background video, unless it is badly coded, such that the coding distortion distracts the viewer from the main scene.

An efficient way of coding the background is to represent the background movement with a global motion model. In section 9.4.5 we saw that motion compensation with the spatial transform was capable of dealing with complex motions such as translation, sheering, zooming etc. The same concept can be used to compensate for the global motion of the background. Here the whole background VOP is transformed to match against the VOP of the previous frame. The transformation parameters then represent the amount of information required to code the background VOP, which corresponds to an extremely low bit rate.

In MPEG-4, the global motion is either represented by six motion parameters (three motion vectors) through the affine transform, defined as:

$$x = \alpha_0 u + \alpha_1 v + \alpha_2 \quad \text{and} \quad y = \alpha_3 u + \alpha_4 v + \alpha_5 \tag{10.13}$$

or, with eight parameters (four motion vectors), using the perspective transform, defined by:

$$x = \frac{\alpha_0 u + \alpha_1 v + \alpha_2}{\alpha_6 u + \alpha_7 v + 1} \quad \text{and} \quad y = \frac{\alpha_3 u + \alpha_4 v + \alpha_5}{\alpha_6 u + \alpha_7 v + 1} \tag{10.14}$$

In both cases, similar to the bilinear transform of section 9.4.5, the coordinates $(u, v)$ in the current frame are matched against the coordinates $(x, y)$ in the previous frame. The best matching parameters then define the motion of the object (in this case, the background VOP).

In MPEG-4, global motion compensation is based on the transmission of a static sprite. A static sprite is a (possibly large) still image, describing panoramic background. For each consecutive image in a sequence, only eight global motion parameters describing camera motion are coded to reconstruct the object. These parameters represent the appropriate perspective transform of the sprite transmitted in the first frame.

Figure 10.22 depicts the basic concept for coding an MPEG-4 video sequence using a sprite panorama image. It is assumed that the foreground object (tennis player, top right image) can be segmented from the background and that the sprite panorama image can be extracted from the sequence prior to coding. (A sprite panorama is a still image that describes the content of the background over all frames in the sequence.) The large panorama sprite image is transmitted to the receiver only once as the first frame of the sequence to describe the background. The sprite is stored in a sprite buffer. In each consecutive frame only the camera parameters relevant for the background are transmitted to the receiver. This allows the receiver to reconstruct the background image for each frame in the sequence based on the sprite. The moving foreground object is transmitted separately as an arbitrary-shape video object. The receiver composes both the foreground and background images to reconstruct each frame (bottom picture in Figure 10.22). For low delay applications it is possible to transmit the sprite in multiple smaller pieces over consecutive frames or to build up progressively the sprite at the decoder.

326  *Standard codecs: image compression to advanced video coding*

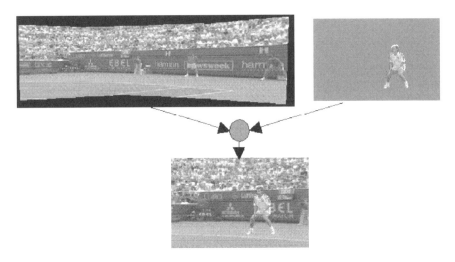

*Figure 10.22   Static sprite of Stefan (courtesy of MPEG-4)*

Global motion compensation can also be applied to the foreground objects. Here, for each VOP, a spatial transform, like the perspective transform, is used to estimate the transformation parameters. These are regarded as the global motion parameters that are used to compensate for the global motion of the foreground object. Globally motion compensated VOP is then coded by motion compensated texture coding, where it is once more motion compensated, but this time with a conventional block matching technique.

## 10.8   Coding of synthetic objects

Synthetic images form a subset of computer graphics, that can be supported by MPEG-4. Of particular interest in synthetic images is the animation of head-and-shoulders or cartoon-like images. Animation of synthetic faces was studied in the first phase of MPEG-4, and the three-dimensional body animation was addressed in the second phase of MPEG-4 development [2].

The animation parameters are derived from a two-dimensional mesh, which is a tessellation of a two-dimensional region into polygonal patches. The vertices of the polygonal patches are referred to as the node points or vertices of the mesh. In coding of the objects, these points are moved according to the movement of the body, head, eyes, lips and changes in the facial expressions. A two-dimensional mesh matched to the Claire image is shown in Figure 10.23a. Since the number of nodes representing the movement can be very small, this method of coding, known as model-based coding, requires a very low bit rate, possibly in the range of 10–100 bit/s [14].

In order to make synthetic images look more natural, the texture of the objects is mapped into the two-dimensional mesh, as shown in Figure 10.23b. For coding of the animated images, triangular patches in the current frame are deformed by the

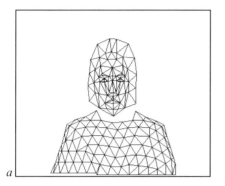 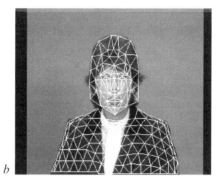

*Figure 10.23*   *a*   a two-dimensional mesh
　　　　　　　　*b*   the mapped texture

movement of the node points to be matched into the triangular patches or facets in the reference frame. The texture inside each patch in the reference frame is thus warped onto the current frame, using a parametric mapping, defined as a function of the node point motion vectors.

For triangular meshes, affine mapping with six parameters (three node points or vertices) is a common choice [15]. Its linear form implies that texture mapping can be accomplished with low computational complexity. This mapping can model a general form of motion including translation, rotation, scaling, reflection and shear, and preserves straight lines. This implies that the original two-dimensional motion field can be compactly represented by the motion of the node points, from which a continuous, piecewise affine motion field can be reconstructed. At the same time, the mesh structure constrains movements of adjacent image patches. Therefore, meshes are well suited to representing mildly deformable but spatially continuous motion fields.

However, if the movement is more complex, like the motion of lips, then affine modelling may fail. For example, Figure 10.24 shows the reconstructed picture of Claire, after nine frames of affine modelling. The accumulated error due to model failure around the lips is very evident.

For a larger complex motion, requiring a more severe patch deformation, one can use quadrilateral mappings with eight degrees of freedom. Bilinear and perspective mappings are these kinds of mapping, which have a better deformation capability over affine mapping [16,17].

## 10.9   Coding of still images

MPEG-4 also supports coding of still images with a high coding efficiency as well as spatial and SNR scalability. The coding principle is based on the discrete wavelet transform, which was described in some length in Chapter 4. The lowest subband

328　*Standard codecs: image compression to advanced video coding*

*Figure 10.24　Reconstructed model-based image with the affine transform*

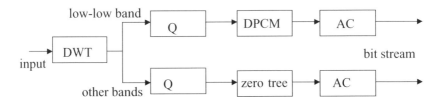

*Figure 10.25　Block diagram of a wavelet-based still image encoder*

after quantisation is coded with a differential pulse code modulation (DPCM) and the higher bands with a variant of embedded zero tree wavelet (EZW) [18]. The quantised DPCM and zero tree data are then entropy coded with an arithmetic encoder. Figure 10.25 shows a block diagram of the still image encoder.

In the following sections each part of the encoder is described.

### 10.9.1　*Coding of the lowest band*

The wavelet coefficients of the lowest band are coded independently from the other bands. These coefficients are DPCM coded with a uniform quantiser. The prediction for coding a wavelet coefficient $w_x$ is taken from its neighbouring coefficients $w_A$ or $w_C$, according to:

$$w_{prd} = w_C, \quad \text{if} |w_A - w_B| < |w_A - w_C|,$$
$$\text{otherwise } w_{prd} = w_A \qquad (10.15)$$

The difference between the actual wavelet coefficient $w_x$ and its predicted value $w_{prd}$ is coded. The positions of the neighbouring pixels are shown in Figure 10.26.

The coefficients after DPCM coding are encoded with an adaptive arithmetic coder. First the minimum value of the coefficient in the band is found. This value,

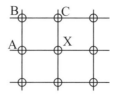

*Figure 10.26   Prediction for coding the lowest band coefficients*

known as *band_offset*, is subtracted from all the coefficients to limit their lower bound to zero. The maximum value of the coefficients as *band_max_value* is also calculated. These two values are included in the bit stream.

For adaptive arithmetic coding [19], the arithmetic coder model is initialised at the start of coding with a uniform distribution in the range of 0 to *band_max_value*. Each quantised and DPCM coded coefficient after arithmetic coding is added to the distribution. Hence, as the encoding progresses, the distribution of the model adapts itself to the distribution of the coded coefficients (adaptive arithmetic coding).

## 10.9.2   Coding of higher bands

For efficient compression of higher bands as well as for a wide range of scalability, the higher order wavelet coefficients are coded with the embedded zero tree wavelet (EZW) algorithm first introduced by Shapiro [18]. Details of this coding technique were given in Chapter 4. Here we show how it is used within the MPEG-4 still image coding algorithm.

Figure 10.27 shows a multiscale zero tree coding algorithm, based on EZW, for coding of higher bands. The wavelet coefficients are first quantised with a quantiser $Q_0$. The quantised coefficients are scanned with the zero tree concept (exploiting similarities among the bands of the same orientation), and then entropy coded with an arithmetic coder (AC). The generated bits comprise the first portion of the bit stream, as the base layer data, $BS_0$. The quantised coefficients of the base layer after inverse quantisation are subtracted from the input wavelet coefficients, and the residual quantisation distortions are requantised by another quantiser, $Q_1$. These are then zero tree scanned (ZTS), entropy coded to represent the second portion of the bit stream, $BS_1$. The procedure is repeated for all the quantisers, $Q_0$ to $Q_N$, to generate $N + 1$ layers of the bit stream.

The quantisers used in each layer are uniform with a dead band zone of twice the quantiser step size of that layer. The quantiser step size in each layer is specified by the encoder in the bit stream. As we are already aware each quantiser is a multilayer and the quantiser step size of a lower layer is several times that of its immediate upper layer. This is because, for a linear quantiser $Q_i$, with a quantiser step size of $q_i$, the maximum residual quantisation distortion is $q_i$ (for those which fall in the dead zone of $Q_i$) and $q_i/2$ (for those which are quantised). Hence for a higher layer quantiser $Q_{i+1}$, with a quantiser step size of $q_{i+1}$ to be efficient, then $q_{i+1}$

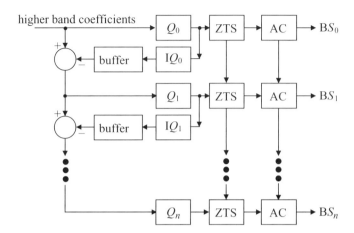

*Figure 10.27  A multiscale encoder of higher bands*

should be several times smaller than $q_i$. If $q_{i+1} = \frac{1}{2}q_i$, then it becomes a bilevel quantiser.

The number of quantisers indicates the number of SNR-scalable layers, and the quantiser step size in each layer determines the granularity of SNR scalability at that layer. For finest granularity of SNR scalability, all the layers can use a bilevel (one bit) quantiser. In this case, for optimum encoding efficiency, the quantiser step size of each layer is exactly twice that of its immediate upper layer. Multistage quantisation in this mode now becomes quantisation by successive approximation or bit plane encoding, described in Chapter 4. Here, the number of quantisers is equal to the number of bit planes, required to represent the wavelet transform coefficients. In this bilevel quantisation, instead of quantiser step sizes, the maximum number of bit planes is specified in the bit stream.

As the Figure shows, the quantised coefficients are zero tree scanned (ZTS) to exploit similarities among the bands of the same orientation. The zero tree takes advantage of the principle that if a wavelet coefficient at a lower frequency band is insignificant, then all the wavelet coefficients of the same orientation at the same spatial location are also likely to be insignificant. A zero tree exists at any node, when a coefficient is zero and all the node's children are zero trees. The wavelet trees are efficiently represented and coded by scanning each tree from the root at the lowest band through the children and assigning symbols to each state of the tree.

If a multilevel quantiser is used, then each node encounters three symbols: zero tree root (ZT), value zero tree root (VZ) and value (V). A zero tree root symbol denotes a coefficient that is the root of a zero tree. When such a symbol is coded, the zero tree does not need to be scanned further, because it is known that all the coefficients in such a tree have zero values. A value zero tree root symbol is a node where the coefficient has a nonzero value, and all its four children are zero tree roots. The scan of this tree can stop at this symbol. A value symbol identifies a coefficient with value either zero or nonzero, but some of the descendents are nonzero. The symbols and the quantised coefficients are then entropy coded with an adaptive arithmetic coder.

When a bilevel quantiser is used, the value of each coefficient is either 0 or 1. Hence, depending on the implementation procedure, different types of symbol can be defined. Since multilayer bilevel quantisation is in fact quantisation by successive approximation, then this mode is exactly the same as coding the symbols at EZW. There we defined four symbols of $+$, $-$, ZT and Z, where ZT is a zero tree root symbol, and Z is an isolated zero within a tree and $+$ and $-$ are the values for refinement (see section 4.5 for details).

In order to achieve both spatial and SNR scalability, two different scanning methods are employed in this scheme. For spatial scalability, the wavelet coefficients are scanned from subband to subband, starting from the lowest frequency band to the highest frequency band. For SNR scalability, the wavelet coefficients are scanned quantiser to quantiser. The scanning method is defined in the bit stream.

### 10.9.3 Shape adaptive wavelet transform

Shape adaptive wavelet (SA-wavelet) coding is used for compression of arbitrary shaped textures. SA-wavelet coding is different from the regular wavelet coding mainly in its treatment of the boundaries of arbitrary shaped texture. The coding ensures that the number of wavelet coefficients to be coded is exactly the same as the number of pixels in the arbitrary shaped region, and coding efficiency at the object boundaries is the same as for the middle of the region. When the object boundary is rectangular, SA-wavelet coding becomes the same as regular wavelet coding.

The shape information of an arbitrary shaped region is used in performing the SA-wavelet transform in the following manner. Within each region, the first row of pixels belonging to that region and the first segment of the consecutive pixels in the row are identified. Depending on whether the starting point in the region has odd or even coordinates, and the number of pixels in the row of the segment is odd or even, the proper arrangements for 2 : 1 downsampling and use of symmetric extensions are made [3].

Coding of the SA-wavelet coefficients is the same as coding of regular wavelet coefficients, except that a modification is needed to handle partial wavelet trees that have wavelet coefficients corresponding to pixels outside the shape boundary. Such wavelet coefficients are called out nodes of the wavelet trees. Coding of the lowest band is the same as that of the regular wavelet, but the out nodes are not coded. For the higher bands, for any wavelet trees without out nodes, the regular zero tree is applied. For a partial tree, a minor modification to the regular zero tree coding is needed to deal with the out nodes. That is, if the entire branch of a partial tree has out nodes only, no coding is needed for this branch, because the shape information is available to the decoder to indicate this case. If a parent node is not an out node, all the children out nodes are set to zero, so that the out nodes do not affect the status of the parent node as the zero tree root or isolated zero. At the decoder, shape information is used to identify such zero values as out nodes. If the parent node is an out node and not all of its children are out nodes, there are two possible cases. The first case is that some of its children are not out nodes, but they are all zeros. This case is treated as a zero tree root and there is no need to go down the tree. The shape information indicates which children are zeros and which are out nodes. The second case is that some of its children are not out nodes and at least one of such nodes is nonzero. In this case, the out node parent is set to zero and the shape information helps the decoder to know that this is an out node, and coding continues further down the tree. There is no need to use a separate symbol for any out nodes.

## 10.10 Video coding with the wavelet transform

The success of the zero tree in efficient coding of wavelet transform coefficients has encouraged researchers to use it for video coding. Although wavelet-based video coding is not part of the standard, there is no reason why wavelet-based video coding cannot be used in the future. This of course depends on the encoding efficiency of wavelet-based coding and its functionalities. For this reason, in this section we look at some video coding scenarios and examine the encoding efficiency.

One way of wavelet-based video coding is to use the generic video encoder of Figure 3.18, but replacing the DCT with the DWT, as shown in Figure 10.28. Here, variable length coding of the quantised coefficients is replaced by the zero tree coding, of either EZW or SPIHT [18,21]. Also, overlapped motion compensation has been found to be very effective with the wavelet transform, as it prevents the motion compensation blocking artefacts from creating spurious vertical and horizontal edges.

Another method that might be used with the wavelet transform is shown in Figure 10.29.

Each frame of the input video is transformed into $n$-band wavelet subbands. The lowest LL band is fed into a DCT-based video encoder, such as MPEG-1 or H.263.

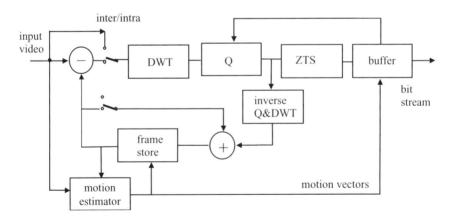

*Figure 10.28  Block diagram of a wavelet video codec*

## Content-based video coding (MPEG-4)  333

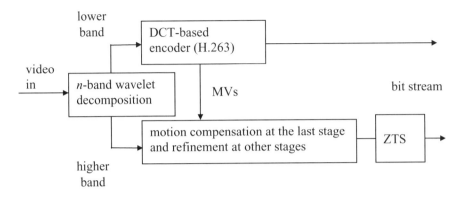

Figure 10.29  *A hybrid H.263/wavelet video coding scheme*

The other bands undergo a hierarchical motion compensation. First, the three high frequency bands of the last stage are motion compensated using the motion vectors from MPEG-1/H.263. The reconstructed picture from these four bands (LL, LH, HL and HH), which is the next level LL band, only requires a $\pm 1$ pixel refinement [20]. The other three bands at this stage are also motion compensated by the same amount. This process is continued for all the bands. Hence at the end all the bands are motion compensated. Now, these motion compensated bands are coded with a zero tree-based coding method (EZW or SPIHT).

### 10.10.1  Virtual zero tree (VZT) algorithm

The lowest band of the wavelet, LL, is a reduced size replica of the original video, and hence can be encoded with an efficient encoder, such as H.263. When zero tree-based coding such as EZW/SPIHT is used along with the standard codecs (e.g. Figure 10.29), it meets some problems. First, the subband decomposition stops when the top level LL band reaches a size of SIF/QSIF or subQSIF. At these levels there will be too many clustered zero tree roots. This is very common for either static parts of the pictures or when motion compensation is very efficient. Even for still images or I-pictures, a large part of the picture may contain only low spatial frequency information. As a result, at the early stages of the quantisation by successive approximation, where the yardstick is large, a vast majority of the wavelet coefficients fall below the yardstick. Secondly, even if the subband decomposition is taken to more stages, such that the top stage LL is a small picture of $16 \times 16$ pixels (e.g. Figure 10.28), it is unlikely that many zero trees can be generated. In other words, with a higher level of wavelet decomposition, the tree structure is bound to break and hence the efficiency of EZW/SPIHT is greatly reduced.

To improve the efficiency of zero tree-based coding, we have devised a version of it called virtual zero tree (VZT) [22]. The idea is to build trees outside the image boundary, hence the word virtual, as an extension to the existing trees that have roots in the top stage, so that the significant map can be represented in a more efficient way.

It can be imagined as replacing the top level LL band with zero value coefficients. These coefficients represent the roots of wavelet trees of several virtual subimages in normal EZW/SPIHT coding, although no decomposition and decimation actually takes place, as demonstrated in Figure 10.30.

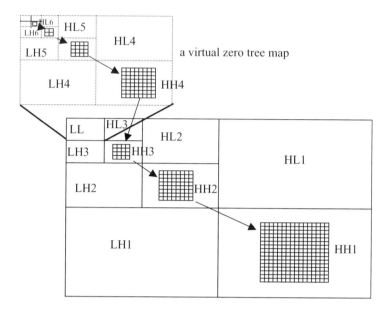

*Figure 10.30  A virtual zero tree*

In this Figure, virtual trees, or a virtual map, are built in the virtual subbands on the high frequency bands of the highest stage. Several wavelet coefficients of the highest stage form a virtual node at the bottom level of the virtual map. Then in the virtual map, four nodes of a lower level are represented by one node of a higher level in the same way as a zero tree is formed in EZW/SPIHT coding. The virtual map has only two symbols: VZT root or nonVZT root. If four nodes of a $2 \times 2$ block on any level of a virtual tree are all VZT roots, the corresponding node on the higher level will also be a VZT root. Otherwise this one node of the higher level will be a nonVZT node. This effectively constructs a long rooted tree of clustered real zero trees. One node on the bottom level of the virtual map is a VZT root only when the four luminance coefficients of a $2 \times 2$ block and their two corresponding chrominance coefficients on the top stage wavelet band are all zero tree roots. Chrominance pictures are also wavelet decomposed and, for a $4:2:0$ image format, four zero tree roots of the luminance and one from each chrominance can be made a composite zero tree root [22].

### 10.10.2  Coding of high resolution video

A wavelet transform can be an ideal tool for coding of high resolution pictures or video. This is because the high correlation among the dense pixels of high resolution

*Content-based video coding (MPEG-4)* 335

pictures is better exploited if a larger area for decorrelation of pixels is used. Unfortunately, in the standard video codecs, the DCT block sizes are fixed at 8 × 8, and hence pixel correlation beyond eight-pixel distances cannot be exploited. On the other hand, in the wavelet transform, increasing the number of decomposition levels means decorrelating the pixels at larger distances if so desired.

We have tested the encoding efficiency of the wavelet transform (for Figure 10.29) for high definition video (HDTV) as well as the super HDTV (SHD). For HDTV, we have used the test sequence Gaynor (courtesy of BBC). For this image sequence, a two-stage (seven-band) wavelet transform is used and the LL band of the SIF size was MPEG-1 coded. Motion compensated higher bands were coded with VZT and EZW. For the SHD video, a park sequence with 2048 × 2048 pixels at 60 Hz (courtesy of NHK Japan [21]) was used. The SHD images after three-stage subband decomposition result in ten bands. The LL band, with picture resolutions of 256 × 256 pixels, is MPEG-1 coded; the remaining nine bands are VZT and EZW coded in two separate experiments.

It appears at first that, by creating virtual nodes, we have increased the number of symbols to be coded, and hence the bit rate tends to increase rather than to decrease. However, these virtual roots will cluster the zero tree roots into a bigger zero tree root, such that instead of coding these roots one by one, at the expense of a large overhead by a simple EZW, we can code the whole cluster by a single VZT with only a few bits. VZT is more powerful at the early stages of encoding, where the vast majority of top stage coefficients are zero tree roots. This can be seen from Table 10.2, where a complete breakdown of the total bit rate required to code a P-picture of the park sequence by both methods is given.

The first row of the Table shows that 171 kbits are used to code the LL band by MPEG-1. The second row shows that 15 kbits is used for the additional ±1 pixel refinement in all bands. For the higher bands the image is scanned in five passes,

*Table 10.2 Breakdown bit rate in coding of a P-picture with VZT and EZW*

|  | VZT (kbits) | | | | | EZW (kbits) | | |
| --- | --- | --- | --- | --- | --- | --- | --- | --- |
|  | virtual pass | real pass | dominant pass | sub- ordinate pass | sum | dominant pass | sub- ordinate pass | sum |
| MPEG | – | – | – | – | 171 | – | – | 171 |
| MV | – | – | – | – | 15 | – | – | 15 |
| Pass-1 | 1.5 | 3.2 | 4.4 | 0.16 | 4.9 | 25 | 0.16 | 25 |
| Pass-2 | 7.7 | 31 | 39 | 1.7 | 41 | 153 | 1.7 | 156 |
| Pass-3 | 18 | 146 | 164 | 11 | 175 | 465 | 11 | 476 |
| Pass-4 | 29 | 371 | 400 | 41 | 441 | 835 | 41 | 896 |
| Pass-5 | 42 | 880 | 992 | 128 | 1050 | 1397 | 128 | 1326 |
| Grand total |  |  |  |  | 1898 |  |  | 3265 |

where the bits used in the dominant and subordinate passes of VZT and EZW are shown. In VZT the dominant pass is made up of two parts: one used in coding of the virtual nodes and the other parts for real data in the actual nine bands. Note that although some bits are used to code the virtual nodes (are not used in EZW), the total bits of the dominant pass in VZT is much less than for EZW. The number of bits in the subordinate passes, which codes the real subordinate data, is the same for both methods. In the Table the grand total is the total number of bits used to code the P-frame under the two coding schemes. It can be seen that VZT requires two-thirds of the bit rate required by EZW.

For HDTV, our results show that, although a good quality video at 18 Mbit/s can be achieved under EZW, VZT only needs 11 Mbit/s [22].

### 10.10.3 Coding of low resolution video

Although coding of low spatial resolution (e.g. QCIF, subQCIF) video may not benefit from wavelet decomposition to the extent that higher resolution video does, nevertheless zero tree coding is efficient enough to produce good results. In fact, the multiresolution property of wavelet-based coding is a bonus that the DCT-based coding most suffers from. In section 8.5.7, we saw that two-layer spatial and SNR scalability of the standard codecs (MPEG, H.263) reduces the compression efficiency by as much as 30 per cent. That is, spatial/SNR coders are required to generate about 30 per cent more bits to be able to produce the same video quality as a single layer coder does. Undoubtedly, increasing the number of layers, or combining them in a hybrid form, will widen this deficiency gap.

On the other hand, with the wavelet transform, as the number of decomposition levels increases, the encoding efficiency increases too. Moreover, with the concept of virtual zero tree a large area of static parts of the picture can be grouped together for efficient coding, as we have seen from the results of higher resolution pictures of the previous section.

To compare the quality of wavelet-coded video (purely wavelet-based video codec of Figure 10.28) against the best of the standard codec, we have coded 10 Hz QCIF size of Akio and Carphone standard video test sequences at various bit rates, as shown in Figure 10.31, for Akio and Figure 10.32 for Carphone.

For the wavelet coder, we have used three levels of wavelet decomposition (ten bands) and two levels of virtual nodes, with an SPIHT-type zero tree coding, called virtual SPHIT [23]. Unlike Figure 10.29, where the lowest subband was coded by an standard DCT codec, here we have coded all the bands, including the LL band with the virtual SPIHT (Figure 10.28). This is because, after three-level wavelet decomposition, the lowest LL band has a dimension of $22 \times 18$ pixels which is not viable, nor practical to be coded with codecs using $16 \times 16$ pixel macroblocks. On the motion compensation, the whole QCIF image was motion compensated with an overlapped block matching algorithm, with half a pixel precision. Motion compensated wavelet coefficients after the zero tree scan were arithmetic coded.

For comparison we have also coded these sequences with single and two-layer SNR scalable H.263 codecs, at various bit rates from 20 kbit/s up to 60 kbit/s. For the

Content-based video coding (MPEG-4) 337

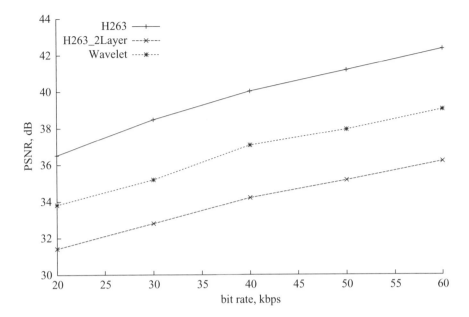

*Figure 10.31*   *Quality of QCIF size Akio sequence coded at various bit rates*

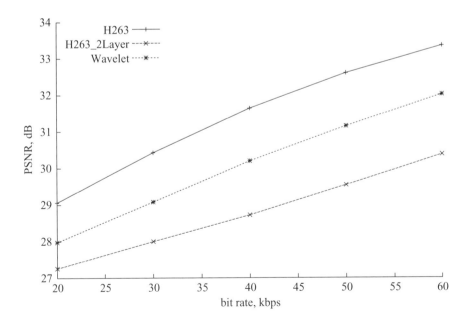

*Figure 10.32*   *Quality of QCIF size Carphone sequence coded at various bit rates*

Figure 10.33  A snap shot of Akio coded at 20 kbit/s, 10 Hz
   a   H.263
   b   SNR-scalable H.263
   c   wavelet SPIHT

Figure 10.34  A snap shot of Carphone coded at 40 kbit/s, 10 Hz
   a   H.263
   b   SNR-scalable H.263
   c   wavelet SPIHT

two-layer H.263, 50 per cent of the total bits were assigned to the base layer. As the PSNR Figures 10.31 and 10.32 show, single-layer H.263 is consistently better than the wavelet which itself is better than the two-layer H.263. However, subjectively wavelet-coded video appears better than single layer H.263, despite being 1–2 dB poorer on the PSNR.

Figures 10.33 and 10.34 show snap shots of Akio and Carphone, coded at 20 kbit/s and 40 kbit/s, respectively. Although these small pictures may not show the subjective quality differences between the two codecs, the fact is that the H.263 picture is blocky but that of the wavelet is not. In the Carphone picture, the wavelet-coded picture is smeared but it is still much sharper than the two-layer H.263. Some nonexperts prefer the smeared picture to the blocky one, but expert viewers have a different opinion. Comparing the wavelet with the two-layer H.263, both subjective and objective results are in favour of the wavelet.

Considering that the above three-level decomposition wavelet transform can be regarded as an $N$-layer video without impairing its compression efficiency, it implies that the wavelet has a high potential for multilayer coding of video. Although this video codec is not a part of any standard, there is no doubt that its high potential for multilayer coding makes it very attractive.

## 10.11 Scalability

We have covered scalability in the standard codecs several times, but due to the nature of content-based coding, scalability in MPEG-4 can be different from the other standard codecs. However, since MPEG-4 is also used as a frame-based video codec, the scalability methods we have discussed so far can also be used in this codec. Thus we introduce two new methods of scalability that are only defined for MPEG-4.

### 10.11.1 Fine granularity scalability

The scalability methods we have seen so far, SNR, spatial and temporal, are normally carried out in two or a number of layers. These create coarse levels of representing various quality, spatial and temporal resolutions. A few spatial and temporal resolutions are quite acceptable for natural scenes, as it is of no practical use to have more resolution levels of these kinds. Hence, for the frame-based video, MPEG-4 also recommends the spatial and temporal scalabilities as we have discussed so far.

On the other hand, the existing SNR scalability has the potential to be represented in more quality levels. The increment in quality is both practical and appreciated by the human observer. Thus in MPEG-4, instead of SNR scalability, a synonymous method named fine granularity scalability (FGS) is recommended. In this method the base layer is coded similar to the base layer of a SNR scalable coder, namely coding of video at a given frame rate and a relatively large quantiser step size. Then the difference between the original DCT coefficients and the quantised coefficients in the base layer (base layer quantisation distortion) rather than being quantised with a finer quantiser step size, as is done in the SNR scalable coder, is represented in bit planes. Starting from the highest bit plane that contains nonzero bits, each bit plane is successively coded using run length coding, on a block by block basis. The codewords for the run lengths can be derived either from Huffman or arithmetic coding. Typically different codebooks are used for different bit planes, because the run length distributions across the bit planes are different.

### 10.11.2 Object-based scalability

In the various scalability methods described so far, including FGS, the scalability operation is applied to the entire frame. In object-based scalability, the scalability operations are applied to the individual objects. Of particular interest to MPEG-4 is object-based temporal scalability (OTS), where the frame rate of a selected object is enhanced, such that it has a smoother motion than the remaining area. That is, the temporal scalability is applied to some objects to increase their frame rates against the other objects in the frames.

MPEG-4 defines two types of OTS. In type-1 the video object layer 0 (VOL0) comprises the background and the object of interest. Higher frame rates for the object of interest are coded at VOL1, as shown in Figure 10.35, where it is predictively coded, or Figure 10.36, if the enhancement layer is bidirectionally coded.

340  *Standard codecs: image compression to advanced video coding*

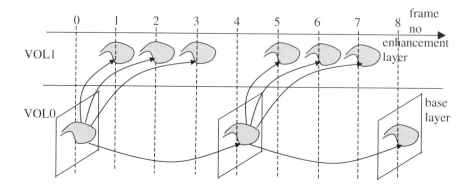

*Figure 10.35*  *OTS enhancement structure of type-1, with predictive coding of VOL*

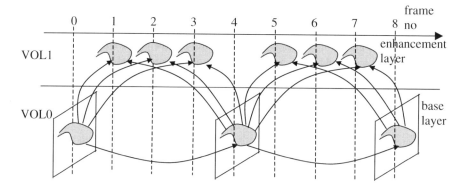

*Figure 10.36*  *OTS enhancement structure of type-1, with bidirectional coding of VOL*

On type-2 OTS, the background is separated from the object of interest, as shown in Figure 10.37. The background, VO0, is sent at a low frame rate without scalability. The object of interest is coded at two frame rates of base and enhancement layers of VOL0 and VOL1, respectively, as shown in the Figure.

## 10.12 MPEG-4 *versus* H.263

There is a strong similarity between the simple profile MPEG-4 and H.263, such that a simple profile MPEG-4 can decode a bit stream generated by the core H.263. In fact most of the annexes introduced into H.263 after the year 2000 were parts of the profiles designed for MPEG-4 in the mid 1990s. However, there is no absolute reverse compatibility between the simplest profile of MPEG-4 and the H.263.

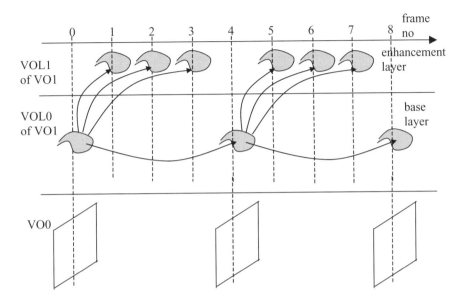

*Figure 10.37   OTS enhancement structure of type-2*

The main source of incompatibility is concerned with the systems function of MPEG-4. Like its predecessor, MPEG-4 uses a packet structure form of transporting compressed audio-visual data. In MPEG-2 we saw that packetised elementary streams (PES) of audio, video and data are multiplexed into a transport packet stream. In MPEG-4, due to the existence of many objects (up to 256 visual objects plus audio and data), the transport stream becomes significantly important. Even if one object is coded (frame-based mode), the same transpost stream should be used. In particular, for transmission of MPEG-4 video over mobile networks, the packetised system has to add extra error resilience to the MPEG-4 bit stream. It is this packetised format of MPEG-4 that makes it nondecodable by an H.263 codec.

In MPEG-4 the elementary objects of the audio-visual media are fed to the synchronisation layer (SL), to generate SL-packets. An SL-packet has a resynchronisation marker, similar to the GOB and slice resynchronisation markers in H.263. However, in MPEG-4 the resynchronisation marker is inserted after a certain number of bits are coded, but in H.263 they are added after several macroblocks. Since the number of bits generated per macroblock is variable, then while distances between the resynchronisation markers in H.263 are variable, those of MPEG-4 are fixed. This helps to reduce the effect of channel errors in MPEG-4, hence making it more error resilient.

A copy of the picture header is repeated in every SL-packet, to make them independent of each other. Each medium may generate more than one SL-packet. For example, if scalability is used, several interrelated packets are generated. Packets are then multiplexed into the output stream. This layer is unaware of the transport or delivery layer. The sync layer is interfaced to the delivery layer through the delivery multimedia integration framework (DMIF) application interface (DAI). The DAI is network independent but demands session setup and stream control functions. It also enables setting up quality of service for each stream.

The transport or delivery layer is delivery aware but unaware of media. MPEG-4 does not define any specific delivery layer. It relies mainly on the existing transport layers, such as RTP for Internet, MPEG-2 transport stream for wired and wireless or ATM for B-ISDN networks.

There are also some small differences on the employment of the coding tools in the two codecs. For example, reversible VLC used in the simple profile of MPEG-4 is not exactly the same as the RVLC used in data partitioning of Annex V of H.263. In the former, RVLC is also used for DCT coefficients, and in the latter it is used for the nonDCT coefficients, e.g. motion vectors, macroblock addresses etc. Although in Chapter 9 we showed that RVLC due to its higher overhead over the conventional VLC is not viable for the DCT-coefficients, nevertheless in the experiments of Chapter 9 macroblocks of the P-pictures were all interframe coded. Had any macroblock been intraframe coded, its retrieval through RVLC would have improved the picture quality.

These differences are significant enough to ask which one of H.263 and MPEG-4 might be suitable for video over mobile networks. In the late 1990s, an international collaboration under the project 3gpp (3rd generation partnership project) set up an extensive investigation to compare the performance of these two codecs [24]. The outcome of one of the experiments is shown in Figure 10.38.

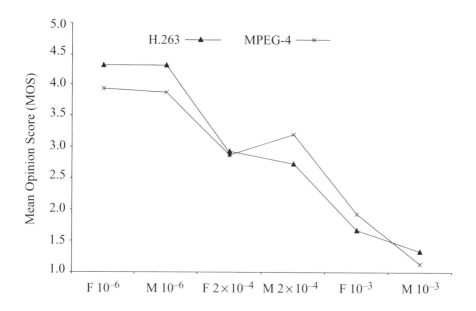

Figure 10.38  Comparison between MPEG-4 and H.263

In this experiment wireless channels of 64 kbit/s were used, of which 7.6 kbit/s were assigned to the speech signal. The remaining bits were the target video bit rates for coding of the overtime video test sequence for the two codecs, including the overheads. The channel errors were set to $10^{-6}$, $2 \times 10^{-4}$ and $10^{-3}$, for both fixed and mobile sets, identified by F and M in the Figure, respectively. The H.263 codec is equipped with annexes D, F, I, J and N (see list of H.263 annexes in Chapter 9 [25]) and that of MPEG-4 was the simple profile (see section 10.1). The performance of the reconstructed pictures after error concealment was subjectively evaluated. The average of the viewers' scores (see section 2.4) as the mean opinion score (MOS) is plotted in Figure 10.38 against the tested error rates.

At the low error rate of $10^{-6}$, H.263 outperforms MPEG-4 for both mobile and fixed environments. Thus it can be concluded that H.263 is a more compression efficient encoder than MPEG-4. Much of this is due to the use of RVLC in the MPEG-4, as we have seen in Table 9.3, and RVLC is less compression efficient than MPEG-4. Other overheads, such as packet header and more frequent resynchronisation markers, add to this overhead. On the other hand, at high error rates, the superior error resilience of MPEG-4 over H.263 compensates for compression inefficiency and in some cases the decoded video under MPEG-4 is perceived to be better than H.263.

Considering the above analysis, one cannot for sure say whether H.263 is better or worse than MPEG-4. In fact what makes a bigger impact on the experimental results is not the basic H.263 or MPEG-4 definitions but the annexes or optionalities of these two codecs. For example, had data partitioning (Annex V) been used in H.263, it would have performed as well as MPEG-4 at high error rates. Unfortunately, data partitioning in H.263 was introduced in 2001, and this study was carried out in 1997. Hence the 3ggp project fails to show the true difference between the H.263 and MPEG-4.

## 10.13 Problems

1  Figure 10.39 shows a part of a binary alpha plane, where shaded pixels are represented by 1 and blank pixels by 0:

   *a*  calculate the context number for pixels A, B and C
   *b*  use the intra table of Appendix D to calculate the probability for coding of alpha pixels A, B and C.

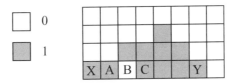

*Figure 10.39*

344  *Standard codecs: image compression to advanced video coding*

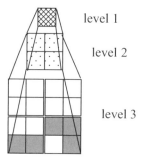

*Figure 10.40*

*Figure 10.41*

2  Draw a part of the contour of the shape of Figure 10.39 starting from point X, and ending at point Y. Use the Huffman Table 10.1 to calculate the number of bits required to differentially chain code the part of the contour from X to Y.

3  Figure 10.40 shows a part of a quad tree binary alpha block (BAB). Assume shaded pixels at level three are binary 1 and blank ones binary 0:

   a  calculate the index of each pixel at level 2
   b  calculate the index of the pixel at level 1.

4  A part of a foreground texture within an $8 \times 8$ pixel block is shown in Figure 10.41. The background pixels are shown blank. The block is DCT coded and the quantised coeffiicients with a step size of $th = q = 8$ are zigzag scanned. Compare the efficeincy of the following two coding methods:

   a  use a shape adaptive DCT to code the block; assume the shape of this object is available to the decoder
   b  use normal DCT, but assume the background pixels are zero.

## 10.14 References

1 KOENEN, R., PEREIRA, F., and CHIARIGLIONE, L.: 'MPEG-4: Context and objectives', *Image Commun. J.*, 1997, **9**:4
2 MPEG-4: 'Testing and evaluation procedures document'. ISO/IEC JTC1/SC29/WG11, N999, July 1995
3 MPEG-4 video verification model version-11, ISO/IEC JTC1/SC29/WG11, N2171, Tokyo, March 1998
4 Special issue on segmentation, *IEEE Trans. Circuits Syst. Video Technol.*, 1998, **8**:5
5 KASS, M., WITKIN, A., and TERZOPOULOS, D.: 'Snakes: active contour models', *Inter. J. Comput. Vis.*, 1987, 1:321–331
6 ROERDINK, J.B.T.M., and MEIJSTER, A.: 'The watershed transform: definitions, algorithms, and parallellization strategies', *Fundamental Informatics*, 2000, 41, pp.187–228
7 IZQUIERDO, E., and GHANBARI, M.: 'Key components for an advanced segmentation systems', *IEEE Trans. Multimedia*, 2002, **4:1**, pp.97–113
8 PERONA, P., and MALIK, J.: 'Scale-space and edge detection using anisotropic diffusion', *IEEE Trans. Pattern Anal. Mach. Intell.*, 1990, **12**:7, pp.629–639
9 SHAFARENKO, L., PETROU, M., and KITTLER, J.: 'Automatic watershed segmentation of randomly textured color images', *IEEE Trans. Image Process.*, 1997, **6:11**, pp.1530–1544
10 WYSZECKI, G., and STILES, W.S.: 'Color science: concepts and methods, quantitative data and formulae' (Wiley, 1982, 2nd edn.)
11 VINCENT, L., and SOILLE, P.: 'Watersheds in digital spaces: an efficient algorithm based on immersion simulations', *IEEE Trans. Pattern Anal. Mach. Intell.*, 1991, **13:6**, pp.583–589
12 MPEG-4: 'Video shape coding'. ISO/IEC JTC1/SC29/WG11, N1584, March 1997
13 GHANBARI, M.: 'Arithmetic coding with limited past history', *Electron. Lett.*, 1991, **27:13**, pp.1157–1159
14 PEARSON, D.E.: 'Developments in model-based video coding', *Proc. IEEE*, 1995, **83:6**, pp.892–906
15 WOLBERG, G.: 'Digital image warping' (IEEE Computer Society Press, Los Alamitos, California, 1990)
16 SEFERIDIS, V., and GHANBARI, M.: 'General approach to block matching motion estimation', *J. Opt. Engineering*, 1993, **37:7**, pp.1464–1474
17 GHANBARI, M., DE FARIA, S., GOH, I.J., and TAN, K.T.: 'Motion compensation for very low bit-rate video', *Signal Process. Image Commun.*, special issue on very low bit rate video, 1995, **7:(4–6)**, pp.567–580
18 SHAPIRO, J.M.: 'Embedded image coding using zero-trees of wavelet coefficients', *IEEE Trans. Signal Process.*, 1993, **4:12**, pp.3445–3462
19 WITTEN, I.H., NEAL, R.M., and CLEARY, J.G.: 'Arithmetic coding for data compression', *Commun. ACM*, 1987, **30:6**, pp.520–540

20 WANG, Q., and GHANBARI, M.: 'Motion-compensation for super high definition video'. Proceedings of IS&T/SPIE symposium, on *Electronic imaging, science and technology, very high resolution and quality imaging*, San Jose, CA, Jan. 27–Feb. 2, 1996
21 SAID, A., and PEARLMAN, W.A.: 'A new, fast and efficient image codec based on set partitioning in hierarchical trees', *IEEE Trans. Circuits Syst. Video Technol.*, 1996, **6:3**, pp.243–250
22 WANG, Q., and GHANBARI, M.: 'Scalable coding of very high resolution video using the virtual zero-tree', *IEEE Trans. Circuits Syst. Video Technol.*, special issue on multimedia, 1997, **7:5**, pp.719–729
23 KHAN, E., and GHANBARI, M.: 'Very low bit rate video coding using virtual SPIHT', *Electron. Lett.*, 2001, **37:1**, pp.40–42
24 3rd Generation Partnership Project (3gpp): TSG-SA codec working group, release 1999, technical report TR 26.912 V 3.0.0

*Chapter 11*

# Content description, search and delivery (MPEG-7 and MPEG-21)

As more and more audio-visual information becomes available in digital form, there is an increasing pressure to make use of it. Before one can use any information, however, it has to be located. Unfortunately, widespread availability of interesting material makes this search extremely difficult.

For textual information, currently many text-based search engines, such as Google, Yahoo, Altavista etc. are available on the worldwide web (www), and they are among the most visited sites. This is an indication of real demand for searching information in the public domain. However, identifying information for audio-visual content is not so trivial, and no generally recognised description of these materials exists. In the mean time, there is no efficient way of searching the www for, say, a piece of video concert by Pavarotti, or improving the user friendliness of interconnected computers *via* the Internet by rich-spoken queries, hand-drawn sketches and image-based queries.

The question of finding content is not restricted to database retrieval applications. For example, TV programme producers may want to search and retrieve famous events stored among the thousands of hours of audio-visual records, in order to collect material for a programme. This will reduce programme time and increase the quality of its content. Another example is the selection of a favourite TV programme from a vast number of available satellite television channels. Currently 6–8 MPEG-2 coded TV programmes can be accommodated in a satellite transponder. Considering that each satellite can have up to 12 transponders, each in horizontal and vertical polarisation mode and satellites can be stationed within two degrees guard band, it is not unrealistic that users may have access to thousands of TV channels. Certainly the current method of printing weekly TV programmes will not be practical (tens of thousands of pages per week!), and a more intelligent computerised way of choosing a TV programme is needed. MPEG-7, under the name of 'Multimedia content-based description standard', aims to address these issues and define how humans expect to interact with computers [1].

The increasing demand for searching multimedia content on the web has opened up new opportunities for creation and delivery of these items on the Internet. Today, many elements exist to build an infrastructure for the delivery and consumption of multimedia content. There is, however, no standard way of describing these elements or relating them to each other. It is hoped that such a standard will be devised by the ISO/IEC MPEG committee under the name of MPEG-21 [2].

The main aim of this standard is to specify how various elements for content creation fit together and, when a gap exists, MPEG-21 will recommend which new standards are required. The MPEG standard will then develop new standards as appropriate while other bodies may develop other relevant standards. These specifications will be integrated into the multimedia framework, through collaboration between MPEG and these bodies. The result is an open framework for multimedia delivery and consumption, with both the content creators and content consumers as the main beneficiaries. The open framework aims to provide content creators and service providers with equal opportunities in the MPEG-21 enabled market. It will also be to the benefit of the content users, providing them access to a large variety of data in an interoperable manner.

In summary, MPEG-7 is about describing and finding contents and MPEG-21 deals with the delivery and consumption of these contents. As we see, none of these standards are about video compression, which is the main subject of this book. However, for the completeness of a book on the standard codecs we briefly describe these two new standards that incidentally are developed by the ISO/IEC MPEG standard bodies.

## 11.1 MPEG-7: multimedia content description interface

The main goal of MPEG-7 is to specify a standard set of descriptors that can be used to describe various types of multimedia information coded with the standard codecs, as well as other databases and even analogue audio-visual information. This will be in the form of defining descriptor schemes or structures and the relationship between various descriptors. The combination of the descriptors and description schemes will be associated with the content itself to allow a fast and efficient searching method for material of user interest. The audio-visual material that has MPEG-7 data associated with it can be indexed and searched for. This material may include still pictures, graphics, three-dimensional models, audio, speech, video and information about how these elements are combined in a multimedia presentation.

Figure 11.1 shows a highly abstract block diagram of the MPEG-7 mission. In this Figure object features are extracted and they are described in a manner which is meaningful to the search engine. As usual, MPEG-7 does not specify how features should be extracted, nor how they should be searched for, but only specifies the order in which features should be described.

### 11.1.1 Description levels

Since the description features must be meaningful in the context of the application, they may be defined in different ways for different applications. Hence a specific

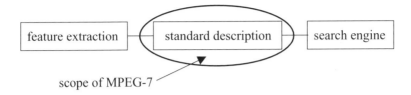

*Figure 11.1   Scope of MPEG-7*

audio-visual event might be described with different sets of features if their applications are different. To describe visual events, they are first described by their lower abstraction level, such as shape, size, texture, colour, movement and their positions inside the picture frame. At this level, the audio material may be defined as key, mood, tempo, tempo changes and position in the sound space.

The high level of the abstraction is then a description of the semantic relation between the above lower level abstractions. For the earlier example of a piece of Pavarotti's concert, the lower level of abstraction for the picture would be: his portrait, a picture of an orchestra, shapes of musical instruments etc. For the audio, this level of abstraction could be, of course, his song, as well as other background music. All these descriptions are of course coded in a way in which they can be searched as efficiently as possible.

The level of abstraction is related to the way the required features are extracted. Many low level features can be extracted in a fully automatic manner. High level features, however, need more human interaction to define the semantic relations between the lower level features.

In addition to the description of contents, it may also be required to include other types of information about the multimedia data. For example:

- *The form*: an example of the form is the coding scheme used (e.g. JPEG, MPEG-2), or the overall data size.
- *Conditions for accessing material*: this would include copyright information, price etc.
- *Classification*: this could include parental rating, and content classification into a number of predefined categories.
- *Links to other relevant material*: this information will help the users to speed up the search operation.
- *The context*: for some recorded events, it is very important to know the occasion of recording (e.g. World Cup 2002, final between Brazil and Germany).

In many cases addition of textual information to the descriptors may be useful. Care must be taken such that the usefulness of the descriptors is as independent as possible from the language. An example of this is giving names of authors, films and places. However, providing text-only documents will not be among the goals of MPEG-7.

The MPEG group has also defined a laboratory reference model for MPEG-7. This time it is called the experimental model (XM), and has the same role as RM, TM and VM in H.261, MPEG-2 and MPEG-4, respectively.

## 11.1.2 Application area

The elements that MPEG-7 standardises will support a broad range of applications. MPEG-7 will also make the web as searchable for multimedia content as it is searchable for text today. This would apply especially to large content archives as well as to multimedia catalogues enabling people to identify content for purchase. The information used for content retrieval may also be used by agents, for the selection and filtering of broadcasted material or for personalised advertising.

All application domains making use of multimedia will benefit from MPEG-7. In the mean time some of these domains that might find MPEG-7 useful are:

- architecture, real estate and interior design (e.g. searching for ideas)
- broadcast multimedia selection (e.g. radio and TV channels)
- cultural service (e.g. history museums, art galleries etc.)
- digital libraries (e.g. image catalogue, musical dictionary, biomedical imaging catalogues, film, video and radio archives)
- e-commerce (e.g. personalised advertising, online catalogues, directories of electronic shops)
- education (e.g. repositories of multimedia courses, multimedia search for support material)
- home entertainment (e.g. systems for the management of personal multimedia collections, including manipulation of content, e.g. home video editing, searching a game)
- investigation services (e.g. human characteristics recognition, forensics)
- journalism (e.g. searching speeches of a certain politician using his name, his voice or his face)
- multimedia directory services (e.g. yellow pages, tourist information, geographical information systems)
- multimedia editing (e.g. personalised electronic news services, media authoring)
- remote sensing (e.g. cartography, ecology, natural resources management)
- shopping (e.g. searching for clothes that you like)
- surveillance (e.g. traffic control, surface transportation, nondestructive testing in a hostile environment)

and many more.

The way MPEG-7 data will be used to answer user queries is outside the scope of the standard. In principle, any type of audio-visual material may be retrieved by means of any type of query material. For example, video material may be queried using video, music, speech etc. It is up to the search engine to match the query data and the MPEG-7 audio-visual description. A few query examples are:

- play a few notes on a keyboard and retrieve a list of musical pieces similar to the required tune
- draw a few lines on a screen and find a set of images containing similar graphics, logos, and ideograms
- sketch objects, including colour patches or textures and retrieve examples among which you select the interesting objects to compose your design

- on a given set of multimedia objects, describe movements and relations between the objects and so search for animations fulfilling the described temporal and spatial relations
- describe actions and get a list of scenarios containing such actions
- using an excerpt of Pavarotti's voice, obtaining a list of Pavarotti's records, video clips where Pavarotti is singing and photographic material portraying Pavarotti.

### 11.1.3 Indexing and query

The current status of research at this stage of MPEG-7 development is concentrated into two interrelated areas, indexing and query. In the former, significant events of video shots are indexed, and in the latter, given a description of an event, the video shot for that event is sought. Figure 11.2 shows how the indices for a video clip can be generated.

In the Figure a video programme (normally 30–90 minutes) is temporally segmented into video shots. A shot is a piece of video clip, where the picture content from one frame to the other does not change significantly, and in general there is no scene cut within a shot. Therefore, a single frame in a shot has a high correlation to all the pictures within the shot. One of these frames is chosen as the key frame. Selection of the key frame is an interesting research issue. An ideal key frame is the one that has maximum similarity with all the pictures within its own shot, but minimum similarity with those of the other shots. The key frame is then spatially segmented into objects with meaningful features. These may include colour, shape, texture, where a semantic relation between these individual features defines an object of interest. As mentioned, depending on the type of application, the same features might be described in a different order. Also, in extracting the features, other information like motion of the objects, background sound or sometimes text might be useful. Here, features are then indexed, and the indexed data along with the key frames is stored in the database, sometimes called metadata.

The query process is the opposite of indexing. In this process, the database is searched for a specific visual content. Depending on how the query is defined to the

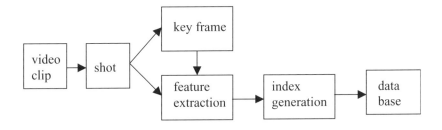

*Figure 11.2  Index generation for a video clip*

search engine, the process can be very complex. For instance in our earlier example of Pavarotti's singing, the simplest form of the query is that a single frame (picture) of him, or a piece of his song is available. This picture (or song) is then matched against all the key frames in the database. If such a picture is found then, due to its index relation with the actual shot and video clip, that piece of video is located. Matching of the query picture with the key frames is under active research, since this is very different from the conventional pixel-to-pixel matching of the pictures. For example, due to motion, obstruction of the objects, shading, shearing etc. the physical dimensions of the objects of interest might change such that pixel-to-pixel matching does not necessarily find the right object. For instance, with the pixel-to-pixel matching, a circle can be more similar to a hexagon of almost the same number of pixels and intensity than to a smaller or larger circle, which is not a desired match.

The extreme complexity in the query is when the event is defined verbally or in text, like the text of Pavarotti's song. Here, this data has to be converted into audio-visual objects, to be matched with the key frames. There is no doubt that most of the future MPEG-7 activity will be focused on this extremely complex audio and image processing task.

In the following some of the description tools used for indexing and retrieval are described. To be consistent we have ignored speech and audio and consider only the visual description tools. Currently there are five visual description tools that can be used for indexing. During the search, any of them or a combination, as well as other data, say from audio description tools, might be used for retrieval.

## 11.1.4 Colour descriptors

Colour is the most important descriptor of an object. MPEG-7 defines seven colour descriptors, to be used in combination for describing an object. These are defined in the following sections.

### 11.1.4.1 Colour space

Colour space is the feature that defines how the colour components are used in the other colour descriptors. For example, R, G, B (red, green, blue), $Y$, $C_r$, $C_b$ (luminance and chrominance), HSV (hue, saturation, value) components or monochrome, are the types describing colour space.

### 11.1.4.2 Colour quantisation

Once the colour space is defined, the colour components of each pixel are quantised to represent them with a small (manageable) number of levels or bins. These bins can then be used to represent the colour histogram of the object. That is, the distribution of the colour components at various levels.

### 11.1.4.3 Dominant colour(s)

This colour descriptor is most suitable for representing local (object or image region) features where a small number of colours is enough to characterise the colour information in the region of interest. Dominant colour can also be defined for the whole image. To define the dominant colour, colour quantisation is used to extract a small number of representing colours in each region or image. The percentage of each quantised colour in the region then shows the degree of the dominance of that colour. A spatial coherency on the entire descriptor is also defined and is used in similarity retrieval (objects having similar dominant colours).

### 11.1.4.4 Scalable colour

The scalable colour descriptor is a colour histogram in HSV colour space, which is encoded by a Haar transform. Its binary representation is scalable in terms of bin numbers and bit representation accuracy over a wide range of data rates. The scalable colour descriptor is useful for image-to-image matching and retrieval based on colour feature. Retrieval accuracy increases with the number of bits used in the representation.

### 11.1.4.5 Colour structure

The colour structure descriptor is a colour feature that captures both colour content (similar to colour histogram) and information about the structure of this content (e.g. colour of the neighbouring regions). The extraction method embeds the colour structure information into the descriptor by taking into account the colours in a local neighbourhood of pixels instead of considering each pixel separately. Its main usage is image-to-image matching and is intended for still image retrieval. The colour structure descriptor provides additional functionality and improved similarity-based image retrieval performance for natural images compared with the ordinary histogram.

### 11.1.4.6 Colour layout

This descriptor specifies the spatial distribution of colours for high speed retrieval and browsing. It can be used not only for image-to-image matching and video-to-video clip matching, but also in layout-based retrieval for colour, such as sketch-to-image matching which is not supported by other colour descriptors. For example, to find an object, one may sketch the object and paint it with the colour of interest. This descriptor can be applied either to a whole image or to any part of it. This descriptor can be applied to arbitrary shaped regions.

### 11.1.4.7 GOP colour

The group of pictures (GOP) colour descriptor extends the scalable colour descriptor that is defined for still images to a video segment or a collection of still images. Before applying the Haar transform, the way the colour histogram is derived should be defined. MPEG-7 considers three ways of defining the colour histogram for GOP colour descriptor, namely average, median and intersection histogram methods.

The average histogram refers to averaging the counter value of each bin across all pictures, which is equivalent to computing the aggregate colour histogram of all pictures with proper normalisation. The median histogram refers to computing the median of the counter value of each bin across all pictures. It is more robust to round-off errors and the presence of outliers in image intensity values compared with the average histogram. The intersection histogram refers to computing the minimum of the counter value of each bin across all pictures to capture the least common colour traits of a group of images. The same similarity/distance measures that are used to compare scalable colour descriptions can be employed to compare GOP colour descriptors.

### 11.1.5 Texture descriptors

Texture is an important structural descriptor of objects. MPEG-7 defines three texture-based descriptors, which are explained in the following sections.

#### 11.1.5.1 Homogeneous texture

Homogeneous texture has emerged as an important primitive for searching and browsing through large collections of similar looking patterns. In this descriptor, the texture features associated with the regions of an image can be used to index the image. Extraction of texture features is done by filtering the image at various scales and orientations. For example, using the wavelet transform with Gabor filters in say six orientations and four levels of decompositions, one can create 24 subimages. Each subimage reflects a particular image pattern at certain frequency and resolution. The mean and the variance of each subimage are then calculated. Finally, the image is indexed with a 48-dimensional vector (24 mean and 24 standard deviations). In image retrieval, the minimum distance between this 48-dimensional vector of the query image and those in the database are calculated. The one that gives the minimum distance is retrieved. The homogeneous texture descriptor provides a precise and quantitative description of a texture that can be used for accurate search and retrieval in this respect.

#### 11.1.5.2 Texture browsing

Texture browsing is useful for representation of homogeneous texture for browsing type applications, and is defined by at most 12 bits. It provides a perceptual characterisation of texture, similar to a human characterisation, in terms of regularity, coarseness and directionality. Derivation of this descriptor is done in a similar way to the homogeneous texture descriptor of section 11.1.5.1. That is, the image is filtered with a bank of orientation and scale-tuned filters, using Gabor functions. From the filtered image, the two dominant texture orientations are selected. Three bits are needed to represent each of the dominant orientations (out of, say, six). This is followed by analysing the filtered image projections along the dominant orientations to determine the regularity (quantified by two bits) and coarseness (two bits $\times 2$). The second dominant orientation and second scale feature are optional. This descriptor, combined with the homogeneous texture descriptor, provides a scalable solution to representing homogeneous texture regions in images.

### 11.1.5.3 Edge histogram

The edge histogram represents the spatial distribution of five types of edge, namely four directional edges and one nondirectional edge. It consists in the distribution of pixel values in each of these directions. Since edges are important in image perception, they can be used to retrieve images with similar semantic meaning. The primary use of this descriptor is image-to-image matching, especially for natural edges with nonuniform edge distribution. The retrieval reliability of this descriptor is increased when it is combined with other descriptors, such as the colour histogram descriptor.

## 11.1.6 Shape descriptors

Humans normally describe objects by their shapes, and hence shape descriptors are very instrumental in finding similar shapes. One important property of shape is its invariance to rotation, scaling and displacement. MPEG-7 identifies three shape descriptors, which are defined as follows.

### 11.1.6.1 Region-based shapes

The shape of an object may consist of either a single region or a set of regions. Since a region-based shape descriptor makes use of all pixels constituting the shape within a picture frame, it can describe shapes of any complexity.

The shape is described as a binary plane, with the black pixel within the object corresponding to 1 and the white background corresponding to 0. To reduce the data required to represent the shape, its size is reduced. MPEG-7 recommends shapes to be described at a fixed size of 17.5 bytes. The feature extraction and matching processes are straightforward enough to have a low order of computational complexity, so as to be suitable for tracking shapes in video data processing.

### 11.1.6.2 Contour-based shapes

The contour-based shape descriptor captures the characteristics of the shapes and is more similar to the human notion of understanding shapes. It is the most popular method for shape-based image retrieval. In section 11.2 we will demonstrate some of its practical applications in image retrieval.

The contour-based shape descriptor is based on the so-called curvature scale space (CSS) representation of the contour. That is, by filtering the shape at various scales (various degrees of the smoothness of the filter), a contour is smoothed at various levels. The smoothed contours are then used for matching. This method has several important properties:

- it captures important characteristics of the shapes, enabling similarity-based retrieval
- it reflects properties of the perception of the human visual system
- it is robust to nonrigid motion or partial occlusion of the shape
- it is robust to various transformations on shapes, such as rotation, scaling, zooming etc.

### 11.1.6.3 Three-dimensional shapes

Advances in multimedia technology have brought three-dimensional contents into today's information systems in the forms of virtual worlds and augmented reality. The three-dimensional objects are normally represented as polygonal meshes, such as those used in MPEG-4 for synthetic images and image rendering. Within the MPEG-7 framework, tools for intelligent content-based access to three-dimensional information are needed. The main applications for three-dimensional shape description are search, retrieval and browsing of three-dimensional model databases.

## *11.1.7  Motion descriptors*

Motion is a feature which discriminates video from still images. It can be used as a descriptor for video segments and the MPEG-7 standard recognises four motion-based descriptors.

### 11.1.7.1  Camera motion

Camera motion is a descriptor that characterises the three-dimensional camera motion parameters. These parameters can be automatically extracted or generated by the capturing devices.

The camera motion descriptor supports the following well known basic camera motion:

- fixed: camera is static
- panning: horizontal rotation
- tracking: horizontal traverse movement, also known as travelling in the film industry
- tilting: vertical rotation
- booming: vertical traverse movement
- zooming: change of the focal length
- dollying: translation along the optical axis
- rolling: rotation around the optical axis.

The subshots for which all frames are characterised by a particular camera motion, which can be single or mixed, determine the building blocks for the camera motion descriptor. Each building block is described by its start time, the duration, the speed of the induced image motion, by the fraction of time of its duration compared with a given temporal window size and the focus-of-expansion or the focus-of-contraction.

### 11.1.7.2  Motion trajectory

The motion trajectory of an object is a simple high level feature defined as the localisation in time and space of one representative point of this object. This descriptor can be useful for content-based retrieval in object-oriented visual databases. If *a priori* knowledge is available, the trajectory motion can be very useful. For example, in surveillance, alarms can be triggered if an object has a trajectory that looks unusual (e.g. passing through a forbidden area).

The descriptor is essentially a list of key points along with a set of optional interpolating functions that describe the path of the object between the key points in terms of acceleration. The key points are specified by their time instant and either two- or three-dimensional Cartesian coordinates, depending on the intended application.

#### 11.1.7.3 Parametric motion

Parametric motion models of affine, perspective etc. have been extensively used in image processing, including motion-based segmentation and estimation, global motion estimation. Some of these we have seen in Chapter 9 for motion estimation and in Chapter 10 for global motion estimation used in the sprite of MPEG-4. Within the MPEG-7 framework, motion is a highly relevant feature, related to the spatio–temporal structure of a video and concerning several MPEG-7 specific applications, such as storage and retrieval of video databases and hyperlinking purposes.

The basic underlying principle consists of describing the motion of an object in video sequences in terms of its model parameters. Specifically, the affine model includes translations, rotations, scaling and a combination of them. Planar perspective models can take into account global deformations associated with perspective projections. More complex movements can be described with the quadratic motion model. Such an approach leads to a very efficient description of several types of motion, including simple translations, rotations and zooming or more complex motions such as combinations of the above mentioned motions.

#### 11.1.7.4 Motion activity

Video scenes are usually classified in terms of their motion activity. For example, sports programmes are highly active and newsreader video shots represent low activity. Therefore motion can be used as a descriptor to express the activity of a given video segment.

An activity descriptor can be useful for applications such as surveillance, fast browsing, dynamic video summarisation and movement-based query. For example, if the activity descriptor shows a high activity, then during the playback the frame rate can be slowed down to make highly active scenes viewable. Another example of an application is finding all the high action shots in a news video programme.

### *11.1.8 Localisation*

This descriptor specifies the position of a query object within the image, and it is defined with two types of description.

#### 11.1.8.1 Region locator

The region locator is a descriptor that enables localisation of regions within images by specifying them with a brief and scalable representation of a box or a polygon.

### 11.1.8.2 Spatio–temporal locator

This descriptor defines the spatio–temporal locations of the regions in a video sequence, such as moving object regions, and provides localisation functionality. The main application of it is hypermedia, which displays the related information when the designated point is inside the object. Another major application is object retrieval by checking whether the object has passed through particular points, which can be used in, say, surveillance. The spatio–temporal locator can describe both spatially connected and nonconnected regions.

## 11.1.9 Others

MPEG-7 is an ongoing process and in the future other descriptors will be added to the above list. In this category, the most notable descriptor is face recognition.

### 11.1.9.1 Face recognition

The face recognition descriptor can be used to retrieve face images. The descriptor represents the projection of a face vector onto a set of basis vectors, which are representative of possible face vectors. The face recognition feature set is extracted from a normalised face image. This normalised face image contains 56 lines, each line with 46 intensity values. At the 24th row, the centres of the two eyes in each face image are located at the 16th and 31st column for the right and left eye, respectively. This normalised image is then used to extract the one-dimensional face vector that consists of the luminance pixel values from the normalised face image arranged into a one-dimensional vector in the scanning direction. The face recognition feature set is then calculated by projecting the one-dimensional face vector onto the space defined by a set of basis vectors.

## 11.2 Practical examples of image retrieval

In this section some of the methods described in the previous sections are used to demonstrate practical visual information retrieval from image databases. I have chosen texture and shape-based retrievals for demonstration purposes, since those of colour and motion-based methods are not easy to demonstrate in black and white pictures and within the limited space of this book.

## 11.2.1 Texture-based image retrieval

Spatial frequency analysis of textures provides an excellent way of classifying them. The complex Gabor wavelet which is a modulated Guassian function to a complex exponential is ideal for this purpose [3]. A two-dimensional complex Gabor wavelet is defined as:

$$g(x, y) = \frac{1}{2\pi \sigma_x \sigma_y} e^{1/2(x^2+y^2)/(\sigma_x^2+\sigma_y^2)} \times e^{j2\pi f_0 x} \tag{11.1}$$

where $\sigma_x$ and $\sigma_y$ are the horizontal and vertical standard deviations of the Gaussian and $f_0$ is the filter bandwidth. Thus its frequency domain representation (Fourier transform) is given by:

$$G(u, v) = e^{1/2(((u-f_0)^2+v^2)/(\sigma_u^2+\sigma_v^2))} \quad (11.2)$$

where $u$ and $v$ are the horizontal and vertical spatial frequencies and $\sigma_u$ and $\sigma_v$ are their respective standard deviations.

The $g(x, y)$ of eqn. 11.1 can be used as the mother wavelet to decompose a signal into various levels and orientations. In Chapter 4 we showed how mother wavelets could be used in the design of discrete wavelet transform filters. The same procedure can be applied to the mother Gabor wavelet.

Figure 11.3 shows the spectrum of the Gabor filter at four levels and six orientations. It is derived by setting the lowest and highest horizontal spatial frequencies to $u_l = 0.05$ and $u_h = 0.4$, respectively. The intermediate frequencies are derived by constraining the bands to touch each other.

The above set of filters can decompose an image into $4 \times 6 = 24$ subimages. As we had seen in Chapter 4, each subimage reflects characteristics of the image at a specific direction and spatial resolution. Hence it can analyse textures of images and describe them at these orientations and resolutions. For the example given above, one can calculate the mean and standard deviation of each subimage, and use it as a 48-dimensional vector to describe it. This is called the feature vector that can be

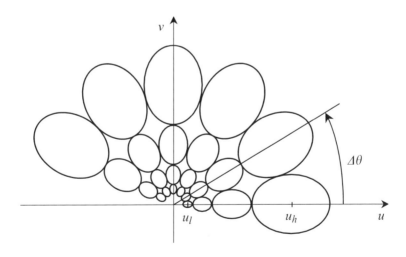

Figure 11.3  Gabor filter spectrum; the contours indicate the half-peak magnitude of the filter responses in the Gabor filter dictionary. The filter parameters used are $u_h = 0.4$, $u_l = 0.05$, $M = 4$ and $L = 6$

360  *Standard codecs: image compression to advanced video coding*

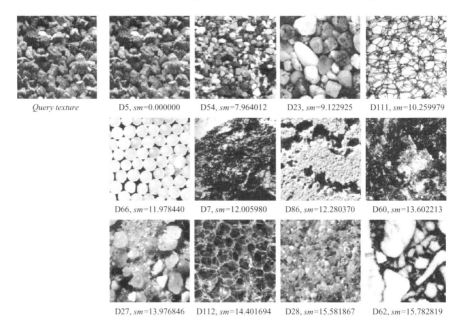

*Figure 11.4    An example of texture-based image retrieval; query texture: D5*

used for indexing the texture of an image and is given by:

$$\text{feature vector} = \big[\mu_{00}\mu_{01}\mu_{02}\mu_{03}\mu_{04}\mu_{05}\mu_{10}\mu_{11}\cdots \\ \mu_{35}\sigma_{00}\sigma_{01}\sigma_{02}\sigma_{03}\sigma_{04}\sigma_{05}\sigma_{10}\sigma_{11}\cdots\sigma_{35}\big]_{1\times 48} \quad (11.3)$$

To retrieve a query texture, its feature vector is compared against the feature vectors of the textures in the database. The similarity measure is the Euclidean distance between the feature vectors, and the one that gives the least distance is the most similar texture.

Figure 11.4 demonstrates retrieving a texture from a texture database of 112 images. In this database, images are identified as D1 to D112. The query image, D5, is shown at the top left, along with the 12 most similar retrieved images, in the order of their similarity measure. As we see, the query texture itself, D5, is found as the closest match, followed by a visually similar texture, D54, and so on. In the Figure the similarity distance, *sm*, of each retrieved candidate texture is also given.

## 11.2.2   Shape-based retrieval

Shapes are best described by the strength of their curvatures along their contours. A useful way of describing this strength is the curvature function, $k(s, \theta)$, defined as the instantaneous rate of change of the angle of the curve (tangent) $\theta$ over its arc length $s$:

$$k(s, \theta) = \frac{d\theta}{ds} \tag{11.4}$$

At sharp edges, where the rate of change of the angle is fast, the curvature function, $k(s, \theta)$, has large values. Hence contours can be described by some values of their curvature functions as feature points. For example, feature points can be defined as the positions of large curvature points or their zero crossings. However, since contours are normally noisy, direct derivation of the curvature function from the contour can lead to false feature points.

To eliminate these unwanted feature points, contours should be denoised through smoothing filters. Care should be taken on the degree of filter smoothness, since heavily filtered contours lose the feature points and lightly filtered ones cannot get rid of the false feature points. Large numbers of feature points also demand more storage and heavy processing for retrieval.

Filtering a contour with a set of Gaussian filters of varying degrees of smoothness is the answer to this question, the so-called scale-space representation of curves [4]. Figure 11.5 shows four smoothed contours of a shape at scaling (smoothing) factors of 16, 64, 256 and 1024. The positions of the curvature extremes on the contour at each scale are also shown. These points exhibit very well the most important structure of each contour.

For indexing and retrieval applications, these feature points can be joined together to approximate each smoothed contour with a polygon [5]. Moving round the contour, the angle of every polygon line with the horizontal is recorded, and this set of angles

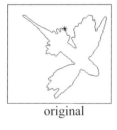

original

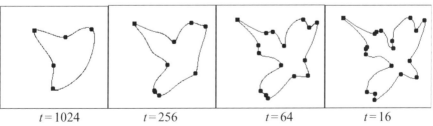

$t = 1024$       $t = 256$       $t = 64$       $t = 16$

*Figure 11.5   A contour and the positions of its curvature extremes at four different scales*

is called the turning function. The turning function is the index of the shape that can be used for retrieval.

The similarity measure is based on the minimisation of the Euclidean distance between the query turning function and the turning functions of the shapes in the database. That is:

$$sm = \min \sum_{i=1}^{N} \left\| \theta_q^i - \theta_j^i \right\| \qquad (11.5)$$

where $\theta_q^i$ is the $i$th angle of the query turning function and $\theta_j^i$ is the $i$th angle of the $j$th shape turning function in the database and $N$ is the number of angles in the turning function (e.g. number of vertices of the polygons or feature points on the contours). Calculation of the Euclidean distance necessitates that all the indices should have an equal number of turning angles in their turning functions. This is done by inserting additional feature points on the contours such that polygons have $N$ vertices.

Insertion of new feature points on the contour should be such that they should also represent important curvature extremes. This is done by inserting points on the contour, one at a time, where the contour has its largest distance from the polygon. Figure 11.6 shows the polygons of Figure 11.5 with the added new vertices. The number of total vertices is chosen such that approximation distortion of the original contour with a polygon is less than an acceptable value.

To show the retrieval efficiency of this method of shape description, Figure 11.7a shows a query shape, to be searched in a database of almost 1100 marine creatures.

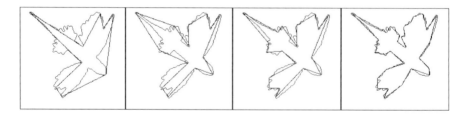

Figure 11.6  New polygons of Figure 11.5 with some added vertices

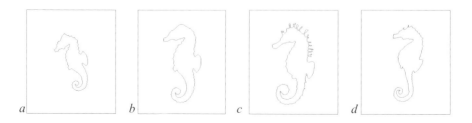

Figure 11.7  a the query shape, b, c and d the three closest shapes in order

The query marine is found as the closest shape, followed by three next closet marine shapes in the order of their closeness as, shown in Figure 11.7.

### 11.2.3 Sketch-based retrieval

In the above two examples of image retrieval, it was assumed that the query image (texture or shape) is available to the user. However, there are occasions when these query images may not be available. Users might have a verbal description of objects of interest, or have in their minds an idea of a visual object.

One way of searching for visual objects without the query images is to sketch the object of interest and submit it to the search engine. Then, based on a set of retrieved objects, the user may iteratively modify his sketch, until he finds the desired object. Assume that the fish in Figure 11.8 C1 is the desired shape in a database. The user first draws a rough sketch of this fish, as he imagines it, like the one shown in Figure 11.8 A0. Based on shape similarity, the three best similar shapes in the order of their similarity to the drawn shape are shown in A1, A2 and A3.

By inspecting these outputs, the user then realises that none of the matched fish has any fins. Adding a dorsal and a ventral fin to the sketched fish, the new query fish of B0 is created. With this new query shape, the new set of best matched shapes in the order of their similarity to the refined sketch become B1, B2 and B3.

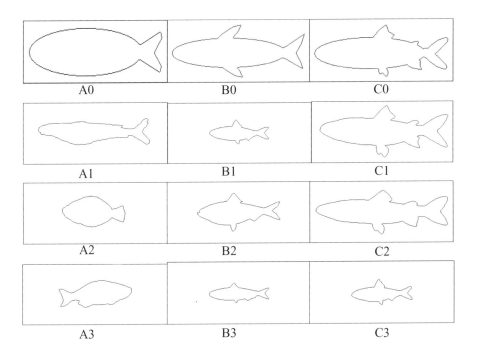

*Figure 11.8   Sketch-based shape retrieval*

Finally, adding an anal fin and a small adipose fin to the refined sketch, a further refined query shape of C0 is created. The new set of retrieved shapes, in the order of their similarity to the refined sketch now become C1, C2 and C3. This is the last iteration step, since the desired shape, C1, comes as the best matched shape.

This technique can also be extended to the other retrieval methods. For example, one might paint or add some texture to the above drawings. This would certainly improve the reliability of the retrieval system

## 11.3 MPEG-21: multimedia framework

Today multimedia technology is so advanced that access to the vast amount of information and services from almost anywhere at any time, through ubiquitous terminals and networks, is possible. However, no complete picture exists of how different communities can best interact with each other in a complex infrastructure. Examples of these communities are the content providers, financial, communication, computer and consumer electronics sectors and their customers. Developing a common multimedia framework will facilitate cooperation between these sectors and support a more efficient implementation and integration of different models, rules, interests and content formats. This is the task given to the multimedia framework project under the name of MPEG-21. The name is chosen to signify the coincidence of the start of the project with the 21st century.

The chain of multimedia content delivery encompasses content creation, production, delivery and consumption. To support this, the content has to be identified, described, managed and protected. The transport and delivery of content will undoubtedly be over a heterogeneous set of terminals and networks. Reliable delivery of contents, management of personal data, financial transactions and user privacy are some of the issues that the multimedia framework should take into account. In the following sections the seven architectural key elements that the multimedia framework considers instrumental in the realisation of the task are explained.

### 11.3.1  Digital item declaration

In multimedia communication to facilitate a wide range of actions involving digital items, there is a strong need for a concrete description for defining exactly what constitutes such an item. Clearly, there are many kinds of content, and nearly as many possible ways of describing it. This presents a strong challenge to lay out a powerful and flexible model for a digital item, from which the content can be described more accurately. Such a model is only useful if it yields a format that can be used to represent any digital item defined within the model unambiguously, and communicate it successfully.

Consider a simple web page as a digital item. This web page typically consists of an HTML (hypertext markup language) document with embedded links or dependencies to various image files (e.g. JPEG images) and possibly some layout information (e.g. style sheet). In this simple case, it is a straightforward exercise to inspect the HTML document and deduce that this digital item consists of the HTML document itself plus all the other resources upon which it depends.

Now let us constrain the above example, such that the web page should be viewed with the JavaScript language. The presence of the language logic now raises the question of what constitutes this digital item and how it can be unambiguously determined. The first problem is that addition of the scripting code changes the declaration of the links, since the links can be determined only by running the embedded script on the specific platform. This could still work as a method of deducing the structure of a digital item, assuming that the author intended each translated version of the web page to be a separate and distinct digital item. This assumption creates a second problem, as it is ambiguous whether the author actually intends for each translation of the page to be a standalone digital item, or whether the intention is for the digital item to consist of the page with the language choice left unresolved. If the latter is the case, it makes it impossible to deduce the exact set of resources that this digital item consists of, which leads back to the first problem. In the course of standardisation MPEG-21 aims to come up with a standard way of defining and declaring digital items.

### 11.3.2 Digital item identification and description

Currently, the majority of content lacks identification and description. Moreover, there is no mechanism to ensure that this identity and description is persistently associated with the content, which hinders any kind of efficient content storage.

However, some identifiers have been successfully implemented and commonly used for several years, but they are defined in a single media type. ISBN, the International Standard Book Number, or URN, Universal Resource Name, are a few examples of how a digital item can be identified. This is just the beginning and in the future we will see more of these.

There are many examples of businesses, which have requirements for the deployment of a unique identification system on a global scale. Proprietary solutions such as labelling and watermarking for insertion, modification and extractions of IDs have emerged in the past. However, no international standard is available today for the deployment of such technologies, and it is the second task of MPEG-21 to identify and describe them.

### 11.3.3 Content handling and usage

The availability and access of content within networks is exponentially increasing over time. With the goal of MPEG-21 to enable transparent use of this content over a variety of networks and devices, it becomes extremely important that standards should exist to facilitate searching, locating, caching, archiving, routing, distributing and using content. In addition, the content has to be relevant to the customer needs and provide a better return of money for the business.

Thus the goal of the MPEG-21 multimedia framework is to provide interfaces and protocols that enable creation, manipulation, search, access, storage, delivery and use of content, across the content creation and consumption chain. The emphasis should be given to improve an interaction model for users with personalisation and content handling.

### 11.3.4 Intellectual property and management

MPEG-21 should provide a uniform framework that enables users to express their rights and interests in, and agreements related to, digital items. They should be assured that those rights, interests and agreements will be persistently and reliably managed and protected across a wide range of networks and devices.

### 11.3.5 Terminal and networks

Accessibility of heterogeneous content is becoming widespread to many network devices. Today we receive a variety of information through set-top boxes for terrestrial/cable/satellite networks, personal digital assistants, mobile phones etc. Additionally, these access devices are used in different locations and environments. This makes it difficult for service providers to ensure that content is available anywhere, anytime and can be used and rendered in a meaningful way.

The goal of MPEG-21 is to enable transparent use of multimedia resources across a wide range of networked devices. This inevitably has an impact on the way network and terminal resources themselves are being dealt with.

Users accessing content should be offered services with a known subjective quality, perhaps at a known or agreed price. They should be shielded from network and terminal installation, management and implementation issues.

From the network point of view, it is desirable that the application serving the user can translate the user requirements into a network quality of service (QOS) contract. This contract, containing a summary of negotiated network parameters, is handled between the user or his agents and the network. It guarantees the delivery of service over the network for a given QOS. However, the actual implementation of network QOS does not fall within the scope of MPEG-21. The intent is to make use of these mechanisms and propose requirements to network QOS functionality extensions to fulfil the overall MPEG-21 QOS demands.

### 11.3.6 Content representation

Content is the most important element of a multimedia framework. Within the framework, content is coded, identified, described, stored, delivered, protected, transacted, consumed etc.

Although MPEG-21 assumes that content is available in digital form, it should be represented in a form to fulfil some requirements. For example, digital video as a digital item needs to be compressed and converted into a format to be stored more economically. Although there are several standards for efficient compression (representation) of image and video, they have been devised for specific purposes. Throughout the book we have seen that JPEG and JPEG2000 are some of the standards for coding of still images. For video H.261, H.263, H.26L, MPEG-1, MPEG-2 are used for frame-based video and MPEG-4 is for coding of arbitrary shaped objects.

But this is not enough for unique and unambiguous representation of digital video items. The same is true for audio.

In fact users are becoming more mobile and have a need to access information on multiple devices in different contexts at different locations. Currently, content providers and authors have to create multiple formats of content and deploy them in a multitude of networks. Also, no satisfactory automated configurable way of delivering and consuming content exists that scales automatically to different network characteristics and device profiles. This is despite the introduction of various scalable coding of video to alleviate some of these problems.

The content representation element of the framework is intended to address the technology needed in order that content is represented in a way adequate for pursuing the general objectives of MPEG-21. In this regard MPEG-21 assumes that content consists of one or a combination of:

(i) content represented by MPEG standards (e.g. MPEG-4 video)
(ii) content used by MPEG but not covered by MPEG standards (e.g. plain text, HTML etc.)
(iii) content that can be represented by (i) and (ii) but is represented by different standards or proprietary specifications
(iv) future standards for other sensory media.

### 11.3.7 Event reporting

Every interaction is an event and there is a necessity that events should be reported. However, there are a number of difficulties in providing an accurate report about the event. Different observers of the event may have different perspectives, needs and focuses. Currently there exists no standardised means of event reporting.

In a multimedia environment there are many events that need reporting. For example, accurate product cost, consumer cost, channel costs or profitability information. This allows users to understand operational processes and simulate dynamics in order to optimise efficiencies and outputs.

However, every industry reports information about its performance to other users. A number of issues make this difficult for the receiving users to process this information. For example, different reporting formats, different standards from country to country, different currencies, languages etc. make event processing difficult. As the last key architectural element, MPEG-21 intends to standardise event reporting, to eliminate these shortfalls.

## 11.4 References

1 ISO/IEC JTC1/SC29/WG11/N4031: 'Overview of the MPEG-7 standard'. Singapore, March 2001
2 ISO/IEC JTC1/SC29/WG11/N4040: 'Study on MPEG-21 (digital audiovisual framework) part 1'. Singapore, March 2001

3 LEE, T.S.: 'Image representation using 2D Gabor wavelets', *IEEE Trans. Pattern Anal. Mach. Intell.*, 1996, **18:10**, pp.959–971
4 IZQUIERDO, E., and GHANBARI, M.: 'Nonlinear Gaussian filtering approach for object segmentation', *IEE Proc. Vis. Image Signal Process.*, 1999, **146:3**, pp.137–143
5 ARKIN, E., CHEW, L., HUTTENLOCHER, D., and MITCHELL, J.: 'An efficiently computable metric for comparing polygonal shapes'. *IEEE Trans. Pattern Anal. Mach. Intell.*, 1991, **PAMI-13**, pp.209–216

## Appendix A
# A C program for the fast discrete cosine transform

```
/*ffdct.c*/
/*fast forward discrete cosine transform in the current
  frame */

#include "config.h"
#include "global.h"

#define W1  2841            /* sqrt(2)cos(π/16)<< 11*/
#define W2  2676            /* sqrt(2)cos(2π/16)<< 11*/
#define W3  2408            /* sqrt(2)cos(3π/16)<< 11*/
#define W5  1609            /* sqrt(2)cos(5π/16)<< 11*/
#define W6  1108            /* sqrt(2)cos(6π/16)<< 11*/
#define W7  565             /* sqrt(2)cos(7π/16)<< 11*/
#define W10 2276            /* W1 - W7 */
#define W11 3406            /* W1 + W7 */
#define W12 4017            /* W3 + W5 */
#define W13 799             /* W3 - W5 */
#define W14 1568            /* W2 - W6 */
#define W15 3784            /* W2 + W6 */

/* global declarations */
void ffdct _ANSI_ARGS_((int *block));

void ffdct(block)
int *block;
{
int s[10],t[10],r[10];
int *p;
int j, temp;
```

```
/*forward transformation in "H" direction*/
p = block;
for(j=0; j<64; j +=8 )
   {
   /* first stage transformation */
   s[0]  =  (*(p)   + *(p+7));
   s[1]  =  (*(p+1) + *(p+6));
   s[2]  =  (*(p+2) + *(p+5));
   s[3]  =  (*(p+3) + *(p+4));
   s[4]  =  (*(p+3) - *(p+4));
   s[5]  =  (*(p+2) - *(p+5));
   s[6]  =  (*(p+1) - *(p+6));
   s[7]  =  (*(p)   - *(p+7));

   /* second stage transformation */
   t[0] = s[0] + s[3];
   t[1] = s[1] + s[2];
   t[2] = s[1] - s[2];
   t[3] = s[0] - s[3];
   t[5] = ((s[6] - s[5]) * 181) >> 8;
   t[6] = ((s[6] + s[5]) * 181) >> 8;

   /* third stage transformation */
   r[4] = s[4] + t[5];
   r[5] = s[4] - t[5];
   r[6] = s[7] - t[6];
   r[7] = s[7] + t[6];

   /* fourth stage transformation */
   block[0+j] = (t[0] + t[1]);
   block[4+j] = (t[0] - t[1]);
   temp = (r[4] + r[7]) * W1;
   block[1+j] = (temp - r[4] * W10) >> 11;
   block[7+j] = (r[7] * W11 - temp) >> 11;
   temp = ( r[5] + r[6]) * W3;
   block[3+j] = (temp - r[5] * W12) >> 11;
   block[5+j] = (temp - r[6] * W13) >> 11;
   temp = (t[2] + t[3]) * W6;
   block[2+j] = (temp + t[3] * W14) >> 11;
   block[6+j] = (temp - t[2] * W15) >> 11;
   p += 8;
   }
```

## A C program for the fast discrete cosine transform

```
/* forward transformation in 'V' direction */
for(j=0; j<8; j++)
   {
   /* first stage transformation */
   s[0] = block[j]    + block[j+56];
   s[1] = block[j+8]  + block[j+48];
   s[2] = block[j+16] + block[j+40];
   s[3] = block[j+24] + block[j+32];
   s[4] = block[j+24] - block[j+32];
   s[5] = block[j+16] - block[j+40];
   s[6] = block[j+8]  - block[j+48];
   s[7] = block[j]    - block[j+56];

   /* second stage transformation */
   t[0] = s[0] + s[3];
   t[1] = s[1] + s[2];
   t[2] = s[1] - s[2];
   t[3] = s[0] - s[3];
   t[5] = ((s[6] - s[5]) * 181) >> 8;
   t[6] = ((s[6] + s[5]) * 181) >> 8;

   /* third stage transformation */
   r[4] = s[4] + t[5];
   r[5] = s[4] - t[5];
   r[6] = s[7] - t[6];
   r[7] = s[7] + t[6];

   /* fourth stage transformation */
   /* transform coefficients */
   /* coefficients are divided by 8 and rounded */
   block[0+j] = ((t[0] + t[1]) + 4) >> 3;
   block[32+j]= ((t[0] - t[1]) + 4) >> 3;
   temp = (r[4] + r[7]) * W1;
   block[8+j] = ((temp - r[4] * W10) + 8192) >> 14;
   block[56+j]= (((r[7] * W11) - temp) + 8192) >> 14;
   temp = (r[5] + r[6]) * W3;
   block[24+j]= ((temp - r[5] * W12) + 8192) >> 14;
   block[40+j]= ((temp - r[6] * W13) + 8192) >> 14;
   temp = (t[2] + t[3]) * W6;
   block[16+j]= ((temp + t[3] * W14) + 8192) >> 14;
   block[48+j]= ((temp - t[2] * W15) + 8192) >> 14;
   }
}
```

## Appendix B

# Huffman tables for the DC and AC coefficients of the JPEG baseline encoder

*Table B.1  DC Huffman coefficients of luminance*

| Category (CAT) | Codeword |
|---|---|
| 0 | 00 |
| 1 | 010 |
| 2 | 011 |
| 3 | 100 |
| 4 | 101 |
| 5 | 110 |
| 6 | 1110 |
| 7 | 11110 |
| 8 | 111110 |
| 9 | 1111110 |
| 10 | 11111110 |
| 11 | 111111110 |

Table B.2  AC Huffman coefficients of luminance

| (RUN,CAT) | Codeword | (RUN,CAT) | Codeword |
|---|---|---|---|
| 0,0 (**EOB**) | 1010 | | |
| 0,1 | 00 | 4,1 | 111011 |
| 0,2 | 01 | 4,2 | 1111111000 |
| 0,3 | 100 | 4,3 | 1111111110010110 |
| 0,4 | 1011 | 4,4 | 1111111110010111 |
| 0,5 | 11010 | 4,5 | 1111111110011000 |
| 0,6 | 1111000 | 4,6 | 1111111110011001 |
| 0,7 | 11111000 | 4,7 | 1111111110011010 |
| 0,8 | 1111110110 | 4,8 | 1111111110011011 |
| 0,9 | 1111111110000010 | 4,9 | 1111111110011100 |
| 0,10 | 1111111110000011 | 4,10 | 1111111110011101 |
| 1,1 | 1100 | 5,1 | 1111010 |
| 1,2 | 11011 | 5,2 | 11111110111 |
| 1,3 | 1111001 | 5,3 | 1111111110011110 |
| 1,4 | 111110110 | 5,4 | 1111111110011111 |
| 1,5 | 11111110110 | 5,5 | 1111111110100000 |
| 1,6 | 1111111110000100 | 5,6 | 1111111110100001 |
| 1,7 | 1111111110000101 | 5,7 | 1111111110100010 |
| 1,8 | 1111111110000110 | 5,8 | 1111111110100011 |
| 1,9 | 1111111110000111 | 5,9 | 1111111110100100 |
| 1,10 | 1111111110001000 | 5,10 | 1111111110100101 |
| 2,1 | 11100 | 6,1 | 1111011 |
| 2,2 | 11111001 | 6,2 | 111111110110 |
| 2,3 | 1111110111 | 6,3 | 1111111110100110 |
| 2,4 | 111111110100 | 6,4 | 1111111110100111 |
| 2,5 | 1111111110001001 | 6,5 | 1111111110101000 |
| 2,6 | 1111111110001010 | 6,6 | 1111111110101001 |
| 2,7 | 1111111110001011 | 6,7 | 1111111110101010 |
| 2,8 | 1111111110001100 | 6,8 | 1111111110101011 |
| 2,9 | 1111111110001101 | 6,9 | 1111111110101100 |
| 2,10 | 1111111110001110 | 6,10 | 1111111110101101 |
| 3,1 | 111010 | 7,1 | 11111010 |
| 3,2 | 111110111 | 7,2 | 111111110111 |
| 3,3 | 111111110101 | 7,3 | 1111111110101110 |
| 3,4 | 1111111110001111 | 7,4 | 1111111110101111 |
| 3,5 | 1111111110010000 | 7,5 | 1111111110110000 |
| 3,6 | 1111111110010001 | 7,6 | 1111111110110001 |
| 3,7 | 1111111110010010 | 7,7 | 1111111110110010 |
| 3,8 | 1111111110010011 | 7,8 | 1111111110110011 |
| 3,9 | 1111111110010100 | 7,9 | 1111111110110100 |
| 3,10 | 1111111110010101 | 7,10 | 1111111110110101 |

*Table B.2   Continued*

| (RUN,CAT) | Codeword | (RUN,CAT) | Codeword |
|---|---|---|---|
| 8,1 | 111111000 | 12,1 | 1111111010 |
| 8,2 | 111111111000000 | 12,2 | 1111111111011001 |
| 8,3 | 1111111110110110 | 12,3 | 1111111111011010 |
| 8,4 | 1111111110110111 | 12,4 | 1111111111011011 |
| 8,5 | 1111111110111000 | 12,5 | 1111111111011100 |
| 8,6 | 1111111110111001 | 12,6 | 1111111111011101 |
| 8,7 | 1111111110111010 | 12,7 | 1111111111011110 |
| 8,8 | 1111111110111011 | 12,8 | 1111111111011111 |
| 8,9 | 1111111110111100 | 12,9 | 1111111111100000 |
| 8,10 | 1111111110111101 | 12,10 | 1111111111100001 |
| 9,1 | 111111001 | 13,1 | 11111111000 |
| 9,2 | 1111111110111110 | 13,2 | 1111111111100010 |
| 9,3 | 1111111110111111 | 13,3 | 1111111111100011 |
| 9,4 | 1111111111000000 | 13,4 | 1111111111100100 |
| 9,5 | 1111111111000001 | 13,5 | 1111111111100101 |
| 9,6 | 1111111111000010 | 13,6 | 1111111111100110 |
| 9,7 | 1111111111000011 | 13,7 | 1111111111100111 |
| 9,8 | 1111111111000100 | 13,8 | 1111111111101000 |
| 9,9 | 1111111111000101 | 13,9 | 1111111111101001 |
| 9,10 | 1111111111000110 | 13,10 | 1111111111101010 |
| 10,1 | 111111010 | 14,1 | 1111111111101011 |
| 10,2 | 1111111111000111 | 14,2 | 1111111111101100 |
| 10,3 | 1111111111001000 | 14,3 | 1111111111101101 |
| 10,4 | 1111111111001001 | 14,4 | 1111111111101110 |
| 10,5 | 1111111111001010 | 14,5 | 1111111111101111 |
| 10,6 | 1111111111001011 | 14,6 | 1111111111110000 |
| 10,7 | 1111111111001100 | 14,7 | 1111111111110001 |
| 10,8 | 1111111111001101 | 14,8 | 1111111111110010 |
| 10,9 | 1111111111001110 | 14,9 | 1111111111110011 |
| 10,10 | 1111111111001111 | 14,10 | 1111111111110100 |
| 11,1 | 1111111001 | 15,1 | 1111111111110101 |
| 11,2 | 1111111111010000 | 15,2 | 1111111111110110 |
| 11,3 | 1111111111010001 | 15,3 | 1111111111110111 |
| 11,4 | 1111111111010010 | 15,4 | 1111111111111000 |
| 11,5 | 1111111111010011 | 15,5 | 1111111111111001 |
| 11,6 | 1111111111010100 | 15,6 | 1111111111111010 |
| 11,7 | 1111111111010101 | 15,7 | 1111111111111011 |
| 11,8 | 1111111111010110 | 15,8 | 1111111111111100 |
| 11,9 | 1111111111010111 | 15,9 | 1111111111111101 |
| 11,10 | 1111111111011000 | 15,10 | 1111111111111110 |
| the special symbol representing 16 zero | | 15,0 (ZRL) | 11111111001 |

# Appendix C
# Huffman tables for quad tree shape coding

Huffman tables for quad tree shape coding. table_012.dat is used at levels 0, 1 and 2 and table_3.dat is used at level 3.

*Table C.1  table_012.dat*

| Index | Code | Index | Code |
|---|---|---|---|
| 0  | –              | 24 | 1111111111111110 |
| 1  | 10011          | 25 | 11111111001      |
| 2  | 11110101       | 26 | 111111010        |
| 3  | 101101         | 27 | 10001            |
| 4  | 0010           | 28 | 1111111010       |
| 5  | 1110110        | 29 | 111111111101     |
| 6  | 111111011      | 30 | 01111            |
| 7  | 1100110        | 31 | 1110101          |
| 8  | 01110          | 32 | 11111111000      |
| 9  | 101111         | 33 | 1101011          |
| 10 | 10010          | 34 | 1110010          |
| 11 | 11110100       | 35 | 11111100         |
| 12 | 1111111111100  | 36 | 0001             |
| 13 | 11110000       | 37 | 1110001          |
| 14 | 1110100        | 38 | 111111111011     |
| 15 | 1111111111111111 | 39 | 11110110       |
| 16 | 111111111010   | 40 | 10101            |
| 17 | 1101110        | 41 | 11110011         |
| 18 | 11111011       | 42 | 11111111010      |
| 19 | 11111000       | 43 | 11110010         |
| 20 | 110010         | 44 | 0101             |
| 21 | 111111111100   | 45 | 1110011          |
| 22 | 1111111000     | 46 | 1101111          |
| 23 | 11111001       | 47 | 11110111         |

Table C.1  Continued

| Index | Code | Index | Code |
|---|---|---|---|
| 48 | 111111111100 | 65 | 111111111000 |
| 49 | 11101111 | 66 | 1101000 |
| 50 | 101110 | 67 | 11110000 |
| 51 | 111111111111101 | 68 | 11111111011 |
| 52 | 111111111001 | 69 | 11101110 |
| 53 | 110001 | 70 | 01101 |
| 54 | 11110001 | 71 | 101100 |
| 55 | 1111111111101 | 72 | 0000 |
| 56 | 111111111111110 | 73 | 1100111 |
| 57 | 110000 | 74 | 1101101 |
| 58 | 1111111001 | 75 | 1101001 |
| 59 | 1111111111110 | 76 | 0011 |
| 60 | 0100 | 77 | 10100 |
| 61 | 1101010 | 78 | 1101100 |
| 62 | 11111010 | 79 | 01100 |
| 63 | 10000 | 80 | – |
| 64 | 1111111011 | | |

Table C.2  table_3.dat

| Index | Code |
|---|---|
| 0 | – |
| 2 | 1101 |
| 6 | 1111110 |
| 8 | 1011 |
| 18 | 11101 |
| 20 | 100 |
| 24 | 11111111 |
| 26 | 111110 |
| 54 | 1100 |
| 56 | 11111110 |
| 60 | 01 |
| 62 | 11110 |
| 72 | 00 |
| 74 | 1010 |
| 78 | 11100 |
| 80 | – |

# Appendix D
# Frequency tables for the CAE encoding of binary shapes

Frequency tables for intra and inter blocks, used in the context-based arithmetic encoding (CAE) method of binary shapes.

Intra_prob[1024] = {
65267,16468,65003,17912,64573,8556,64252,5653,40174,3932,29789,277,45152,1140,32768,2043,
4499,80,6554,1144,21065,465,32768,799,5482,183,7282,264,5336,99,6554,563,
54784,30201,58254,9879,54613,3069,32768,58495,32768,32768,32768,2849,58982,54613,32768,12892,
31006,1332,49152,3287,60075,350,32768,712,39322,760,32768,354,52659,432,61854,150,
64999,28362,65323,42521,63572,32768,63677,18319,4910,32768,64238,434,53248,32768,61865,13590,
16384,32768,13107,333,32768,32768,32768,32768,32768,32768,1074,780,25058,5461,6697,233,
62949,30247,63702,24638,59578,32768,32768,42257,32768,32768,49152,546,62557,32768,54613,19258,
62405,32569,64600,865,60495,10923,32768,898,34193,24576,64111,341,47492,5231,55474,591,
65114,60075,64080,5334,65448,61882,64543,13209,54906,16384,35289,4933,48645,9614,55351,7318,
49807,54613,32768,32768,50972,32768,32768,32768,15159,1928,2048,171,3093,8,6096,74,
32768,60855,32768,32768,32768,32768,32768,32768,32768,32768,32768,32768,32768,55454,32768,57672,
32768,16384,32768,21845,32768,32768,32768,32768,32768,32768,32768,5041,28440,91,32768,45,
65124,10923,64874,5041,65429,57344,63435,48060,61440,32768,63488,24887,59688,3277,63918,14021,
32768,32768,32768,32768,32768,32768,32768,32768,690,32768,32768,1456,32768,32768,8192,728,
32768,32768,58982,17944,65237,54613,32768,2242,32768,32768,32768,42130,49152,57344,58254,16740,
32768,10923,54613,182,32768,32768,32768,7282,49152,32768,32768,5041,63295,1394,55188,77,
63672,6554,54613,49152,64558,32768,32768,5461,64142,32768,32768,32768,62415,32768,32768,16384,
1481,438,19661,840,33654,3121,64425,6554,4178,2048,32768,2260,5226,1680,32768,565,
60075,32768,32768,32768,32768,32768,32768,32768,32768,32768,32768,32768,32768,32768,32768,32768,
16384,261,32768,412,16384,636,32768,4369,23406,4328,32768,524,15604,560,32768,676,
49152,32768,49152,32768,32768,32768,64572,32768,32768,32768,32768,54613,32768,32768,32768,32768,
4681,32768,5617,851,32768,32768,59578,32768,32768,32768,3121,3121,49152,32768,6554,10923,
32768,32768,54613,14043,32768,32768,32768,3449,32768,32768,32768,32768,32768,32768,32768,32768,
57344,32768,57344,3449,32768,32768,32768,3855,58982,10923,32768,239,62259,32768,49152,85,
58778,23831,62888,20922,64311,8192,60075,575,59714,32768,57344,40960,62107,4096,61943,3921,
39862,15338,32768,1524,45123,5958,32768,58982,6669,930,1170,1043,7385,44,8813,5011,
59578,29789,54613,32768,32768,32768,32768,32768,32768,32768,32768,32768,58254,56174,32768,32768,
64080,25891,49152,22528,32768,2731,32768,10923,10923,3283,32768,1748,17827,77,32768,108,
62805,32768,62013,42612,32768,32768,61681,16384,58982,60075,62313,58982,65279,58982,62694,62174,
32768,32768,10923,950,32768,32768,32768,32768,5958,32768,38551,1092,11012,39322,13705,2072,
54613,32768,32768,11398,32768,32768,32768,145,32768,32768,32768,29789,60855,32768,61681,54792,
32768,32768,32768,17348,32768,32768,32768,8192,57344,16384,32768,3582,52581,580,24030,303,

380   *Standard codecs: image compression to advanced video coding*

62673,37266,65374,6197,62017,32768,49152,299,54613,32768,32768,32768,35234,119,32768,3855,
31949,32768,32768,49152,16384,32768,32768,32768,24576,32768,49152,32768,17476,32768,32768,57445,
51200,50864,54613,27949,60075,20480,32768,57344,32768,32768,32768,32768,32768,45875,32768,32768,
11498,3244,24576,482,16384,1150,32768,16384,7992,215,32768,1150,23593,927,32768,993,
65353,32768,65465,46741,41870,32768,64596,59578,62087,32768,12619,23406,11833,32768,47720,17476,
32768,32768,2621,6554,32768,32768,32768,32768,32768,5041,32768,16384,32768,4096,2731,
63212,43526,65442,47124,65410,35747,60304,55858,60855,58982,60075,19859,35747,63015,64470,25432,
58689,1118,64717,1339,24576,32768,32768,1257,53297,1928,32768,33,52067,3511,62861,453,
64613,32768,32768,32768,64558,32768,32768,2731,49152,32768,32768,32768,61534,32768,32768,35747,
32768,32768,32768,32768,13107,32768,32768,32768,32768,32768,32768,20480,32768,32768,32768,
32768,32768,32768,54613,40960,5041,32768,32768,32768,32768,3277,64263,57592,32768,3121,
32768,32768,32768,32768,10923,32768,32768,32768,8192,32768,32768,5461,6899,32768,1725,
63351,3855,63608,29127,62415,7282,64626,60855,32768,32768,60075,5958,44961,32768,61866,53718,
32768,32768,32768,32768,32768,32768,6554,32768,32768,32768,32768,32768,2521,978,32768,1489,
58254,32768,58982,61745,21845,32768,54613,58655,60075,32768,49152,16274,50412,64344,61643,43987,
32768,32768,32768,1638,32768,32768,32768,24966,54613,32768,32768,2427,46951,32768,17970,654,
65385,27307,60075,26472,64479,32768,32768,4681,61895,32768,32768,16384,58254,32768,32768,6554,
37630,3277,54613,6554,4965,5958,4681,32768,42765,16384,32768,21845,22827,16384,32768,6554,
65297,64769,60855,12743,63195,16384,32768,37942,32768,32768,32768,60075,32768,62087,54613,
41764,2161,21845,1836,17284,5424,10923,1680,11019,555,32768,431,39819,907,32768,171,
65480,32768,64435,33803,2595,32768,57041,32768,61167,32768,32768,32768,32768,32768,32768,1796,
60855,32768,17246,978,32768,32768,8192,32768,32768,32768,14043,2849,32768,2979,6554,6554,
65507,62415,65384,61891,65273,58982,65461,55097,32768,32768,32768,55606,32768,2979,3745,16913,
61885,13827,60893,12196,60855,53248,51493,11243,56656,783,55563,143,63432,7106,52429,445,
65485,1031,65020,1380,65180,57344,65162,36536,61154,6554,26569,2341,63593,3449,65102,533,
47827,2913,57344,3449,35688,1337,32768,22938,25012,910,7944,1008,29319,607,64466,4202,
64549,57301,49152,20025,63351,61167,32768,45542,58982,14564,32768,9362,61895,44840,32768,26385,
59664,17135,60855,13291,40050,12252,32768,7816,25798,1850,60495,2662,18707,122,52538,231,
65332,32768,65210,21693,65113,6554,65141,39667,62259,32768,22258,1337,63636,32768,64255,52429,
60362,32768,6780,819,16384,32768,16384,4681,49152,32768,8985,2521,24410,683,21535,16585,
65416,46091,65292,58328,64626,32768,65016,39897,62687,47332,62805,28948,64284,53620,52870,49567,
65032,31174,63022,28312,64299,46811,48009,31453,61207,7077,50299,1514,60047,2634,46488,235
};

Inter_prob[512] = {
65532,62970,65148,54613,62470,8192,62577,8937,65480,64335,65195,53248,65322,62518,62891,38312,
65075,53405,63980,58982,32768,32768,54613,32768,65238,60009,60075,32768,59294,19661,61203,13107,
63000,9830,62566,58982,11565,32768,25215,3277,53620,50972,63109,43691,54613,32768,39671,17129,
59788,6068,43336,27913,6554,32768,12178,1771,56174,49152,60075,43691,58254,16384,49152,9930,
23130,7282,40960,32768,10923,32768,32768,27307,32768,32768,32768,32768,32768,32768,32768,32768,
36285,12511,10923,32768,45875,16384,32768,32768,16384,23831,4369,32768,8192,10923,32768,32768,
10175,2979,18978,10923,54613,32768,6242,6554,1820,10923,32768,32768,32768,32768,32768,5461,
28459,593,11886,2030,3121,4681,1292,112,42130,23831,49152,29127,32768,6554,5461,2048,
65331,64600,63811,63314,42130,19661,49152,32768,65417,64609,62415,64617,64276,44256,61068,36713,
64887,57525,53620,61375,32768,8192,57344,6554,63608,49809,49152,62623,32768,15851,58982,34162,
55454,51739,64406,64047,32768,32768,7282,32768,32768,49152,58756,62805,64990,32768,14895,16384,19418,
57929,24966,58689,31832,32768,16384,10923,6554,54613,42882,57344,64238,58982,10082,20165,20339,
62687,15061,32768,10923,32768,10923,32768,16384,59578,34427,32768,16384,32768,7825,32768,7282,
58052,23400,32768,5041,32768,2849,32768,32768,47663,15073,57344,4096,32768,1176,32768,1320,
24858,410,24576,923,32768,16384,16384,5461,16384,1365,32768,5461,32768,5699,8192,13107,
46884,2361,23559,424,19661,712,655,182,58637,2094,49152,9362,8192,85,32768,1228,
65486,49152,65186,49152,61320,32768,57088,25206,65352,63047,62623,49152,64641,62165,58986,18304,
64171,16384,60855,54613,42130,32768,61335,32768,58254,58982,49152,32768,60985,35289,64520,31554,
51067,32768,64074,32768,40330,32768,34526,4096,60855,32768,63109,58254,57672,16384,31009,2567,
23406,32768,44620,10923,32768,32768,32099,10923,49152,49152,54613,60075,63422,54613,64388,39719,
58982,32768,54613,32768,14247,32768,22938,5041,32768,49152,32768,32768,25321,6144,29127,10999,
41263,32768,46811,32768,267,4096,426,16384,32768,19275,49152,32768,1008,1437,5767,11275,

5595,5461,37493,6554,4681,32768,6147,1560,38229,10923,32768,40960,35747,2521,5999,312,
17052,2521,18808,3641,213,2427,574,32,51493,42130,42130,53053,11155,312,2069,106,
64406,45197,58982,32768,32768,16384,40960,36864,65336,64244,60075,61681,65269,50748,60340,20515,
58982,23406,57344,32768,6554,16384,19661,61564,60855,47480,32768,54613,46811,21701,54909,37826,
32768,58982,60855,60855,32768,32768,39322,49152,57344,45875,60855,55706,32768,24576,62313,25038,
54613,8192,49152,10923,32768,32768,32768,32768,32768,19661,16384,51493,32768,14043,40050,44651,
59578,5174,32768,6554,32768,5461,23593,5461,63608,51825,32768,23831,58887,24032,57170,3298,
39322,12971,16384,49152,1872,618,13107,2114,58982,25705,32768,60075,28913,949,18312,1815,
48188,114,51493,1542,5461,3855,11360,1163,58982,7215,54613,21487,49152,4590,48430,1421,
28944,1319,6868,324,1456,232,820,7,61681,1864,60855,9922,4369,315,6589,14
};

# Appendix E
# Channel error/packet loss model

Digital channels and packet networks can be modelled as a discrete two-state Elliot–Gilbert model, as shown in Figure E.1.

When the model is at the bad state, b, a bit is corrupted or a packet is received erroneously, and can be regarded as lost. The model is run for every bit (or packet) to be transmitted, and the average number of times that the model stays at the bad position is the average bit error rate, or packet loss rate, $P$. Consequently, the average number of times it stays at the good state is $1 - P$. In the following, we show how the average bit error or packet loss rate is related to the transition probabilities $\alpha$ and $\beta$. We analyse on packet loss rate, and the same procedure can be used for the bit error rate. The model has also been accepted by the ITU-T for the evaluation of ATM networks.[1]

*Calculation of mean packet loss rate*

At each run of the model, a packet is lost in two ways. First, before the run it was at the bad state, but after the run it is also at the bad state. Second, it was in a good state, but after the run is at the bad state. Thus the probability of being at the bad

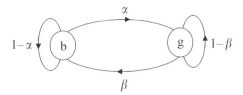

*Figure E.1   An Elliot–Gilbert two-level error model*

---

[1] ITU SGXV working party XV/I, Experts Group for ATM video coding, working document AVC-205, January 1992

state (loss), $P$, is:

$$P = P(1 - \alpha) + (1 - P)\beta$$

and rearranging the equation, the average packet loss rate $P$ is:

$$P = \frac{\beta}{\alpha + \beta} \quad (E.1)$$

## Calculation of mean burst length

A burst of lost packets is defined as a sequence of consecutive packets, all of which are marked as lost. The burst starts when a packet is lost after one or more packets have not been lost. Thus, the probability of a burst length of one is being at the bad state and then going to the good state. This probability is $\alpha$. Similarly, the probability of a burst length of two is the probability of being at the bad state, but coming to this state at the next run. This probability is $(1 - \alpha)\alpha$. Thus in general the probability of a burst length of $k$ packets is to be at the bad state for $k - 1$ times and the next run to go to the good state, which is $(1 - \alpha)^{(k-1)}\alpha$.

The mean bust length $B$ is then:

$$B = \alpha + 2(1 - \alpha)\alpha + 3(1 - \alpha)^2\alpha + 4(1 - \alpha)^3\alpha + \cdots + k(1 - \alpha)^{k-1}\alpha + \cdots$$

summing this series, leads to:

$$B = \frac{1}{\alpha} \quad (E.2)$$

Rearranging eqn. E.2 gives:

$$\alpha = \frac{1}{B} \quad (E.3)$$

and rearranging eqn. E.1 and substituting for $\alpha$ from eqn. E.3 gives:

$$\beta = \frac{P}{B(1 - P)} \quad (E.4)$$

## Simulation of packet loss

For a given mean packet loss rate $P$ and average burst length $B$, the transition probabilities $\alpha$ and $\beta$ can be calculated from eqns E.3 and E.4. The model is run for as many packets as are to be served, and at each run it is decided whether the packet should be marked as lost or not. To do this, each run is equivalent to running a random number between 0 and 1, and if it is less than the relevant transition probabilities, the packet is marked lost, otherwise it is received safely.

A pseudo code to perform packet loss is given below:

```
PreviousPacketLost=FALSE;
Readln(P,B);
```
$\alpha = (1/B);$
$\beta = P/(B \times (1-P));$

```
   FOR (number of packets to be transmitted)
     BEGIN
     CASE PreviousPacketLoss OF
           TRUE: IF random <1-α THEN PacketLoss=TRUE;
                                ELSE Packetloss=FALSE;
           FALSE: IF random <β  THEN PacketLoss=TRUE;
                                ELSE Packetloss=FALSE;
     END
```

*Appendix F*

# Solutions to the problems

## Chapter 2

1. 
|   | r | g | b | y | c | m | w |
|---|---|---|---|---|---|---|---|
| R | 1 | 0 | 0 | 1 | 0 | 1 | 1 |
| G | 0 | 1 | 0 | 1 | 1 | 0 | 1 |
| B | 0 | 0 | 1 | 0 | 1 | 1 | 1 |

2. 
|   | r | g | b | y | c | m | w |
|---|---|---|---|---|---|---|---|
| $Y$ | 82 | 145 | 41 | 210 | 170 | 107 | 235 |
| $C_b$ | 90 | 54 | 240 | 16 | 166 | 202 | 128 |
| $C_r$ | 240 | 34 | 110 | 146 | 16 | 222 | 128 |

3. $T = \dfrac{1}{25 \times 625} = 64\ \mu s$ and $T = \dfrac{1}{30 \times 525} = 63.5\ \mu s$

4. $\dfrac{13.5 \times 10^6}{25 \times 625} = 864$ pixels; $864 - 720 = 144$ pixels; $\dfrac{144}{13.5 \times 10^6} = 10.67\ \mu s$

5. 857 pixels; $857 - 720 = 137$ pixels; 10 μs

6. 
   a. $720 \times 576 \times 25 \times 2 \times 8 = 166$ Mbit/s
   b. $720 \times 480 \times 30 \times 2 \times 8 = 166$ Mbit/s
   c. $360 \times 288 \times 25 \times 1.5 \times 8 = 31$ Mbit/s
   d. $360 \times 240 \times 30 \times 1.5 \times 8 = 31$ Mbit/s
   e. 37 Mbit/s
   f. 31 Mbit/s
   g. 31 Mbit/s
   h. 4.7 Mbit/s
   i. 1.4 Mbit/s

7. 94  73  194  184  50  204  207
8. 94  82  73  132  194  201  184  109  50  121  204  222  207
   PSNR = 20.2 dB

9  a  A sinusoid with amplitude $A$ has a peak-to-peak $2A = 2^n \Delta \Rightarrow \Delta = A\, 2^{1-n}$

$$SNR = 10 \log_{10}\left(\frac{A^2/2}{\Delta^2/12}\right) = 10 \log(3 \times 2^{2n-1}) = 1.78 + 6n \text{ dB}$$

   b  Peak-to-peak power of the sinusoid is

$$10 \log_{10}\left(\frac{(2A)^2}{A^2/2}\right) = 9 \text{ dB}$$

   higher than its mean power $\Rightarrow$ PSNR $= 10.78 + 6n$

   c  $10.78 + 6n \geq 58 \Rightarrow n \geq 8$ bits

# Chapter 3

1  a  $|x| \leq 16;\ y = 0 \quad 16 < |x| \leq 32;\ y = \pm 24 \quad 32 < |x| \leq 48;\ y = \pm 40$  etc.
   b  $|x| \leq 16;\ y = \pm 8 \quad 16 < |x| \leq 32;\ y = \pm 24 \quad 32 < |x| \leq 48;\ y = \pm 40$  etc.
2  12  16  28  240  196  32     PSNR $= 43.4$ dB
3  6  12  27  77  127  77     PSNR $= 10.7$ dB
4  15 21 19 21 19 69 119 169 219 234 232 230 232 230

   ⟵⟶  ⟵⟶  ⟵⟶
   G-noise   slope overload   G-noise

5  $T = \dfrac{1}{2} \begin{vmatrix} 0.71 & 0.71 & 0.71 & 0.71 & 0.71 & 0.71 & 0.71 & 0.71 \\ 0.98 & 0.83 & 0.56 & 0.2 & -0.2 & -0.56 & -0.83 & -0.98 \\ 0.92 & 0.38 & -0.38 & -0.92 & -0.92 & -0.38 & 0.38 & 0.92 \\ 0.83 & -0.2 & -0.98 & -0.56 & 0.56 & 0.98 & 0.2 & -0.83 \\ 0.71 & -0.71 & -0.71 & 0.71 & 0.71 & -0.71 & -0.71 & 0.71 \\ 0.56 & -0.98 & 0.2 & 0.83 & -0.83 & -0.2 & 0.98 & -0.56 \\ 0.38 & -0.92 & 0.92 & -0.38 & -0.38 & 0.92 & -0.92 & 0.38 \\ 0.2 & -0.56 & 0.83 & -0.98 & 0.98 & -0.83 & 0.56 & -0.2 \end{vmatrix}$

$T^{-1} = T^T$

$= \dfrac{1}{2} \begin{vmatrix} 0.71 & 0.98 & 0.92 & 0.83 & 0.71 & 0.56 & 0.38 & 0.2 \\ 0.71 & 0.83 & 0.38 & -0.2 & -0.71 & -0.98 & -0.92 & -0.56 \\ 0.71 & 0.56 & -0.38 & -0.98 & -0.71 & 0.2 & 0.92 & 0.83 \\ 0.71 & 0.2 & -0.92 & -0.56 & 0.71 & 0.83 & -0.38 & -0.98 \\ 0.71 & -0.2 & -0.92 & 0.56 & 0.71 & -0.83 & -0.38 & 0.98 \\ 0.71 & -0.56 & -0.38 & 0.98 & -0.71 & -0.2 & 0.92 & -0.83 \\ 0.71 & -0.83 & 0.38 & 0.2 & -0.71 & 0.98 & -0.92 & 0.56 \\ 0.71 & -0.98 & 0.92 & -0.83 & 0.71 & -0.56 & 0.38 & -0.2 \end{vmatrix}$

6  a  364  15  $-211$  $-26$  $-5$  38  $-2$  $-1$
   b  the basis vector of the second AC coefficient matches the input pixels

Solutions to the problems 389

    c  35 82 190 250 200 150 101 23. Due to mismatch (approximating the cosine elements) some of the input pixels cannot be reconstructed, e.g. 81/82 and 100/101.

7  quantised coefficients:   360   0   −216   −24   0   40   0   0
    reconstructed pixels:   30   70   185   250   204   153   104   27
    PSNR = 30 dB

8  a  reconstructed pixels:   128   128   128   128   128   128   128   128
       PSNR = 10.4 dB
   b  reconstructed pixels:   28   87   169   227   227   169   87   28
       PSNR = 23.4 dB

9  a  mv(−1, −1)
   b  mv(−1, −1)

10  a  169
    b  (i)   multiplications $= 256 \times 169$, additions $= 511 \times 169$
       (ii)  multiplications $= 0$, additions $= 511 \times 169$

11  type   operations   multiplications   additions
    TDL   23              $23 \times 256$       $23 \times 511$
    TSS   25              $25 \times 256$       $25 \times 511$
    CSA   17              $17 \times 256$       $17 \times 511$
    OSA   13              $13 \times 256$       $13 \times 511$

12  **a** = 010, **b** = 1, **c** = 00, **d** = 011, av bits = 1.8, entropy = 1.72

13  a  **cbdad** = 001011010011
    b  (i)   1st bit in error decoded string = **babbad**
       (ii)  3rd bit in error, decoded string = **ccbbad**
       (iii) 5th bit in error, decoded string = **cbcbad**

14  lower value $= 0.83875$, upper value $= 0.841875$

15  a  the first three symbols = **cbc**
    b  the first five symbols = **cbcab**

16  the same as 15

17  11010110111

18  the same as 17

19  a  first bit in error $= 0.01010110111 = 2^{-2} + 2^{-4} + 2^{-6} + 2^{-7} + 2^{-9} + 2^{-10} + 2^{-11} = 0.33935546875$ which is decoded to string **bbacb**
    b  similarly, with the third bit in error the decimal number would be 0.96435546875, decoded to **dbacb**
    c  with the fifth bit in error, the decimal number is 0.87060546875 and it is decoded to string **cbacb**.

## Chapter 4

1  $P(z) = H_0(z)H_1(-z)$ should be factorised into two terms. The given $P(z)$ is zero at $z^{-1} = -1$, hence it is divisible by $1 + z^{-1}$. Divide as many times as possible, that gives:

    (i)   $\frac{1}{16}(1 + z^{-1})(-1 + z^{-1} + 8z^{-2} + 8z^{-3} + z^{-4} - z^{-5})$

(ii) $\frac{1}{16}(1+z^{-1})^2(-1+2z^{-1}+6z^{-2}+2z^{-3}-z^{-4})$
(iii) $\frac{1}{16}(1+z^{-1})^{-3}(-1+3z^{-1}+3z^{-2}-z^{-3})$
(iv) $\frac{1}{16}(1+z^{-1})^{-4}(-1+4z^{-1}-z^{-2})$

Thus in (i) the lowpass analysis filter will be:

$$H_0(z) = \frac{1}{16}(-1+z^{-1}+8z^{-2}+8z^{-3}+z^{-4}-z^{-5})$$

and the high pass analysis filter is:

$$H_1(z) = 1-z^{-1}$$

In (ii) $H_0(z) = \frac{1}{8}(-1+2z^{-1}+6z^{-2}+2z^{-3}-z^{-4})$ and the highpass:

$$H_1(z) = \frac{1}{2}(1-z^{-1})^2 = \frac{1}{2}(1-2z^{-1}+z^{-2})$$

which are the (5,3) subband filter pairs.

In (iii) $H_0(z) = \frac{1}{4}(-1+3z^{-1}+3z^{-2}-z^{-3})$ and the highpass

$$H_1(z) = \frac{1}{4}(1-z^{-1})^{-3} = \frac{1}{4}(1-3z^{-1}+3z^{-2}-z^{-3})$$

which gives the second set of (4,4) subband filters.

In (iv) $H_0(z) = \frac{1}{2}(-1+4z^{-1}-z^{-2})$ and

$$H_1(z) = \frac{1}{8}(1-z^{-1})^4 = \frac{1}{2}(1-4z^{-1}+6z^{-2}-4z^{-3}+z^{-4})$$

Any other combinations may be used, as desired.

2 (i) $P(z) = H_0(z) \times H_1(-z) = \frac{1}{2}(1+z^{-1})^2 = \frac{1}{2}+z^{-1}+\frac{1}{2}z^{-2}$. Thus with $P(z)-P(-z) = 2z^{-1}$, results in one sample delay.

(ii) $G_0(z) = H_1(-z) = \frac{1}{\sqrt{2}}(1-z^{-1})$

$G_1(z) = -H_0(-z) = -\frac{1}{\sqrt{2}}(1-z^{-1})$

3 (i) With a weighting factor of $k$, $P(z) = k(1+z^{-1})^{-4}(-1+4z^{-1}-z^{-2})$, and using $P(z)-P(-z) = 2z^{-m}$, gives $k = 1/16$ and $m = 3$.

(ii) The factor for the other set will be $k = 3/256$ and $m = 5$ samples delay.

4 In problem 3, $P(z)$ is in fact type (iv) of problem 1. Hence it leads not only to the two sets of (5,3) and (4,4) filter pairs, but also to two new types of filter, given in (i) and (iv) of problem 1.

5 With

$$P(z) = \frac{3}{256}(1+z^{-1})^6 \left(1-6z^{-1}+\frac{38}{3}z^{-2}-6z^{-3}+z^{-4}\right)$$

retaining $H_1(-z) = (1+z^{-1})^2 = 1+2z^{-1}+z^{-2}$, that gives the three-tap highpass filter of $H_1(z) = 1-2z^{-1}+z^{-2}$ and the remaining parts give the nine-tap lowpass filter

$$H_0(z) = \frac{3}{256}(1+z^{-1})^4\left(1-6z^{-1}+\frac{38}{3}z^{-2}-6z^{-3}+z^{-4}\right)$$

or

$$H_0(z) = \frac{1}{25}[3 - 6z^{-1} - 16z^{-2} + 38z^{-3} + 90z^{-4} + 38z^{-5} - 16z^{-6} - 6z^{-7} - 3z^{-8}]$$

Had we divided the highpass filter coefficients, $H_1(z)$, by $2\sqrt{2}$, and hence multiplying those of lowpass $H_0(z)$ by this amount, we get the 9/3 tap filter coefficients of Table 4.2.

6. Use $G_0(z) = H_1(-z)$ and $G_1(z) = -H_0(-z)$ to derive the synthesis filters
7. See pages 84 and 88
8. 33 bits
9. 29 bits

## Chapter 5

1. a  multiply all the luminance matrix elements by $\alpha = 50/25 = 2$
   b  $\alpha = 2 - \frac{2 \times 99}{100} = 0.02$, and small elements of the matrix will be 1 and larger ones become 2.
   c  $\alpha = 0$, hence all the matrix elements will be 1.
2. the same as problem 1.
3. a  

| 62 | 0  | 3  | 1 | 0 | 0 | 1 | 0 |
|----|----|----|---|---|---|---|---|
| 0  | 1  | −1 | 0 | 1 | 0 | 0 | 0 |
| 0  | 0  | 0  | 0 | 1 | 0 | 0 | 0 |
| −1 | 0  | 0  | 0 | 0 | 0 | 0 | 0 |
| 0  | 0  | 1  | 0 | 0 | 0 | 0 | 0 |
| 0  | 0  | 0  | 0 | 0 | 0 | 0 | 0 |
| 0  | 0  | 0  | 0 | 0 | 0 | 0 | 0 |
| 0  | 0  | 0  | 0 | 0 | 0 | 0 | 3 |

   b  

| 31 | 0 | 1 | 0 | 0 | 0 | 0 | 0 |
|----|---|---|---|---|---|---|---|
| 0  | 0 | 0 | 0 | 0 | 0 | 0 | 0 |
| 0  | 0 | 0 | 0 | 0 | 0 | 0 | 0 |
| 0  | 0 | 0 | 0 | 0 | 0 | 0 | 0 |
| 0  | 0 | 0 | 0 | 0 | 0 | 0 | 0 |
| 0  | 0 | 0 | 0 | 0 | 0 | 0 | 0 |
| 0  | 0 | 0 | 0 | 0 | 0 | 0 | 0 |
| 0  | 0 | 0 | 0 | 0 | 0 | 0 | 1 |

4. for 50% quality: DIFF = 62 − 50 = 12, symbol_1 = 4; symbol_2 = 12
   scanned pairs (3,1)(0,3)(0,1)(0,−1)(1,−1)(6,1)(6,1)(1,1)(1,1)(35,3) and the resultant events:
   (3,1)(0,2)(0,1)(0,1)(1,1)(6,1)(6,1)(1,1)(1,1)(15,0)(15,0)(3,3)
   for 25% quality:   DIFF = 31−50 = −19,   symbol_1 = 5,   symbol_2 = −19−1 = −20
   scanned pairs: (4,1)(57,1)
   events: (4,1)(15,0)(15,0)(15,0)(9,1)

5  for DC: DIFF= −19 ⇒ CAT = 5; DIFF−1 = −20
   VLC for CAT = 5 is **110** and −20 in binary is 111**01100**, hence the VLC for the DC coefficient is 11001100
   for AC, using the AC VLC tables:
   for each (15,0) the VLC is 11111111001
   and for (9,1) the VLC is 111111001
   total number of bits: $8 + 3 \times 11 + 9 = 50$ bits.
6  At bit-plane 6 coefficient 65 at clean-up pass. At bit-plane 5 coefficient 65 at all passes and coefficient 50 at clean-up pass.

## Chapter 6

1  a  33, 198
   b  396, 2376
2  $CIF: \frac{1}{30 \times 396} = 84.2$ μs; $QCIF: \frac{1}{10 \times 99} = 1$ ms
3  a  MC
   b  NOMC
   c  NOMC
4  due to motion vector overhead
5  a  inter
   b  intra
   c  inter
6  For small values in intra mode DC still needs eight bits, while in inter mode it is less.
7  a  63
   b  60
   c  3
8  
   | 83 | 0 | 2 | 1 | 0 | 0 | 4 | 0 |
   |---|---|---|---|---|---|---|---|
   | 0 | 0 | −1 | 0 | 3 | 0 | 0 | 0 |
   | 0 | 0 | 0 | 0 | 3 | −1 | 0 | 0 |
   | −1 | 1 | 0 | 0 | 0 | 0 | 0 | 0 |
   | 0 | 0 | 3 | 0 | 2 | 0 | 0 | 0 |
   | 0 | 0 | 0 | 0 | 0 | 0 | 0 | 1 |
   | −1 | 4 | 0 | 0 | 0 | 0 | 0 | 0 |
   | 2 | 0 | 0 | 0 | 0 | 0 | 0 | 31 |

   events: $(0, 83)(4, 2)(0, 1)(0, −1)(1, −1)(1, 1)(4, 3)(4, −1)(1, 3)(1, 3)(1, 4)(2, −1)$ $(3, 4)(0, 2)\ (3, 2)(20, 1)(2, 31)$
   number of bits (including the sign bit): $20 + 9 + 5 + 5 + 6 + 6 + 11 + 8 + 8 + 8 + 9 + 7 + 11 + 5 + 8 + 20 + 20 = 166$
   no EOB is used, as the last coefficient is coded
9  159   149   170
   105   113   133
10 @384 kbit/s $P = 6$ and with $q = 62$ and $P = 6$, the buffer content is 36 kbits, left over capacity $= 5000 \times 8 − 36000 = 4000$ bits

# Chapter 7

1. a  each operation $= (2 \times 15 + 1)^2 + 8 = 969$, total operations $= 3 \times 969 = 2907$
   b  $\omega = 3 \times 15 = 45$, total no of operations $= (2 \times 45 + 1)^2 + 8 = 8289$
2. a  the first B-picture is closer to its forward prediction than its backward prediction picture.
   b  for the second B-picture, FWD $= 9$ and BWD $= 5$
   c  for P-picture $(2 \times 13 + 1)^2 + 8 = 737$, and for each B-picture $(2 \times 5 + 1)^2 + 8 + (2 \times 9 + 1)^2 + 8 = 498$
3. Average $Q = \frac{10+16}{2} = 13$ and the complexity index is $50 \times 1000 \times 13 = 65 \times 10^4$.
4. $(8 \times 7) + (3 \times 10) + 20 = 106$
   for I-pictures:

   $$\frac{1.2 \times 10^6 \times 12 \times 20}{25 \times 106} = 108.7$$

   for $P = 54.3$ kbits and for $B = 38$ kbits.
5. The new index ratio for B becomes $7/1.4 = 5$
   $(8 \times 5) + (3 \times 10) + 20 = 90$, and bits for $I = 128$ kbits, for $P = 64$ and for $B = 32$ kbits
6. $20 + 10 + 10 + 20 + 8 \times 7 = 116$
   for P, the average index of $(10 + 10 + 20)/3 = 13.3$ should be used, hence the target bit rates for $I = 99.3$ kbits, for $P = 66.2$ kbits and for $B = 34.75$ kbits.
7. a  all equal to 48 kbit/s
   b  $60 + 3 \times 20 + 8 \times 15 = 240$
   For $I = \frac{1.2 \times 10^6 \times 12 \times 60}{25 \times 240} = 144$ kbits, for $P = 48$ kbits and for $B = 36$ kbits

# Chapter 8

2. (i)   L
   (ii)  L
   (iii) P
   (iv)  L

3. 

| 4  | −1 | −5 | 0 | 2  | 0 | 0 | −1 |
|----|----|----|---|----|---|---|----|
| 7  | 0  | 2  | 1 | −1 | 0 | 0 | −1 |
| 0  | 1  | 0  | 1 | −1 | 0 | 0 | 0  |
| 0  | −4 | 0  | 0 | 0  | 0 | 0 | 0  |
| −2 | 0  | 0  | 0 | 1  | 0 | 0 | 0  |
| 0  | 0  | 1  | 0 | 0  | 0 | 0 | 0  |
| 0  | 0  | 0  | 0 | 0  | 0 | 0 | 0  |
| 0  | 0  | 0  | 0 | 0  | 0 | 0 | 0  |

2D-events: (0, 4)(0, −1)(0, 7)(2, −5)(1, 2) (0, 1)(1, −2) (0, −4)(1, 1)(0, 2)(1, −1) (0, 1)(7, −1)(2, −1)(4, 1)(5, 1) (2, −1)EOB (using Figure 6.12, the bits including the sign bit)

$8+5+11+20+7+5+7+8+6+5+6+5+9+6+8+8+6+2 = 132$ bits

4   The base layer events: $(0,4)(0,-1)(0,7)(2,-5)(1,2)(0,1)(1,-2)(0,4)$ + PBP the bits: $8+5+11+20+7+5+7+8+6 = 77$ bits
The enhancement layer events: $(1,1)(0,2)(1,-1)(0,1)(7,-1)(2,-1)(4,1)(5,1)(2,-1)$ + EOB the bits: $6+5+6+5+9+6+8+8+6+2 = 61$
Total bits $77+61 = 138$ bits, about 4.5 per cent extra over one-layer coding

5   base:

|   |   |   |   |   |   |   |   |
|---|---|---|---|---|---|---|---|
| 2 | 0 | 2 | 0 | 1 | 0 | 0 | 0 |
| 4 | 0 | 1 | 0 | 0 | 0 | 0 | 0 |
| 0 | 0 | 0 | 0 | -1| 0 | 0 | 0 |
| 0 | -2| 0 | 0 | 0 | 0 | 0 | 0 |
| -1| 0 | 0 | 0 | 0 | 0 | 0 | 0 |
| 0 | 0 | 0 | 0 | 0 | 0 | 0 | 0 |
| 0 | 0 | 0 | 0 | 0 | 0 | 0 | 0 |
| 0 | 0 | 0 | 0 | 0 | 0 | 0 | 0 |

enhancement:

|   |   |   |   |   |   |   |   |
|---|---|---|---|---|---|---|---|
| 0 | -1| 0 | 0 | 0 | 0 | 0 | -1|
| 0 | 0 | 0 | 1 | -1| 0 | 0 | -1|
| 0 | 1 | 0 | 1 | 0 | 0 | 0 | 0 |
| 0 | 0 | 0 | 0 | 0 | 0 | 0 | 0 |
| 0 | 0 | 0 | 0 | 1 | 0 | 0 | 0 |
| 0 | 0 | 0 | 0 | 0 | 0 | 0 | 0 |
| 0 | 0 | 0 | 0 | 0 | 0 | 0 | 0 |
| 0 | 0 | 0 | 0 | 0 | 0 | 0 | 0 |

Base layer events: $(0, 2)(1, 4)(2, -2)(1, 1)(2, -1)(0, -2)(2, 1)(10, -1)$EOB
Bits: $5+9+7+6+6+5+6+10+2 = 56$ bits
Enhancement layer events: $(1, -1)(6, 1)(4, 1)(2, -1)(0, 1)(10, -1)(10, 1)(2, -1)$EOB
Bits: $6+9+8+6+5+10+10+6+2 = 62$ bits
Total bits $= 56+62 = 118$
Note: the overall bit rate is less than the one layer, but the distortion will be larger.

6   $\frac{8-2}{1.25} \log_2 64 = 28.8$ Mbit/s, $28.8/8 = 3.5$, thus three TV programmes.

7   (i)   for $B = 1, \alpha = 1$ and $\beta = 10^{-5}$
    (ii)  for $B = 5, \alpha = 0.2$ and $\beta = 2 \times 10^{-6}$

8   (i)   34 980 Mbits = 4.3725 Gbytes
    (ii)  mean bit rate = $34\,980/(90 \times 60) = 6.478$ Mbit/s
    (iii) CBR = $90 \times 60 \times 20/8 = 13.5$ Gbytes, peak-to-mean = 3.09

9   (i)   if error in any video bits, packet is in error; $P = 47 \times 8 \times 10^{-7} = 3.76 \times 10^{-7}$
    (ii)  if error in the header, packet is lost; $P = 5 \times 8 \times 10^{-7} = 4 \times 10^{-6}$

10  available link rate = $50 \times 30 \times 10^{-2} = 15$ Mbit/s
    total data to be sent = $34\,980 \times 53/47 = 39\,445.5$ Mbits
    time required = $39\,445.5/15$ Mbits/s = 43 min 50 s

11  $25 \times 4 = 100$ Mbits/s, load $\rho = \frac{100 \times 53}{155 \times 47} = 0.7275$ and the error rate is $P = 10^{-10(1-0.7275^2)} = 2 \times 10^{-6}$.

Solutions to the problems 395

12 With SNR scalability, assume 30 per cent more load, then total load $= 100 \times 1.3 = 130$ Mbit/s, of which 65 Mbit/s is assigned to the base layer. Since the base layer has an absolute priority, then network load for the base layer:

$$\rho = \frac{65 \times 53}{155 \times 47} = 0.4729$$

and at this load $P = 1.72 \times 10^{-8}$.

For the enhancement layer, it sees the whole load, of 130 Mbits/s, thus the load will be:

$$\rho = \frac{130 \times 53}{155 \times 47} = 0.9457$$

and the error rate will be $P = 0.088$.

13 In data partitioning, with 4 per cent extra bits over one layer and 50 per cent to the base layer, the base layer load will be:

$$\rho = \tfrac{1}{2} \times 0.7275 \times 1.04 = 0.3783$$

and the error rate $P = 2.7 \times 10^{-9}$. The enhancement layer has a load of $2 \times 3.783 = 0.7566$ that leads to an error rate of $P = 5.3 \times 10^{-5}$.

14 With spatial scalability, assuming 50 per cent more bits over one layer and 50 per cent assigned to the base layer, then the allocated bits to the base layer will be 75 Mbit/s. Base layer load is:

$$\rho = \frac{75 \times 53}{155 \times 47} = 0.5456$$

leading to $P = 9.5 \times 10^{-8}$. For the enhancement layer, the load will be more than 100 per cent and the loss probability will be $P = 1$!

## Chapter 9

1 Prepend 0 to all events of problem 3 of Chapter 8, except the last event, where 1 should be appended, and no need for EOB, e.g. first event $(0, 4, 0)$ and the last event $(1, 2, -1)$
2 For $x$, the median of $(3, 4, -1)$ is 3 and for $y$, the median of $(-3, 3, 1)$ is 1. Hence the prediction vector is $(3, 1)$ and MVD $= (2 - 3 = -1; 1 - 1 = 0) = (-1, 0)$
3 (i) $d = \frac{300 - 120° + 920 - 150}{16} = -8.125$, $d_1 = -(\max(0, 8.125 - \max(0, 2 \times 8.125 - 16))) = 8$
   thus $B_1 = 150 - 8 = 142$ and $C_1 = 115 + 8 = 123$
   (ii) $d = -31.25$, and $d_1 = 0$, hence $B$ and $C$ do not change.
4 (i) $(3, 4)$, (ii) $(0, -3)$, (iii) $(1, 0.5)$, (iv) $(3, 2.6)$ (v) $(-1, -1)$.
5 In order for a matrix to be orthonormal, multiplying each row by itself should be 1. Hence in row 1 and 3 (basis vectors 0 and 2), their values are 4, hence they should

be divided by $\sqrt{4} = 2$. In rows 2 and 4 their products give: $4 + 1 + 1 + 4 = 10$, hence their values should be divided by $\sqrt{10}$.

Thus the forward $4 \times 4$ integer transform becomes

$$T = \begin{bmatrix} \frac{1}{2} \\ \frac{1}{\sqrt{10}} \\ \frac{1}{2} \\ \frac{1}{\sqrt{10}} \end{bmatrix} \begin{bmatrix} 1 & 1 & 1 & 1 \\ 2 & 1 & -1 & -2 \\ 1 & -1 & -1 & 1 \\ 1 & -2 & 2 & -1 \end{bmatrix}$$

And the inverse transform is its transpose

$$T^{-1} = T^T = \begin{bmatrix} \frac{1}{2} & \frac{2}{\sqrt{10}} & \frac{1}{2} & \frac{1}{\sqrt{10}} \\ \frac{1}{2} & \frac{1}{\sqrt{10}} & -\frac{1}{2} & -\frac{2}{\sqrt{10}} \\ \frac{1}{2} & -\frac{1}{\sqrt{10}} & -\frac{1}{2} & \frac{2}{\sqrt{10}} \\ \frac{1}{2} & -\frac{2}{\sqrt{10}} & \frac{1}{2} & -\frac{1}{\sqrt{10}} \end{bmatrix}$$

As can be tested, this inverse transform is orthornormal, e.g.:

$$\left(\frac{1}{2}\right)^2 + \left(\frac{2}{\sqrt{10}}\right)^2 + \left(\frac{1}{2}\right)^2 + \left(\frac{1}{\sqrt{10}}\right)^2 = \frac{1}{4} + \frac{4}{10} + \frac{1}{4} + \frac{1}{10} = 1$$

6 With the integer transform of problem 5, the two-dimensional transform coefficients will be

$$\begin{array}{rrrr} 431 & -156 & 91 & -15 \\ 43 & 52 & 30 & 1 \\ -6 & -46 & -26 & -7 \\ -13 & 28 & -19 & 14 \end{array}$$

The reconstructed pixels with the retained coefficients are; for $N = 10$:

$$\begin{array}{rrrr} 105 & 121 & 69 & 21 \\ 69 & 85 & 62 & 44 \\ 102 & 100 & 98 & 119 \\ 196 & 175 & 164 & 195 \end{array}$$

which gives an MSE error of 128.75, or PSNR of 27.03 dB. The reconstructed pixels with the retained 6 and 3 coefficients give PSNR of 22.90 and 18 dB, respectively.

With $4 \times 4$ DCT, these values are 26.7, 23.05 and 17.24 dB, respectively.

As we see the integer transform has the same performance as the DCT. If we see it is even better for some, this is due to the approximation of cosine elements.

7  index-0 = $QP$

   (i)   index-8 = $2QP$
   (ii)  index-16 = $4QP$
   (iii) index-24 = $8QP$,...index-40 = $32QP$
   (iv)  index-48 = $64QP$
   (v)   index-51 = $(64 + \frac{51-48}{8} \times 64)QP = 88QP$

8  Compared with H.263, at lower indices H.263 is coarser, e.g. at index-8 the quantiser parameter for H.263 is $8QP$, but for H.26L is $2QP$ etc.

   At higher indices, the largest quantiser parameter for H.263 is $31QP$, but that of H.26L is $88QP$, hence at larger indices H.26L has a coarser quantiser.

## Chapter 10

1  For A: $c0 = c2 = c3 = 1$, index $= 1 + 4 + 8 = 13$, the given frequency table in Appendix D is for prob(0), hence prob(0) = 29789, but since A is a 1 pixel, then its probability prob(1) = 65535 − 29789 = 35746 out of 65535.

   For B: $c0 = c1 = c2 = c3 = c4 = 1$, and index $= 1 + 2 + 4 + 8 + 16 = 31$, prob(0) = 6554. As we see this odd pixel of 0 among the 1s has a lower probability.

   For C: $c1 = c2 = c3 = c4 = c5 = c7 = 1$, and the index becomes 190. Like pixel A the prob(0) = 91, but its prob(1) = 65535 − 91 = 65444 out of 65535, which is as expected.

2  The chain code is 0, 1, 0, 1, 7, 7. The differential chain code will be: 0, 1, −1, 1, −2, 0 with bits $1 + 2 + 3 + 2 + 5 + 1 = 14$ bits

3  At level 2 the indices (without swapping) will be 0 for the two blank blocks and index $= 27 \times 2 + 9 \times 2 + 0 + 2 = 74$ and index $= 0 + 0 + 3 \times 2 + 2 = 8$ for the two blocks. At the upper level the index is: index $= 0 + 0 + 3 \times 1 + 1 = 4$

4  The coefficients of the shape adaptive DCT will be

$$\begin{array}{rrr} 127 & -40 & 1 \\ 10 & -15 & \\ 23 & -7 & \\ -6 & & \end{array}$$

confined to the top left corner, while for those of the normal DCT with padded zeros, the significant coefficients are scattered all over the $8 \times 8$ area.

*Appendix G*
# Glossary of acronyms

| | |
|---|---|
| 2D-VLC | two-dimensional variable length code |
| 3D-VLC | three-dimensional variable length code |
| AAL | ATM adaptation layer |
| ABT | adaptive block transform |
| ATM | asynchronous transfer mode |
| BD | boundary difference |
| BISDN | broadband ISDN |
| BMA | block matching algorithm |
| CABAC | context-based adaptive arithmetic coding |
| CAT | category |
| CBP | coded block pattern |
| CBR | constant bit rate |
| $C_b, C_r(u, v)$ | chrominance components |
| CCF | cross correlation function |
| CCIR | International Radio Consultative Committee (now called ITU) |
| CIF | common intermediate format |
| DCT | discrete cosine transform |
| DIFF | difference between DC coefficients |
| DPCM | differential pulse code modulation |
| DSCQS | double stimulus continuous quality score |
| DSIS | double stimulus impairment scale |
| DSM | digital storage media |
| DTS | decoding time stamp |
| DVD | digital versatile (video) disc |
| DWT | discrete wavelet transform |
| EBCOT | embedded block coding with optimised truncation |
| ELNUM | enhancement layer number |
| EOB | end of block |
| EZW | embedded zero tree |
| FFT | fast Fourier transform |

| | |
|---|---|
| FLC | fixed length code |
| FR-TV | full reference TV |
| FSM | full search method |
| GOB | group of blocks |
| GOP | group of pictures |
| HBMA | hierarchical block matching algorithm |
| HDTV | high definition television |
| HVS | human visual system |
| ICT | irreversible colour transform |
| IEC | International Electrotechnical Commission |
| INTER | interframe |
| INTRA | intraframe |
| IDCT | inverse DCT |
| IP | internet protocol |
| I$Q$, $Q^{-1}$ | inverse quantiser |
| ISDN | integrated services digital network |
| ISO | International Standards Organisation |
| ITC, TC$^{-1}$ | inverse transform |
| ITU-T | International Telecommunication Union (telegraphy section) |
| JPEG | Joint Photographic Experts Group |
| L (Y) | luminance component |
| LAN | local area network |
| LIP | list of insignificant pixels |
| LIS | list of insignificant set |
| LPF | lowpass filter |
| LSB | least significant bit |
| LSP | list of significant pixels |
| MAE | mean absolute error |
| MB | macroblock |
| MC | motion compensation |
| MCBPC | combined macroblock and coded block pattern |
| MCU | multipoint control unit |
| ME | motion estimation |
| MOS | mean opinion score |
| MPEG | Motion Picture Experts Group |
| MSB | most significant bit |
| MSE | mean-squared error |
| MV | motion vector |
| MVD | motion vector data |
| MVDB | motion vector data for B-pictures |
| MVDS | motion vector data for shape |
| NAL | network adaptation layer |
| NR-TV | no reference TV |
| NTSC | National Television System Committee |

| | |
|---|---|
| OFDM | orthogonal frequency division multiplex |
| PAL | phase alternate line |
| PBP | priority break point |
| PCRD | post compression rate distortion |
| PES | packetised elementary stream |
| PSNR | peak signal-to-noise ratio |
| PSTN | public switched telephone network |
| PTS | presentation time reference |
| Q | quantiser |
| QCIF | quarter of CIF |
| RCT | reversible colour transform |
| RGB | red, green and blue colour primaries |
| RLNUM | reference layer number |
| RM | reference model |
| ROI | region of interest |
| RR-TV | reduced reference TV |
| RTP | real time transport protocol |
| RVLC | reversible variable length code |
| SAC | syntax-based arithmetic coding |
| SECAM | sequential couleur avec memoire |
| SIF | source input format |
| SOT | special orientation tree |
| SPIHT | set partitioning in hierarchical tree |
| SSCQE | single stimulus continuous quality evaluation |
| ST | statistical table |
| STC | systems time clock |
| subQCIF | one ninth of CIF |
| TC | transform coding |
| TM | test model |
| UBR | unspecified bit rate |
| VBR | variable bit rate |
| VCL | video coding layer |
| VCR | video cassette recorder |
| VLC | variable length code |
| VLD | variable length decode |
| WT | weighting table |

# Subject index

Active contours (snakes) 302
Addressing
　blocks 139
　macroblocks 138, 158
　motion vectors 139, 165
Advanced
　intra coding 254
　inter coding 256
　inter/intra VLC 253
　prediction 235
　video coding 280
　very low bit rate coding 227, 280
　VLC 251
Affine transform 325, 327
Aliasing 11, 66, 208
Alpha blocks/planes 309
Analogue video 9
Annexes of H.263 231, 234, 294
Arithmetic coding 43
　adaptive 53
　binary 46
　conditional 90
　context-based 53, 91, 316
　interval testing loop 49
　models (adaptive, fixed) 43
　principles 44
　syntax-based 251
Asynchronous transfer mode (ATM) 57, 216, 220, 290
ATM adaptation layer (AAL) 182, 220

Back channel 258
Base layer 126, 185, 194
Basis functions (vectors) 28, 63, 69, 323
Bilinear interpolation 243
Bilinear transform 242
Binary alpha plane 309
Binary significance state 90

Bit plane encoding 89, 112
Bit stream switching 287
　random access 289
Block matching algorithm (BMA) 36
Block of pixels 132, 159
Blocking artefacts 30, 63, 144, 165, 198, 218
Boundary macroblock 322
Broken link 178
Buffer regulation 57, 161, 278
Buffer size and delay 172

CCIR-601 11
Cell loss priority 220
Chrominance 10
Code block 89, 122
Coded block pattern (CBP) 132, 139
Coding for videoconferencing 131
Coding of motion vectors 228
Coding of objects 299
Colour
　components 10
　edge detection 304
　similarity 307
　transformation 10, 118
Common intermediate format (CIF) 16
Complexity index 173
Compression efficiency of various codecs 239
Concealment motion vectors 194
　bidirectional 270
　majority 268
　mean 268
　vector mean 268
Conversion from film 19
Constant bit rate (CBR) 57, 146, 171, 220
Context-based arithmetic encoding (CAE) 53, 91, 316
Content-based video coding 295

404  *Index*

Data partitioning 197, 259
DC level shifting 105, 118
DCT mismatch control 30
Deblocking filter 240, 285
Decoder
   generic 57
   MPEG-1 175
   multilayer 203
Differential motion vector correction 191
Differential pulse code modulation (DPCM) 25
Digital storage media 151, 184, 219
Digital versatile disc (DVD) 219
Digital video 11
Discrete cosine transform (DCT) 29
Discrete wavelet transform (DWT) 71, 120, 327

Editing 177
Embedded block coding with optimised truncation (EBCOT) 88
Embedded zero tree wavelet (EZW) 81, 330
Encoder
   generic 55
   H.261 133
   H.263 254
   MPEG-1 151
   MPEG-4 295
   wavelet 119, 332
End of block (EoB) 109, 142
Enhancement layer 126, 185, 194
Entropy coding 25, 41, 111, 121, 286
Error
   concealment 266
   correction 257
   detection 263
   intraframe concealment 266
   interframe concealment 267
   resilience 127
Escape code 142
Experimental model (XM) 349
Extensions of H.263 232

Fast DCT 31
Fast forward 176
Fast motion estimation 37
Field/frame 10
Filtering
   analysis/synthesis 65
   biorthogonal 76, 120
   filter banks 64, 74
   filter design 66, 76
   loop filter 144
   product filter 68, 77
   temporal resampling 19
   up/downsampling 16
Flicker 9
Forced update 137, 258
Forward error correction 257
Fractional bit plane coding 91, 122
   clean up pass 95, 122
   magnitude refinement pass 95, 122
   significant propagation pass 93, 122
Frame-based coding 298, 341
Freeze picture 232
Functionality 295

Group of blocks (GOB) 133, 157
Group of pictures (GOP) 155

H.120 1
H.245 231
H.261 131
H.261/H.263 differences 228
H.261/MPEG-1 differences 159, 169
H.262 *see* MPEG-2
H.263 227
H.263/MPEG-1 differences 228
H.263+/H.263++ 233
H.263/MPEG-4 differences 340
H.264 *see* H.26L
H.26L 234, 280
H.26L/H.263 differences 280
HDTV 18
Huffman coding 41
   modified Huffman 43, 142
   2D Huffman 108, 140
   3D Huffman 228
Hypothetical reference decoder 147

Image format 12, 132, 230
   (CIF, CCIR-601 QCIF, QSIF, SIF 4:4:4, 4:2:2, 4:2:0, 4:1:1)
Image descriptors
   colour 352
   motion 356
   shape 355
   texture 354
Image gradient 303
Image indexing 351
Image retrieval 358
Improved PB frames 249
Independent segment decoding 253
Intra/inter decision 136

Intraframe coding 25, 136, 155, 282
Interactivity 295
Interframe coding 25, 35, 55, 284
Internet 290

JPEG 103
    JPEG2000 116
    AC coefficients 109
    baseline 105
    DC coefficients 108
    extended DCT 112
    hierarchical transmission 113
    interleaving 115
    lossless compression 104
    lossy compression 105
    progressive transmission 112
    quality factor 107
    run length coding 108
    sequential transmission 112

Key frame 351

Layered coding 194
Layering *versus* scalability 195
Level and profile 185, 296
Loop filter 144
Loss concealment 271
Lossless compression 26, 104
Lossy compression 26, 105
Low bit rate coding 227
Luminance 10

Macroblock 132, 158
    addressing 138
    boundary MB 241, 253
    type 137, 167, 247
Metadata 351
Model-based coding 326
Modified Huffman 43, 142
Motion compensated prediction
    $16 \times 8$ 193
    dual prime 191
    field 189
    frame 189
Motion compensation decision 135
Motion estimation/compensation 35, 162
    advanced motion estimation/
        compensation 234
    advanced prediction 235
    bidirectional search 166
    block matching algorithm (BMA) 36

    fast methods 37
    four motion vectors 236
    full search method 37
    global motion compensation 325
    half pixel search 164
    importance of motion vectors 239
    hierarchical 39
    mesh-based 245, 326
    overlapped motion compensation 236
    region motion 307
    search range 163, 166
    sixteen motion vectors 284
    spatial transforms 242
    telescopic search 163
    unrestricted motion vectors 235
Motion JPEG 103
Motion vector data 139, 236
Motion vectors in PB frames 248
Motion vectors for chrominance 194, 230
MPEG-1 151
    multiplexing 153
    synchronisation 153
    systems layer 152
    fast play 176
    editing 177
    pause 176
    reverse play 176
MPEG-1/MPEG-2 differences 187
MPEG-2 181
    applications 181
    systems 182
MPEG-4 295
    motion estimation 321
    texture coding 322
MPEG-4/H.263 differences 340
MPEG-7 347-8
MPEG-21 347, 364
Multimedia 295, 364
    description 348
    framework 364
Multipoint control unit 232
Multiple reference prediction 258, 285
Multiresolution signal analysis 71

Network adaptation layer (NAL) 289
Noise
    aliasing 11, 65, 208
    blockiness 30, 63, 144
    coding distortions 20
    contouring noise 34
    granular noise 34

Noise (contd.)
 jerkiness 258
 mosquito noise 144, 198
 quantisation noise 34
 slope overload 35
Nonlinear diffusion 303

Object-based coding 295
Object-based scalability 339
Object mask creation 307
OFDM 219
Optional modes in H.263 234
Orthonormality 28, 77
Overflow/underflow 57, 177
Overhead due to scalability 213
Overlapped motion compensation 236

Packetisation 220
Packet networks 216, 280
PAL 10
PB frames 247
 improved PB frames 249
Perspective transform 246, 325
Picture drift 198, 203
Picture freeze 231
Picture layer 231
Picture quality 20
Picture reordering 154
Picture type
 B-pictures 169
 D-pictures 170
 I-pictures 167
 P-pictures 168
 S-pictures 287
Pixel 12
Post compression rate distortion optimisation (PCRD) 92
Postprocessing 123, 177
Precinct 123
Prediction loop 55, 135
Predictive coding 25
Preprocessing 118, 123, 153
Priority break point 197
Product filter 68, 77
Protection against error 256
PSNR 21
Pulldown 19

Quantisation 31, 121, 140, 285
 adaptive 34, 173
 basic step size 121
 bilevel 330

bit plane 81, 90, 330
B-pictures 250
dead zone 32
nonlinear 35
parameter 34, 137, 241, 278
step size 31
successive approximation 80, 121
weighting matrix 105, 161
Quadtree representation 81
Quality factor 107
Quality packets 124
Query
 colour-based 352
 motion-based 356
 shape-based 355, 360
 sketched-based 363
 texture-based 354, 358

Rate control 99, 146, 173, 278, 317
Real time transport protocol (RTP) 290
Redundancy reduction
 spatial 25
 statistical 41
 temporal 35
Reference model (RM) codec 131
Region of interest 117, 124
Regularity of wavelets 76
Resynchronisation marker 128, 252
Reversible VLC (RVLC) 252, 261
Run length coding 108, 140, 228

Scalability 124, 194, 273, 339
 applications 216
 data partitioning 197
 fine granularity (FGS) 339
 hybrid 211
 multilayer 275
 object-based 339
 picture transmission order 277
 SNR 126, 200, 274
 spatial 126, 205, 275
 temporal 208, 274
Scanning 9, 285
 alternate 188, 254
 horizontal/vertical 254
 interlaced/progressive 9
 stripe scan 91
 zero tree scan (ZTS) 82, 329
 zigzag 107, 188
Scene complexity 218
Segmentation 301
 automatic 302
 semiautomatic 302

Index    407

Set partitioning in hierarchical trees (SPIHT) 84
Set top box 218
Shape 309, 355
Shape adaptive DCT 323
Shape adaptive wavelet transform 331
Shape coding 309
    chain code 310
    context-based arithmetic encoding 316
    greyscale 321
    MMR 314
    quad tree 311
    size conversion 317
Slice 157
Slice start code 157
Slice structure mode 253
Source input format (SIF) 12
Spatial orientation tree (SOT) 85
Spatial transforms 242
    affine 246, 325
    bilinear 242
    perspective 246, 325
Spectral selection 112
Sprite 325
Statistical multiplexing 218
Still image coding 103, 327
Subband coding 64
Switched multipoint 231
Switching pictures 287
Synthetic objects 326

Temporal reference (TR) 248, 258
Temporal resampling 19
Test model (TM) encoder 189
Texture coding 322
Tiling 118
Three-dimensional VLC 228
Transform coding 27
    adaptive block transform 282
    cosine 29
    Haar 353
    Hadamard 284
    integer 281
    Karhunen Loeve 28
    lapped orthogonal 63
    wavelet 69
Two-dimensional VLC 108, 140

Unrestricted motion vectors 235
Upconversion 15, 178
User data packet (UDP) 290

Vanishing moments 76
Variable bit rate (VBR) 57
Variable length code (VLC) 41
    advanced VLC 251
    advanced inter/intra VLC 253
    alternative inter VLC 256
    reversible VLC (RVLC) 252
Variable length decoding (VLD) 57
Verification model (VM) 295
Very low bit rate coding 280
Video broadcasting 218
Video buffer verifier 171
Video coding with wavelet transform 332
    higher resolution 334
    lower resolution 336
Video coding layer (VCL) 280, 289
Video object plane (VOP) 297
    VOP encoding 299
    VOP formation 300
Video over networks
    ATM 220
    IP 290
    mobile 227, 282
    satellite 219
    terrestrial 219
Video quality metrics 20
Video streaming
    bit rate switching 287
    random access 289
Virtual zero tree 333

Watermarking 117
Watershed transform 305
Wavelet coding 332
Wavelet transform 69, 120
    Haar wavelet 73
    mother wavelet 70
    multiband 75
    similarities among the bands 81
    symmetric extension 79

Z transform 65
Zero tree coding 82, 329